Random Processes with Independent Increments

Mathematics and Its Applications (*Soviet Series*)

Volume 47

Random Processes with Independent Increments

by

A. V. SKOROHOD
Mathematics Institute,
Kiev State University,
U.S.S.R.

KLUWER ACADEMIC PUBLISHERS
DORDRECHT / BOSTON / LONDON

ISBN 0-7923-0340-7

Published by Kluwer Academic Publishers,
P.O. Box 17, 3300 AA Dordrecht, The Netherlands.

Kluwer Academic Publishers incorporates
the publishing programmes of
D. Reidel, Martinus Nijhoff, Dr W. Junk and MTP Press.

Sold and distributed in the U.S.A. and Canada
by Kluwer Academic Publishers,
101 Philip Drive, Norwell, MA 02061, U.S.A.

In all other countries, sold and distributed
by Kluwer Academic Publishers Group,
P.O. Box 322, 3300 AH Dordrecht, The Netherlands.

Printed on acid-free paper

Translated from the Russian by P.V. Malyshev

This is the translation of the original work
СЛУЧАЙНЫЕ ПРОЦЕССЫ
С НЕЗАВИСИМЫМИ ПРИРАЩЕНИЯМИ
Published by Nauka Publishers, Moscow, © 1986.

Printed in the Netherlands

SERIES EDITOR'S PREFACE

'Et moi, ..., si j'avait su comment en revenir, je n'y serais point allé.'

Jules Verne

The series is divergent; therefore we may be able to do something with it.

O. Heaviside

One service mathematics has rendered the human race. It has put common sense back where it belongs, on the topmost shelf next to the dusty canister labelled 'discarded non-sense'.

Eric T. Bell

Mathematics is a tool for thought. A highly necessary tool in a world where both feedback and non-linearities abound. Similarly, all kinds of parts of mathematics serve as tools for other parts and for other sciences.

Applying a simple rewriting rule to the quote on the right above one finds such statements as: 'One service topology has rendered mathematical physics ...'; 'One service logic has rendered computer science ...'; 'One service category theory has rendered mathematics ...'. All arguably true. And all statements obtainable this way form part of the raison d'être of this series.

This series, *Mathematics and Its Applications*, started in 1977. Now that over one hundred volumes have appeared it seems opportune to reexamine its scope. At the time I wrote

> "Growing specialization and diversification have brought a host of monographs and textbooks on increasingly specialized topics. However, the 'tree' of knowledge of mathematics and related fields does not grow only by putting forth new branches. It also happens, quite often in fact, that branches which were thought to be completely disparate are suddenly seen to be related. Further, the kind and level of sophistication of mathematics applied in various sciences has changed drastically in recent years: measure theory is used (non-trivially) in regional and theoretical economics; algebraic geometry interacts with physics; the Minkowsky lemma, coding theory and the structure of water meet one another in packing and covering theory; quantum fields, crystal defects and mathematical programming profit from homotopy theory; Lie algebras are relevant to filtering; and prediction and electrical engineering can use Stein spaces. And in addition to this there are such new emerging subdisciplines as 'experimental mathematics', 'CFD', 'completely integrable systems', 'chaos, synergetics and large-scale order', which are almost impossible to fit into the existing classification schemes. They draw upon widely different sections of mathematics."

By and large, all this still applies today. It is still true that at first sight mathematics seems rather fragmented and that to find, see, and exploit the deeper underlying interrelations more effort is needed and so are books that can help mathematicians and scientists do so. Accordingly MIA will continue to try to make such books available.

If anything, the description I gave in 1977 is now an understatement. To the examples of interaction areas one should add string theory where Riemann surfaces, algebraic geometry, modular functions, knots, quantum field theory, Kac-Moody algebras, monstrous moonshine (and more) all come together. And to the examples of things which can be usefully applied let me add the topic 'finite geometry'; a combination of words which sounds like it might not even exist, let alone be applicable. And yet it is being applied: to statistics via designs, to radar/sonar detection arrays (via finite projective planes), and to bus connections of VLSI chips (via difference sets). There seems to be no part of (so-called pure) mathematics that is not in immediate danger of being applied. And, accordingly, the applied mathematician needs to be aware of much more. Besides analysis and numerics, the traditional workhorses, he may need all kinds of combinatorics, algebra, probability, and so on.

In addition, the applied scientist needs to cope increasingly with the nonlinear world and the extra mathematical sophistication that this requires. For that is where the rewards are. Linear models are honest and a bit sad and depressing: proportional efforts and results. It is in the non-linear world that infinitesimal inputs may result in macroscopic outputs (or vice versa). To appreciate what I am hinting at: if electronics were linear we would have no fun with transistors and computers; we would have no TV; in fact you would not be reading these lines.

There is also no safety in ignoring such outlandish things as nonstandard analysis, superspace and anticommuting integration, p-adic and ultrametric space. All three have applications in both electrical engineering and physics. Once, complex numbers were equally outlandish, but they frequently proved the shortest path between 'real' results. Similarly, the first two topics named have already provided a number of 'wormhole' paths. There is no telling where all this is leading - fortunately.

Thus the original scope of the series, which for various (sound) reasons now comprises five subseries: white (Japan), yellow (China), red (USSR), blue (Eastern Europe), and green (everything else), still applies. It has been enlarged a bit to include books treating of the tools from one subdiscipline which are used in others. Thus the series still aims at books dealing with:

- a central concept which plays an important role in several different mathematical and/or scientific specialization areas;
- new applications of the results and ideas from one area of scientific endeavour into another;
- influences which the results, problems and concepts of one field of enquiry have, and have had, on the development of another.

The first edition of this book was published in 1964 (in Russian). At that time stochastic processes with independent increments were already of prime importance. Indeed the most often used processes in both theory and applications such as Poisson processes, Wiener processes, and compound Poisson processes have independent increments. There are many more, these are just the simplest examples. Also these processes which can be handled well mathematically serve as a touchstone in the development of probability theory in general. It is here that independence, i.e. probability as opposed to analysis, plays a dominant role, and independence of one sort or another is crucial for applications of probability and statistics.

Since 1964 very much has happened; enough to make this a practically new book, both because of all the new material and because the author has taken the opportunity to insert original proofs of older results.

One striking new aspect is the central (and amount of) place given to random measures which are used as a basis of building up the theory of random processes with independent increments; another, the sections on such processes with values in general linear groups, a cornerstone for non-commutative and quantum probability.

This is the third book by this world famous author in this series, and I must say that I am rather pleased about that.

The shortest path between two truths in the real domain passes through the complex domain.

J. Hadamard

La physique ne nous donne pas seulement l'occasion de résoudre des problèmes ... elle nous fait pressentir la solution.

H. Poincaré

Never lend books, for no one ever returns them; the only books I have in my library are books that other folk have lent me.

Anatole France

The function of an expert is not to be more right than other people, but to be wrong for more sophisticated reasons.

David Butler

Amsterdam, 6 March 1991

Michiel Hazewinkel

Contents

Preface to the Second Russian Edition

Processes with independent increments are the oldest part of the relatively young theory of random processes. Construction of the general theory of random processes was started just by studying these processes. Brownian motion, Wiener process, Poisson process—these random processes arose when concrete and important physical problems were treated. Up to now, the study of them is not completely finished and opportunities of applications are not exhausted; and they are the simplest examples of the processes with independent increments. Thus, investigation of such processes is also useful for practical applications. On the other hand, it is this theory where the notion of independence, by which probability theory differs from mathematical analysis, is used most completely. In addition, there is a rich enough collection of concrete random processes in this theory. They can form the sapid material by means of which general concepts of the theory of random processes are illustrated and developed. Without this material, they remain purely speculative.

The first edition of this book was published more than 20 years ago. Since then, the theory of random processes has been enriched substantially. The same happened to the theory of random processes with independent increments. Thus, the book has undergone extensive revisions. Excluded were sections that covered general questions, such as limit theorems for random processes and absolute continuity of measures corresponding to random processes. The material dealing with the random processes with discrete time (random walks) has been expanded. Included are chapters about random measures with independent values, as well as multiplicative processes (generalizing the notion of the process with independent increments onto processes on groups). The chapters dealing with the investigation of processes with independent increments in R are also expanded. In the revised text, interesting new results that are not very complicated in proof are provided where possible.

I am grateful to Mrs. L. V. Lobanova and Mrs. N. F. Ryabova, whose help was essential to preparing the manuscript of this book.

List of Notations

R	real line;
R^m	m-dimensional Euclidean space;
$L(R^m)$	space of linear operators in R^m;
$L_+(R^m)$	set of nonnegative symmetric operators in R^m;
$(\cdot,\cdot)$	scalar product;
$\lvert\cdot\rvert$	modulus of a real (complex) number, norm of a vector;
$\lVert\cdot\rVert$	norm of an operator;
$\mathrm{sp}\, A$	trace of the operator A;
$\mathcal{A}$, $\mathcal{B}$, $\mathcal{F}$	σ-algebras;
$\sigma(\xi_s, s \in T)$	σ-algebra generated by the random variables $\xi_s, s \in T$;
$\vee_\alpha \mathcal{A}_\alpha$	minimal α-algebra containing all σ- algebras $\mathcal{A}_\sigma$;
$\mathbf{M}\xi$, $\mathbf{D}\xi$	mathematical expectation and dispersion of the variable ξ;
	if φ is a function from R^d to R, then $\phi'(x)$ is its derivative;
	$\varphi'(x) \in R^d, \quad (\varphi'(x), a) = \frac{d}{dt}\varphi(x+ta)\big\rvert_{t=a},$
	$\varphi''(x) \in L(R^d), \quad (\varphi''(x)a, b) = \frac{\partial^2}{\partial t \partial s}\varphi(x+ta+sb)\big\rvert_{t=0,\, s=0},$
$\square$	sign marking the end of the proof;
r.h.s. (l.h.s.)	right- (left)-hand side;
iff	if and only if;
s.m.	stopping moment.

CHAPTER 0

PRELIMINARY INFORMATION

§0.1 Probability Space

When probabilistic objects are considered, a certain *probability space*, namely a triple $\{\Omega, \mathcal{F}, \mathbf{P}\}$, is usually assumed to be fixed. Here Ω is a space of elementary events, $\mathcal{F}$ is a σ-algebra of subsets of Ω, the elements of $\mathcal{F}$ are called *events* (*random events*), $\mathbf{P}$ is a normalized measure on $\mathcal{F}$, i.e., a nonnegative countably additive function of the set for which $\mathbf{P}(\Omega) = 1$, and $\mathbf{P}(A)$ for $A \in \mathcal{F}$ is called the *probability of event* A.

The $\mathcal{F}$-measurable function $\xi = \xi(\omega)$ from Ω to R is called a *random (real-valued) variable* on the probability space $\{\Omega, \mathcal{F}, \mathbf{P}\}$. The function $F(x) = \mathbf{P}(\{\omega : \xi(\omega) < x\})$ is called a distribution function of the random variable $\xi(\omega)$.

If $\xi_1, \ldots, \xi_n$ are random variables, then the following function on R^n

$$F(x_1, \ldots, x_n) = \mathbf{P}(\bigcap_{k=1}^{n} \{\omega : \xi_k(\omega) < x_k\}) \tag{0.1}$$

is called a *joint distribution function* of the variables $\xi_1, \ldots, \xi_n$.

Let X be a certain set and $\mathcal{B}$ be a σ-algebra of its subsets, then a pair $(X, \mathcal{B})$ is called a *measurable space*. The measurable mapping $x(\omega)$ of the measurable space $(\Omega, \mathcal{F})$ into the measurable space $(X, \mathcal{B})$ is called a random element in X (measurability of $x(\omega)$ means that $\{\omega : x(\omega) \in B\} \in \mathcal{F}$ for all $B \in \mathcal{B}$). A measure on $\mathcal{B}$ defined by the equality

$$\mu_x(B) = \mathbf{P}(x^{-1}(B)) = \mathbf{P}(\{\omega : x(\omega) \in B\}) \tag{0.2}$$

is called a *distribution of the random element* $x(\omega)$. If X is a finite-dimensional vector space and $\mathcal{B}$ is a σ-algebra of its Borel subsets, then $x(\omega)$ is called a *random vector*. In particular, each set of n random variables $\xi_1, \ldots, \xi_n$ defines the random vector $(\xi_1, \ldots, \xi_n)$ in R^n. The distribution of this vector is called a *joint distribution* of the variables $\xi_1, \dots, \xi_n$ (not to be confused with the joint distribution function: (0.1) defines the function on R^n, whereas (0.2) defines the measure on Borel subsets of R^n). The joint distribution of the variables $\xi_1, \dots, \xi_n$ can be defined by using the joint distribution function that determines it in a unique way.

Along with finite-dimensional vector spaces, we shall consider random elements in separable Hilbert and Banach spaces.

Let $\xi(\omega)$ be a real-valued random variable. If it is integrable with respect to measure $\mathbf{P}$, then

$$\int \xi(\omega)\mathbf{P}(d\omega) = \mathbf{M}\xi(\omega) \tag{0.3}$$

(on the r.h.s., there is a notation) is called the *mathematical expectation* of the random variable ξ. For the nonnegative random variable ξ, the mathematical expectation $\mathbf{M}\xi$ is always considered to be defined; it may take the value $+\infty$. If $\xi \geq 0$, the *Chebyshev inequality* holds, for $c > 0$,

$$\mathbf{P}\{\xi > c\} \leq \frac{1}{c}\mathbf{M}\xi. \tag{0.4}$$

Let X be a separable complete metric space with a distance $r(x, y)$, and $\mathcal{B}$ be a σ-algebra of its Borel subsets. Consider a sequence of random elements $x_n(\omega)$ with values in X; $x_n(\omega)$, $n \geq 1$, converges to $x_0(\omega)$:

(1) *in probability*, if for every $\varepsilon > 0$,

$$\lim_{n\to\infty} \mathbf{P}\{r(x_n(\omega), x_0(\omega) > \varepsilon)\} = 0;$$

(2) *in mean of the power* α, if

$$\lim_{n\to\infty} \mathbf{M}r^\alpha\{(x_n(\omega), x_0(\omega) > \varepsilon)\} = 0 \tag{0.5}$$

(convergence for $\alpha = 1$ is called *convergence in mean* and, for $\alpha = 2$, *mean square convergence*);

(3) *with probability 1*, if an event $\{\omega : \lim_{n\to\infty} r(x_n(\omega), x_0(\omega)) = 0\}$ has probability 1.

The convergences (2) and (3) imply (1); if X is bounded, (1) and (2) are equivalent.

The space of all random elements with values in X [we denote it by $X(\Omega)$] and with convergence in probability is metrizable: we let

$$\rho(x_1, x_2) = \mathbf{M}(1 - \exp\{-r(x_1(\omega), x_2(\omega))\}),$$
$$x_1(\omega), x_2(\omega) \in X(\Omega);$$

with such a metric, it is complete. Convergence (2) is also complete: If for a sequence $x_n(\omega)$ the condition $\lim_{n,m\to\infty} \mathbf{M}r^\alpha(x_n(\omega), x_m(\omega)) = 0$ holds, then there exists the random element $x_0(\omega)$ in X for which (0.5) is valid.

In exactly the same way, if $\mathbf{P}\{\lim_{n,m\to\infty} r(x_n(\omega), x_m(\omega)) = 0\} = 1$, the sequence $x_n(\omega)$ converges with probability 1 to some element $x_0(\omega)$.

We now give a statement that ascertains a relation between the convergence with probability 1 and convergence in probability.

1°. Theorem. *The sequence $x_n(\omega)$ converges to $x_0(\omega)$ in probability iff from each sequence n_k a subsequence n'_k can be chosen such that $x_{n'_k}(\omega)$ converges to $x_0(\omega)$ with probability 1.*

Proof. Note, that if $\varepsilon_k \to 0$ and n_k is an arbitrary subsequence for which $\sum_k \mathbf{P}\{r(x_{n_k}(\omega), x_0(\omega)) > \varepsilon_k\} < \infty$, then only a finite number of events $\{r(x_{n_k}(\omega), x_0(\omega)) > \varepsilon_k\}$ will occur with probability 1. (We use the Borel–Cantelli lemma, which is formulated completely in 24°). Thus for almost all ω and sufficiently large k, $r(x_n(\omega), x_0(\omega)) \leq \varepsilon_k$, i.e., $x_{n_k} \to x_0(\omega)$ with probability 1. On the other hand, if $x_n(\omega)$ doesn't converge to $x_0(\omega)$, then one can find such $\varepsilon > 0$, $\delta > 0$ and a subsequence n_k, for which $\mathbf{P}\{r(x_{n_k}(\omega), x_0(\omega)) > \varepsilon\} \geq \delta$. But in this case, no subsequence for which $x_{n_k}(\omega) \to x_0(\omega)$ with probability 1 can be extracted from the subsequence n_k.

□

Weak Convergence of Measures

Let X be a complete separable metric space with the distance $r(x, y)$, C_X be a space of bounded continuous real-valued functions on X, and $\mathcal{B}$ be a σ-algebra of Borel subsets of X. The sequence of bounded measures μ_n on $\mathcal{B}$ converges weakly to a measure μ, if for every function $f \in C_X$,

$$\lim_{n\to\infty} \int f(x)\mu_n(dx) = \int f(x)\mu(dx). \tag{0.6}$$

2°. *If a sequence of random elements $x_n(\omega)$ in X converges in probability to the element $x_0(\omega)$, μ_0 is the distribution of $x_0(\omega)$, and μ_n is the distribution of $x_n(\omega)$, then μ_n converges weakly to μ.*

Proof.

$$\int f(x)\mu_m(dx) = \mathbf{M}f(x_n(\omega)), \quad n = 0, 1, \ldots.$$

Let us show that $f(x_n(\omega))$ converges in probability to $f(x_0(\omega))$. It is sufficient to use Theorem 1° and to note that if $x_{n_k}(\omega) \to x_0(\omega)$ with probability 1, then $f(x_{n_k}(\omega)) \to f(x_0(\omega))$ also with probability 1.

However, from the boundedness of f and from the convergence of $f(x_n(\omega))$ to $f(x_0(\omega))$ in probability, it follows (according to the Lebesgue theorem about limit transition under the integral sign) that

$$\mathbf{M}f(x_0, (\omega)) = \lim_{n\to\infty} \mathbf{M}f(x_n(\omega)).$$

□

Let $\mathbf{M} = \{\mu_\theta, \theta \in \Theta\}$ be a certain set of bounded measures on $\mathcal{B}$. It is called *weakly compact* if every sequence of measures $\mu_n \in \mathbf{M}$ contains a weakly convergent subsequence.

Now we shall give, without proofs, conditions of weak compactness and of weak convergence of measures.

3°. A. *A set of bounded measures* $\mathbf{M}$ *is weakly compact iff the following conditions hold:* (a) $\sup_{\mu\in\mathbf{M}}\mu(X)<\infty$; (b) *for every* $\varepsilon>0$, *there exists a compactum* $K\subset X$ *such that for all* $\mu\in\mathbf{M}$, *the following inequality* $\mu(X)-\mu(K)<\varepsilon$ *is valid.*

B. *If* X *is a finite-dimensional Euclidean space, the following condition [along with condition* (a)*]*

$$\lim_{c\to 0}\sup_{\mu}\mathbf{P}(\{x:|x|\geq c\})=0$$

is necessary and sufficient for compactness of the set of measures.

Let X be a finite-dimensional Euclidean space; we associate with the measure μ its Fourier transform (a characteristic function),

$$f_\mu(z)=\int e^{i(z,x)}\mu(dx),\quad z\in X,$$

where $(\cdot,\cdot)$ is a scalar product in X.

Let us formulate the conditions of weak compactness of measures in terms of Fourier transforms.

C. *The set* M *in a finite-dimensional Euclidean space is weakly compact iff*
(a) $\sup_{\mu\in M}f_\mu(0)<\infty$
(b) $\lim_{\delta\to 0}\sup_{\mu\in M}\sup_{|z|<\delta}|f_\mu(0)-f_\mu(z)|=0$.

To formulate the conditions of weak convergence of measures, we shall need the conception of a total set.

Definition. A set $\mathcal{T}\subset C_X$ is called *total* if from the equality

$$\int f(x)\mu_1(dx)=\int f(x)\mu_2(dx)\quad\forall f\in\mathcal{T},$$

where μ_1,μ_2 are finite measures on $\mathcal{B}$, it follows that $\mu_1=\mu_2$.

D. *A sequence of finite measures* μ_n *on* $\mathcal{B}$ *converges weakly to a measure* μ *iff it is weakly compact and for all* $f\in\mathcal{T}$ *the equality* (0.6) *holds. If the sequence* μ_n *is weakly compact and if for* $f\in\mathcal{T}$ *there exists a limit* $\lim_{n\to\infty}\int f(x)\mu_n(dx)$, *then* μ_n *converges weakly to some finite measure* μ *on* $\mathcal{B}$.

E. *Let* X *be a finite-dimensional Euclidean space. The sequence of finite measures* μ_n *on* $\mathcal{B}$ *converges weakly iff their Fourier transforms* $f_n(z)$ *converge for all* z *to some continuous function* $f(z)$. *Moreover,* $f(z)$ *is the Fourier transform of a limit measure, and* $f_n(z)$ *converges to* $f(z)$ *uniformly on each bounded set.*

Let X be a locally compact space. A measure μ given on $\mathcal{B}$ is called *locally finite* if the measure of each bounded set is finite. Let us denote by C_X^f a set of functions from C_X with compact supports (i.e., the functions $f(x)$ such that $\{x : |f(x)| > 0\}$ is bounded). A sequence of locally finite measures μ_n converges on finite functions to a locally finite measure μ, if for all $f \in C_X^f$, the relation (0.6) holds. A set M of locally finite measures is called *compact with respect to the convergence on finite functions* if from each sequence of measures $\mu_n \in M$, one can extract a weakly convergent subsequence on finite functions.

4°. Theorem. *The set of locally finite measures M is compact with respect to the convergence on finite functions iff $\sup_{\mu \in M} \mu(K) < \infty$ for every compact set $K \subset X$. If the sequence μ_n is compact with respect to the convergence on finite functions and if there exists the limit $\lim \int f(x)\mu_n(dx)$ for some total subset $T \subset C_X^f$ of functions f, then μ_n converges on finite functions to some locally bounded measure μ.*

Proof. Whatever compact set K_1 is taken, one can find a compact set K_2 such that K_1 lies inside K_2. Let $g_1(x) = 1$ on K_1 and $g_1(x) = 0$ for $x \overline{\in} K_2$, and also $g_1(x) \in C_X$, $0 \le g_1(x) \le 1$. If μ_n is a sequence of measures from M, the sequence of measures $\tilde{\mu}_n(B) = \int_B g_1(x)\mu_n(dx)$ is weakly compact by virtue of 3°A, since $\tilde{\mu}_n(X) \le \mu_n(K_2)$, $\tilde{\mu}_n(X) = \tilde{\mu}_n(K_2)$. Thus one can extract the subsequence n_k such that the limit $\lim \int f(x)\tilde{\mu}_{n_k}(dx)$ exists. If the support of f belongs to K_1, then

$$\lim_{k\to\infty} \int f(x)\tilde{\mu}_{n_k}(dx) = \lim_{k\to\infty} \int f(x)\mu_{n_k}(dx)$$

Now let (K_m) be a sequence of compact sets such that K_m belongs to K_{m+1} and $\bigcup_m K_m = X$. If $g_m(x) \in C_X$, $g_m(x) = 1$ for $x \in K_m$, $g_m(x) = 0$ for $x \overline{\in} K_{m+1}$, $0 \le g_m(x) \le 1$, and $\mu_n^{(m)}(B) = \int_B g_m(x)\mu_n(dx)$, then all the sequences $\{\mu_n^{(m)}, n = 1, 2, \dots\}$ are weakly compact and one can extract a subsequence n_k such that the $\mu_{n_k}^{(m)}$ converge weakly for all m. Then μ_{n_k} obviously converges on finite functions, since for every $f \in C_X^f$ there exists m such that $g_m f = f$.

The second statement follows from 3°E.

□

σ-Algebras

Recall that the collection $\mathcal{A}$ of the subsets of a certain set X is called an *algebra* if $X \in \mathcal{A}$ and, together with two arbitrary sets $A, B \in \mathcal{A}$, also $A \backslash B \in \mathcal{A}$. Then $A \cap B \in \mathcal{A}$, $A \cup B \in \mathcal{A}$, $X \backslash A \in \mathcal{A}$, too. The algebra of the sets $\mathcal{A}$ is called a *σ-algebra*, if together with the arbitrary sequence, $A_n \in \mathcal{A}$, also $\bigcup_n A_n \in \mathcal{A}$.

A collection of subsets $\mathcal{M}$ is called a *monotonic class* if for any monotonic sequence, $A_n \in \mathcal{M}$ (monotonicity means that for all n, either $A_n \subset A_{n+1}$—this is an increasing sequence—or $A_n \supset A_{n+1}$—this is a decreasing one), and

$\lim A_n \in \mathcal{M}$ (for the increasing sequence, $\lim A_n = \bigcup A_n$, whereas for the decreasing one, $\lim A_n = \bigcap A_n$).

5°. Theorem on monotonic class. *The least monotonic class, which contains the algebra $\mathcal{A}$, coincides with the least σ-algebra containing $\mathcal{A}$.*

Proof. For any algebra of the sets $\mathcal{A}$, we denote by $L(\mathcal{A})$ the least algebra containing all the limits of the monotonic sequences of the sets from $\mathcal{A}$. Clearly, if $\mathcal{A} \subset \mathcal{M}$, where $\mathcal{M}$ is the monotonic class, then $L(\mathcal{A}) \subset \mathcal{M}$, too. If $\mathcal{A}_\theta$, $\theta \in \Theta$ is a collection of algebras such that for all θ_1 there exists θ_2 for which $\mathcal{A}_{\theta_1} \subset \mathcal{A}_{\theta_2}$, then $\bigcup_\theta \mathcal{A}_\theta$ is also an algebra. (We call such a collection of algebras *ordered by inclusion.*) Moreover, if all $\mathcal{A}_\theta \subset \mathcal{M}$, then $\bigcup_\theta \mathcal{A}_\theta \subset \mathcal{M}$.

Let us now construct a transfinite sequence of algebras $\mathcal{A}_\theta$, where θ is an ordinal number, $\mathcal{A}_0 = \mathcal{A}$; if θ is an ordinal number and if, for all $\theta < \theta_1$, the algebras $\mathcal{A}_\theta$ have already been built, then $\mathcal{A}_{\theta_1} = L(\mathcal{A}_{\theta_0})$, where θ_0 precedes θ_1 if θ_1 has the preceding number, and $\mathcal{A}_{\theta_1} = \bigcup_{\theta<\theta_1} \mathcal{A}_\theta$ if there is no preceding number. All $\mathcal{A}_\theta \in \mathcal{M}$. Let us write $\mathcal{A}^1 = \bigcup \mathcal{A}_\theta$. Then $L(\mathcal{A}^1) \subset \mathcal{A}^1$, and therefore $\mathcal{A}^1$ is a σ-algebra, $\mathcal{A}^1 \subset \mathcal{M}$. It remains to point out that every σ-algebra is a monotonic class.

□

We shall use the following notations: If $\mathcal{K}$ is some class of sets, then by $\sigma(\mathcal{K})$ we shall denote the least σ-algebra containing $\mathcal{K}$. We shall call it the *σ-algebra generated by the class $\mathcal{K}$.*

Let $\{F = \{f_\theta, \theta \in \Theta\}\}$ be a certain set of real-valued functions. By $\sigma(F) = \sigma(\{f_\theta, \theta \in \Theta\})$ we shall denote the least σ-algebra with respect to which all functions f_θ are measurable. Let $\mathcal{K}$ consist of the sets $\{x : f_\theta(x) < \lambda\}$, $\theta \in \Theta$, $\lambda \in R$. Then $\sigma(F) = \sigma(\mathcal{K})$. If f_θ are functions from X into a certain measurable space $(Y, \mathcal{C})$, then $\sigma(\{f_\theta, \theta \in \Theta\}) = \sigma(\{f_\theta(x) \in C, \theta \in \Theta, C \in \mathcal{C}\})$.

Definition. Let $\{X, \mathcal{B}\}$ be some measurable space. The measure μ on the σ-algebra $\mathcal{B}$ is called *complete* if $\mathcal{B}$ contains all A for which there exists $B \in \mathcal{B}$ such that $A \subset B$ and $\mu(B) = 0$ (then $\mu(A) = 0$, too). In this case, the σ-algebra $\mathcal{B}$ is called *complete* (with respect to the measure μ).

Let us define an operation of "completion" of a measure and of a σ-algebra, i.e., an expansion of the initial σ-algebra and an extension of the measure onto the expanded σ-algebra.

Let us denote by $\widehat{\mathcal{C}}$ a σ-ring of the sets A for which there exists $B \in \mathcal{B}$ such that $\mu(B) = 0$ and $A \subset B$, and by $\widehat{\mathcal{B}}$ a σ-algebra of the sets B for which there exists $A \in \mathcal{B}$ such that $(A \backslash B) \cup (B \backslash A) \in \widehat{\mathcal{C}}$. If we define a measure $\hat{\mu}$ on $\widehat{\mathcal{B}}$ by the equality $\hat{\mu}(B) = \mu(A)$ for $A \in \mathcal{B}$ and $(A \backslash B) \cup (B \backslash A) \in \widehat{\mathcal{C}}$, then the measure $\hat{\mu}$ is the extension of μ onto $\widehat{\mathcal{B}}$ (it is unique); $\hat{\mu}$ is called a *completion of the measure* μ ($\hat{\mu}$ is complete), and $\widehat{\mathcal{B}}$ is called a *completion of the σ-algebra* $\mathcal{B}$ in the measure μ.

Let $\{\mathcal{A}_\theta, \theta \in \Theta\}$ be a certain set of σ-algebras of subsets of the same set. Then $\bigcap_\theta \mathcal{A}_\theta$ is, obviously, a σ-algebra. By $\bigvee_\theta \mathcal{A}_\theta$ we denote a σ-algebra $\sigma\{\bigcup_\theta \mathcal{A}_\theta\}$, i.e., the least one containing all the σ-algebras $\mathcal{A}_\theta$.

If $\{\xi_n, n \geq 1\}$ is a sequence of random variables, then

$$\sigma\{\xi_n, n \geq 1\} = \bigvee_n \sigma\{\xi_n, \ldots, \xi_n\} = \bigvee_n \sigma\{\xi_n\}.$$

Let $(X_1, \mathcal{B}_1)$ and $(X_2, \mathcal{B}_2)$ be two measurable spaces, and let $X_1 \times X_2$ be the Cartesian product of the sets X_1 and X_2. Then the least σ-algebra containing all the sets $B_1 \times B_2$, where $B_i \in \mathcal{B}_i$ is called the ***product*** $\mathcal{B}_1 \otimes \mathcal{B}_2$ ***of the σ-algebras*** $\mathcal{B}_1$ ***and*** $\mathcal{B}_2$. We have

$$\mathcal{B}_1 \otimes \mathcal{B}_2 = \sigma(\{B_1 \times B_2, B_1 \in \mathcal{B}_1, B_2 \in \mathcal{B}_2\}).$$

Let $\mathcal{A}_1$ and $\mathcal{A}_2$ be σ-algebras, $\mathcal{A}_1 \times \mathcal{A}_2$ be their product, i.e., an algebra of the sets $\bigcup_{k=1}^n A_k^{(1)} \times A_k^{(2)}$, where $A_{(k)}^{(i)} \in \mathcal{A}_i$, $i = 1, 2$. Then

$$\mathcal{A}_1 \otimes \mathcal{A}_2 = \sigma(\mathcal{A}_1 \times \mathcal{A}_2).$$

Let $(X_k, \mathcal{B}_k)$ be a sequence of measurable spaces, $\prod_{k=1}^n X_k$ and $\prod_{k=1}^\infty X_k$ be the finite or countable Cartesian product of X_k. Let us define $\otimes_{k=1}^n \mathcal{B}_k$ and $\otimes_{k=1}^\infty \mathcal{B}_k$ as σ-algebras in $\prod_{k=1}^n X_k$ and $\prod_{k=1}^\infty X_k$, respectively, in the following way:

$$\otimes_{k=1}^n \mathcal{B}_k = \sigma(\{B_1 \times \cdots \times B_n, \quad B_k \subset \mathcal{B}_k, \quad k = 1, \ldots, n\})$$

(if all $\mathcal{B}_k$ coincide, we write $\mathcal{B}^{\otimes n}$),

$$\otimes_{k=1}^\infty \mathcal{B}_k = \sigma(\{B_1 \times \cdots \times B_n \times \prod_{k=n+1}^{\infty} X_k, \quad B_k \subset \mathcal{B}_k,$$
$$k = 1, \ldots, n, \quad n = 1, 2, \ldots\}).$$

The sets of the form $B_1 \times \cdots \times B_n \times \prod_{k=n+1}^\infty X_k$, where $B_k \in \mathcal{B}_k$, $k = 1, 2, \ldots, n$ are called *cylindrical sets*, and an algebra $\mathcal{C}$ of the sets of the form $\bigcup_{k=1}^m C_k$, where C_k are cylindrical sets, is called a *cylindrical algebra*. Hence

$$\otimes_{k=1}^\infty \mathcal{B}_k = \sigma(\mathcal{C}).$$

Consider now σ-algebras in functional spaces. Let Θ be a certain ("parametrical") set, and $(X, \mathcal{B})$ be a measurable space. By X^Θ we denote the set of all functions from Θ into X: $\{x_\theta, \theta \in \Theta\}$. By $\otimes_{\theta \in \Theta} \mathcal{B} = \mathcal{B}^{\otimes \Theta}$ we denote the σ-algebra generated by the collection of all cylindrical sets; the definition of the cylindrical set is given below.

For $\theta_1 \in \Theta$ and $B_1 \in \mathcal{B}$, let us denote by $C_{\theta_1}(B_1)$ a subset of X^Θ defined by the equality

$$C_{\theta_1}(B_1) = \{x_\theta : x_{\theta_1} \in B_1\}. \tag{0.7}$$

A set as in (0.7) is called a *one-dimensional cylindrical set.* Sets of the form

$$\bigcap_{k=1}^{n} C_{\theta_k}(B_k), \quad \theta_k \in \Theta, \quad B_k \in \mathcal{B}, \quad k = 1,, \ldots, n,$$

are called *n-dimensional cylindrical sets.* Let $\Lambda \subset \Theta$. We denote

$$\mathcal{B}^{\Lambda} = \sigma(\{C_\theta(B), \quad \theta \in \Lambda, \quad B \in \mathcal{B}\}).$$

If $N(\Theta)$ is a set of all countable subsets of Θ, then

$$\otimes_{\theta \in \Theta} \mathcal{B} = \bigcup_{\Lambda \in N(\Theta)} \mathcal{B}^{\otimes \Lambda}.$$

Let X be a topological separable space. The σ-algebra generated by open sets is called a *σ-algebra of Borel sets*; we denote it by $\mathcal{B}_X$.

If R^n is an n-dimensional space, then

$$\mathcal{B}_{R^n} = \mathcal{B}_R^{\otimes n}.$$

For a separable Banach space X with a space of linear functionals X^*,

$$\mathcal{B}_X = \sigma(\{x : x^*(x) < \alpha\}, \quad x^* \in L, \quad \alpha \in R),$$

where $L \subset X^*$ is weakly sequentially dense in X^* (i.e., for all $x^* \in X^*$, there exists a sequence $x_n^* \subset L$ such that $x^*(x) = \lim_{n\to\infty} x_n^*(x)$ for all x).

In a separable complete metric space X, one can also define $\mathcal{B}_X$ as a σ-algebra generated by spheres or as the least σ-algebra, with respect to which all continuous (uniformly continuous) real-valued functions are measurable.

A metric space X is called *Borel* if X is a Borel subset of its own completion $\overline{X}$, which is already a complete metric space.

§0.2 Random Functions and Processes

Let $(X, \mathcal{B})$ be a measurable space, Θ be a set, and $\{\Omega, \mathcal{F}, \mathbf{P}\}$ be a probability space. The family of random elements $\{x(\omega, \theta), \theta \in \Theta\}$ given on $\{\Omega, \mathcal{F}, \mathbf{P}\}$ and taking values on $(X, \mathcal{B})$ is called a *random function*; $(X, \mathcal{B})$ is a *phase space* of this random function, and Θ is a *parametric set.* If Θ is a subset of R, the random function is called a *random process*, and the parameter $t \in \Theta$ is interpreted as time.

The collection of functions

$$\begin{aligned} \{F_{\theta_1,\ldots,\theta_k}(A_1, \ldots, A_k) = \mathbf{P}\{x(\omega, \theta_1) \in A_1, \ldots, x(\omega, \theta_k) \in A_k\} \\ k = 1, 2, \ldots, \quad \theta_i \in \Theta, \quad A_i \in \mathcal{B}\} \end{aligned} \tag{0.8}$$

is called *distributions* of the random function $x(\omega, \theta)$. For fixed k, the functions $F_{\theta_1,\ldots,\theta_k}(A_1, \ldots, A_k)$ are called *k-dimensional distributions of the process*, the collection (0.8) is also called *partial (finite-dimensional) distributions of the process.*

Finite-dimensional distributions of a process satisfy the following conditions.

(I) For all k, $F_{\theta_1,\dots,\theta_k}(A_1,\dots,A_k)$ is a joint distribution of k variables with values in $(X,\mathcal{B})$.

(II) If $i_1,\dots,i_k$ is a permutation of numbers $1,\dots,k$, then

$$F_{\theta_1,\dots,\theta_k}(A_1,\dots,A_k)=F_{\theta_{i_1},\dots,\theta_{i_k}}(A_{i_1},\dots,A_{i_k}).$$

(III) Conditions I–III are called *compatibility conditions.*

We now give, without a proof, Kolmogorov's theorem about the existence of a random function with given finite-dimensional distributions.

6°. Theorem. *Let X be a complete separable metric space, $\mathcal{B}=\mathcal{B}_X$ then each collection of functions $\{F_{\theta_1,\dots,\theta_k}(A_1,\dots,A_k),\ k=1,2,\dots,\theta_i\in\Theta, A_i\in\mathcal{B}\}$ satisfying the compatibility conditions is a set of finite-dimensional distributions of some random function $x(\omega,\theta),\theta\in\Theta$.*

This process is constructed on a probability space $\{X^{\Theta},\mathcal{B}_X^{\otimes\Theta},\mathbf{P}\}$, where the probability measure $\mathbf{P}$ is given on the cylindrical sets

$$\mathbf{P}\{\bigcap_{i=1}^{k}C_{\theta_i}(B_i)\}=F_{\theta_1,\dots,\theta_k}(B_1,\dots,B_k). \tag{0.9}$$

The measure $\mathbf{P}$ is defined by (0.9) in a unique way.

Two random functions $x_1(\omega,\theta)$ and $x_2(\omega,\theta)$ given on Θ with the same probability $\{\Omega,\mathcal{F},\mathbf{P}\}$ and phase $(X,\mathcal{B})$ spaces are called *stochastically equivalent* if

$$\mathbf{P}\{x_1(\omega,\theta)=x_2(\omega,\theta)\}=1,\quad \theta\in\Theta. \tag{0.10}$$

In this case, $x_1(\omega,\theta)$ and $x_2(\omega,\theta)$ are called *modifications* (for each other).

Let Θ be a separable metric space and X be a complete metric space. The random function $x(\omega,\theta)$ is called a *separable function* if there exists a countable set Λ dense in Θ and a set $N\in\mathcal{F}$ of zero measure such that for every open $U\subset\Theta$ and closed $A\subset X$ sets,

$$\bigcap_{\theta\in U\cap\Lambda}\{\omega:x(\omega,\theta)\in A\}\setminus\bigcap_{\theta\in U}\{\omega:x(\omega,\theta)\in A\}\subset N.$$

We shall now give (also without proof) the useful Doob theorem.

7°. Theorem. *If X is a compact metric space, $\mathcal{B}=\mathcal{B}_X$, then every random function has its separable modification.*

Measurability of Random Functions

Let $(\Theta, \mathcal{C})$ be a measurable space. Let us consider the random functions $x(\omega, \theta)$, given on the probability space $\{\Omega, \mathcal{F}, \mathbf{P}\}$ and on Θ, with the measurable phase space $(X, \mathcal{B})$. The random function $x(\omega, \theta)$ is called *measurable* if it is measurable with respect to $\mathcal{C} \otimes \mathcal{F}$, i.e.,

$$\{(\omega, \theta) : x(\omega, \theta) \in A\} \in \mathcal{C} \otimes \mathcal{F} \quad \text{for all} \quad A \in \mathcal{B}.$$

Assume that Θ is a separable metric space with a metric $\rho(\theta_1, \theta_2)$ and X is a metric space with a metric $r(x_1, x_2)$. The random function $x(\omega, \theta)$ is called *stochastically continuous* if $x(\omega, \theta_n)$ converges in probability to $x(\omega, \theta)$ when $\theta_n \to \theta$ for all $\theta \in \Theta$.

8°. *If $x(\omega, \theta)$ is a stochastically continuous process, then it has measurable modification.*

Further on we shall use the theorem about measurability of "projection."

9°. *Let $x(\omega, \theta)$ be a measurable process, given on the complete probability space $\{\Omega, \mathcal{F}, \mathbf{P}\}$, for which the measure $\mathbf{P}$ is complete and Θ is a separable Borel metric space, and $C = \mathcal{B}_\Theta$. Then for all $A \in \mathcal{B}m$,*

$$\{\omega : \{\theta : x(\omega, \theta) \in A\} = \varnothing\} \in \mathcal{F}. \tag{0.11}$$

Corollary 1. *Let $X = R$ and $\mathcal{B} = \mathcal{B}_R$. Then $\sup_{\theta \in C} x(\omega, \theta)$ is the random variable for all $C \in \mathcal{B}_\theta$.*

In fact, the random function $x(\omega, \theta)$, $\theta \in C$, is also measurable on the Borel space $(C, \mathcal{B}_C)$.

$$\{\sup_{\theta \in C} x(\omega, \theta) < \lambda\} = \bigcup_k \{\omega : \{\theta \in C : x(\omega, \theta) \geq \lambda - 1/k\} = \varnothing\}.$$

□

Sets in $\Theta \times \Omega$ that are measurable (with respect to $\mathcal{C} \otimes \mathcal{F}$) are called *random sets.* For $\widetilde{C} \in \mathcal{C} \otimes \mathcal{F}$, a set

$$\pi_\Omega(\widetilde{C}) = \{\omega : \{\theta : (\omega, \theta) \in \widetilde{C}\} \neq \varnothing\}$$

is called a *projection* of $\widetilde{C}$ onto Ω.

Corollary 2. *If $\mathcal{F}$ is a complete σ-algebra, then $\pi_\Omega(\widetilde{C}) \in \mathcal{F}$.*

§0.3 Conditional Probabilities

The elementary formula for the conditional probability of the event A with the hypothesis B

$$\mathbf{P}(A|B) = \frac{\mathbf{P}(A \cap B)}{\mathbf{P}(B)}, \quad \mathbf{P}(B) > 0,$$

allows such generalization.

Let $\mathcal{A}$ be a finite σ-algebra, E_n be its atoms, and $\mathbf{P}(E_n) > 0$. If A is an arbitrary event from $\mathcal{A}$, we set

$$\mathbf{P}\{A|\mathcal{A}\} = \sum_n \mathbf{P}(A|E_n) I_{E_n}, \tag{0.12}$$

where I_{E_n} is an indicator of E_n.

Let us introduce a notion of the conditional mathematical expectation $\mathbf{M}\{\xi|\mathcal{A}\}$ of a random variable ξ. We naturally set

$$\mathbf{M}\{\xi|\mathcal{A}\} = \int_\Omega \xi \mathbf{P}(d\omega|\mathcal{A})$$

or

$$\mathbf{M}\{\xi|\mathcal{A}\} = \sum_n \frac{I_{E_n}}{\mathbf{P}(E_n)} \int_{E_n} \xi d\mathbf{P}. \tag{0.13}$$

Here we assume that $\mathbf{M}\xi$ is finite. If we set $\xi - I_A$ in the last formula, we obtain the conditional probability of the event A:

$$\mathbf{M}\{I_A|\mathcal{A}\} = \sum_n I_{E_n} \frac{\mathbf{P}(E_n \cap A)}{\mathbf{P}(E_n)} = \mathbf{P}\{A|\mathcal{A}\},$$

i.e., the conditional probability is a partial case of the conditional mathematical expectation.

Let η be an arbitrary bounded $\mathcal{A}$-measurable random variable. Then $\eta = \sum_n a_n I_{E_n}$. Multiplying the equality (0.13) by η, we obtain

$$\mathbf{M}(\eta \mathbf{M}\{\xi|\mathcal{A}\}) = \sum_n a_n \int_{E_n} \xi d\mathbf{P} = \sum_n \int_{E_n} \eta \xi d\mathbf{P},$$

or

$$\mathbf{M}(\eta \mathbf{M}\{\xi|\mathcal{A}\}) = \mathbf{M}\eta\xi. \tag{0.14}$$

The last relation defines the conditional mathematical expectation of the random variable ξ completely. In fact, let $\check{\xi}$ be an $\mathcal{A}$-measurable random variable such that for any bounded $\mathcal{A}$-measurable random variable η, we have

$\mathbf{M}(\eta\check{\xi}) = \mathbf{M}(\eta\xi)$. The variable $\check{\xi}$ is constant on E_n, and thus $\check{\xi} = C_n$ for $\omega \in E_n$. Let $\eta = I_{E_n}$. Then

$$\mathbf{M}(\eta\check{\xi}) = c_n \mathbf{M} I_{E_n} = c_n \mathbf{P}(E_n) = \int_{E_n} \xi \, d\mathbf{P},$$

i.e., $\check{\xi}$ coincides with the r.h.s. of (0.13).

Now we use (0.14) to define the conditional mathematical expectation in the general case. If we set $\eta = I_F$ in (0.14), it takes the form

$$\int_F \mathbf{M}\{\xi|\mathcal{A}\} d\mathbf{P} = \int_F \xi \, d\mathbf{P}. \tag{0.15}$$

Definition. The random $\mathcal{A}$-measurable variable $\mathbf{M}\{\xi|\mathcal{A}\}$ is called a *conditional mathematical expectation* of the random variable ξ (for which $\mathbf{M}|\xi| < \infty$) with respect to the σ-algebra $\mathcal{A}$, if it satisfies the equality (0.15) for any $F \in \mathcal{A}$.

Theorem. *For any random variable ξ possessing the property $\mathbf{M}|\xi| < \infty$, the conditional mathematical expectation exists and* (mod $\mathbf{P}$) *is unique.*

Definition. The random variable

$$\mathbf{P}\{A|\mathcal{A}\} = \mathbf{M}\{I_A|\mathcal{A}\}.$$

is called the *conditional probability of the event A with respect to the σ-algebra $\mathcal{A}$.*

$\mathbf{P}\{A|\mathcal{A}\}$ can be directly defined as follows.

The random $\mathcal{A}$-measurable variable $\mathbf{P}\{A|\mathcal{A}\}$ is called a *conditional probability* if for every $F \in \mathcal{A}$ it satisfies the equality

$$\int_F \mathbf{P}\{A|\mathcal{A}\} d\mathbf{P} = \mathbf{P}(A \cap F). \tag{0.16}$$

It follows from the theorem that $\mathbf{P}\{A|\mathcal{A}\}$ exists and is uniquely determined (mod $\mathbf{P}$) for each $A (A \in \mathcal{F})$.

Clearly, the equality (0.14) holds for each $\mathcal{A}$-measurable random variable if the variable $\eta\xi$ is integrable.

We shall now describe some properties of conditional mathematical expectations. We fix all random variables considered below to have finite mathematical expectations, and all equalities that we write are to be understood as equalities with probability 1.

11°. *If the random variable ξ is $\mathcal{A}$-measurable, then*

$$\mathbf{M}\{\xi|\mathcal{A}\} = \xi. \tag{0.17}$$

12°. *If ξ_1, ξ_2 take values of the same sign or have finite mathematical expectations, then*

$$\mathbf{M}\{\xi_1 + \xi_2|\mathcal{A}\} = \mathbf{M}\{\xi_1|\mathcal{A}\} + \mathbf{M}\{\xi_2|\mathcal{A}\}.$$

Corollary. *If $A \cap B = \varnothing$, then*

$$\mathbf{P}\{A \cup B|\mathcal{A}\} = \mathbf{P}\{A|\mathcal{A}\} + \mathbf{P}\{B|\mathcal{A}\} \ (\text{mod}\,\mathbf{P}).$$

13°. *If ξ_n, $n = 1, 2, \ldots$, is a monotonically nondecreasing sequence of nonnegative random variables, then*

$$\lim \mathbf{M}\{\xi_n|\mathcal{A}\} = \mathbf{M}\{\lim \xi_n|\mathcal{A}\}.$$

Corollary. *If A_n, $n = 1, 2, \ldots$, are mutually disjoint, then*

$$\mathbf{P}\{\bigcup_1^\infty A_n|\mathcal{A}\} = \sum_{n=1}^\infty \mathbf{P}\{A_n|\mathcal{A}\}. \tag{0.18}$$

14°. *If the mathematical expectations of ξ and $\alpha\xi$ are defined (here α is an $\mathcal{A}$-measurable random variable), then*

$$\mathbf{M}\{\alpha\xi|\mathcal{A}\} = \alpha\mathbf{M}\{\xi|\mathcal{A}\}. \tag{0.19}$$

Corollary. *If $F \in \mathcal{A}$, then*

$$\mathbf{P}\{A \cap F|\mathcal{A}\} = I_F\mathbf{P}\{A|\mathcal{A}\}. \tag{0.20}$$

Repeated calculation of the conditional mathematical expectation has the important and widely used property of "absorption."

15°. Theorem. *If $\mathcal{A}_1 \subset \mathcal{A}_2$, then*

$$\mathbf{M}\{\mathbf{M}\{\xi|\mathcal{A}_2\}|\mathcal{A}_1\} = \mathbf{M}\{\xi|\mathcal{A}_i\}.$$

Proof. It follows from $F \in \mathcal{A}_1$ that $F \in \mathcal{A}_2$, and thus

$$\int_F \mathbf{M}\{\mathbf{M}\{\xi|\mathcal{A}_2\}|\mathcal{A}_1\}d\mathbf{P}$$
$$= \int_F \mathbf{M}\{\xi|\mathcal{A}_2\}d\mathbf{P} = \int_F \xi d\mathbf{P} = \int_F \mathbf{M}\{\xi|\mathcal{A}_1\}d\mathbf{P}.$$

□

Consider some experiment described by the random element ζ, $\zeta = g(\omega)$, with values in $\{X, \mathcal{B}\}$. The conditional mathematical expectation $\mathbf{M}\{\xi|\zeta\}$ of a random variable ξ with respect to the random element ζ is just the average value of ξ for fixed ζ.

Definition. $\mathbf{M}\{\xi|\zeta\} = \mathbf{M}\{\xi|\mathcal{F}_\zeta\}$, where $\mathcal{F}_\zeta$ is the σ-algebra generated by the random element ζ.

Proceeding from the initial definition of the conditional mathematical expectation, we can find this definition to be equivalent to the following one:

$\mathbf{M}\{\xi|\zeta\}$ is the $\mathcal{F}_\zeta$-measurable random variable satisfying the relation

$$\int_{g^{-1}(B)} \mathbf{M}\{\xi|\zeta\} d\mathbf{P} = \int_{g^{-1}(B)} \xi d\mathbf{P} \tag{0.21}$$

for every $B \in \mathcal{B}$.

16°. Theorem. *The conditional mathematical expectation* $\mathbf{M}\{\xi|\zeta\}$ *is a* $\mathcal{B}$-*measurable function of* ζ, *i.e., one can find a real* $\mathcal{B}$-*measurable function* $h(x)$, $x \in X$, *such that* $\mathbf{M}\{\xi|\zeta\} = h(\zeta)$ *and* $(\forall B \in \mathcal{B})$

$$\int_B h(x)\mathbf{P}g^{-1}(dx) = \int_{g^{-1}} \xi d\mathbf{P}.$$

Let us describe some properties of conditional mathematical expectations with respect to random variables, which follow directly from the above:

(a) if $\xi = h(\zeta)$, where $h(x)$ is a $\mathcal{B}$-measurable function, then $\mathbf{M}\{\xi|\zeta\} = \xi$;

(b) if ζ_i are random elements in $\{X_i, \mathcal{B}_i\}$, $i = 1, 2$, and (ζ_1, ζ_2) is their direct product, then $\mathbf{M}\{\mathbf{M}\{\xi|(\zeta_1, \zeta_2)\}|\zeta_1\} = \mathbf{M}\{\xi|\zeta_1\}$.

Regular Conditional Probabilities

We set $\mathbf{P}\{A|\mathcal{A}\} = \mathbf{P}^{\mathcal{A}}(A, \omega)$. For every $A \in \mathcal{F}$, the conditional probability $\mathbf{P}^{\mathcal{A}}(A, \omega)$ is defined uniquely, however, with probability 1 only.

Definition. If there exists the function $p(A, \omega)$, $A \in \mathcal{F}$, $\omega \in \Omega$, such that:

(a) as a function of the set A, $p(A, \omega)$ is a probability measure for almost all ω;

(b) $p(A, \omega)$ is $\mathcal{A}$-measurable, and for fixed A,

$$p(A, \omega) = \mathbf{P}^{\mathcal{A}}(A, \omega) \pmod{\mathbf{P}},$$

then $p(A, \omega)$ is called a *regular conditional probability.*

We can give examples when regular conditional probabilities do not exist. But if they exist, conditional mathematical expectations can be expressed in terms of these probabilities by means of integration.

17°. Theorem. *If $p(A,\omega) = \mathbf{P}^{\mathcal{A}}(A,\omega)$ is a regular conditional probability, then*

$$\mathbf{M}\{\xi|\mathcal{A}\} = \int f(\omega')p(d\omega',\omega). \tag{0.22}$$

Since regular conditional probabilities do not exist in some cases, we introduce a modification of this notion that is sufficient for a series of problems arising in applications.

Let $\{X,\mathcal{B}\}$ be a measurable space, ζ be a random element in $\{X,\mathcal{B}\}$, and $\mathcal{A}$ be a σ-algebra $\mathcal{A} \subset \mathcal{F}$.

Definition. If there exists the function $Q(B,\omega)$ defined on $\mathcal{B} \times \Omega$ such that:

(a) it is $\mathcal{A}$-measurable for fixed $B \in \mathcal{B}$;

(b) for fixed ω, it is a probability measure on $\mathcal{B}$ (with probability 1);

(c) for every $B \in \mathcal{B}$, we have $Q(B,\omega) = \mathbf{P}\{(\zeta \in B)|\mathcal{A}\}$ (mod $\mathbf{P}$), then $Q(B,\omega)$ is called a *regular conditional distribution of the random element ζ with respect to the σ-algebra $\mathcal{A}$.*

Condition (c) is equivalent to the requirement that for every $F \in \mathcal{A}$,

$$\int_F Q(B,\omega)\mathbf{P}(d\omega) = \mathbf{P}(\{\xi \in B\} \cap F).$$

Theorem. *Let X be a complete separable metric space, $\mathcal{B}$ be a σ-algebra of Borel sets in X, and ζ be a random element in $\{X,\mathcal{B}\}$. Then ζ has the regular conditional distribution with respect to an arbitrary σ-algebra $\mathcal{A}$, $\mathcal{A} \subset \mathcal{F}$.*

Consider the random elements ζ_1 and ζ_2 in $\{Y_1,\mathcal{B}_1\}$ and $\{Y_2,\mathcal{B}_2\}$, respectively, where $\{Y_i,\mathcal{B}_i\}$ satisfy the conditions of this theorem. Let us write $Y^{(1,2)} = Y_1 \times Y_2$, $\mathcal{B}^{(1,2)} = \sigma\{\mathcal{B}_k, k = 1,2\}$. The sequence $\zeta^{(1,2)} = (\zeta_1,\zeta_2)$ can be regarded as a random element in $\{Y^{(1,2)},\mathcal{B}^{(1,2)}\}$ and $Y^{(1,2)}$ as a complete metric separable space. Denote by q_i a distribution of ζ_i ($i = 1,2$), by $q^{(1,2)}$ a distribution of $\zeta^{(1,2)}$, and by $q^{(2|1)}$ a regular conditional distribution of ζ_2 with respect to the σ-algebra $\mathcal{A}_{\zeta_1}$ generated by the random element ζ_1. Since $q^{(2|1)}$ is an $\mathcal{A}_{\zeta_1}$-measurable function, we have $q^{(2|1)}(B_2,\omega) = q(B_2,\zeta_1)$, where $B_2 \in \mathcal{B}_2$, and the function $q(B_2,y)$ is $\mathcal{B}_1$-measurable. The definition of conditional probabilities yields

$$\int_{g_1^{-1}(B_1)} q(B_2,\zeta_1)d\mathbf{P} = q^{(1,2)}(B_1 \times B_2),$$

where B_1 is an arbitrary set from $\mathcal{B}_1$ and $\zeta_1 = g_1(\omega)$. In accordance with the rule of change of variables, this equality can be written in the form

$$q^{(1,2)}(B_1 \times B_2) = \int_{B_1} q(B_2,y_1)q_1(dy_1)$$

or

$$q^{(1,2)}(B_1 \times B_2) = \int_{Y_1} \chi^{(1)}(B_1,y_1) \int_{Y_2} \chi^{(2)}(B_2,y_2)q(dy_2,y_1)q_1(dy_1),$$

where $\chi^{(i)}$ are indicators of sets in the space Y_i. The next theorem follows from the last formula.

19°. Theorem. *For arbitrary $\mathcal{B}^{(1,2)}$-measurable nonnegative functions $f(y_1, y_2)$,*

$$\int_{Y^{(1,2)}} f(y_1, y_2) dq^{(1,2)} = \int_{Y_1} \left(\int_{Y_2} f(y_1, y_2) q(dy_2, y_1) \right) q_1(dy_1). \tag{0.23}$$

Note that (0.23) also holds for functions with alternating signs (if one of the sides of (0.23) is meaningful).

Corollary.

$$\mathbf{M}\{f(\zeta_1, \zeta_2) | \mathcal{A}_{\zeta_1}\} = \int_{Y_2} f(\zeta_1, y_2) q(dy_2, \zeta_1). \tag{0.24}$$

This follows from (0.23).

The results obtained can be formulated in the following (more general) form. Let ζ_k be random elements in $\{Y_k, \mathcal{B}_k\}$, and Y_k be complete separable metric spaces. We put

$$Y^{(1,s)} = \prod_{k=1}^{s} Y_k, \qquad \mathcal{B}^{(1,s)} = \sigma\{\mathcal{B}_k, \quad k = 1, \ldots, s\},$$

$\eta_s = (\zeta_1, \zeta_2, \ldots, \zeta_s)$, q_k is a distribution of the element ζ_k in $\{Y_k, \mathcal{B}_k\}$, and $q^{(s)} = q^{(s)}(B_s, \zeta_1, \ldots, \zeta_{s-1})$ is a regular conditional distribution of the element ζ_s with respect to the σ-algebra $\mathcal{A}_{\eta_{s-1}} = \mathcal{A}(\zeta_1, \zeta_2, \ldots, \zeta_{s-1})$. By repeatedly using the relation (0.24), we obtain from

$$\mathbf{M}\{f | \mathcal{A}_{\zeta_1}\} = \mathbf{M}\{\ldots \{\mathbf{M}\{f | \mathcal{A}_{\eta_{n-1}}\} | \mathcal{A}_{\eta_{n-2}}\} \cdots | \mathcal{A}_{\eta_1}\}$$

the following equalities

$$\begin{aligned} &\mathbf{M}\{f(\zeta_1, \ldots, \zeta_n) | \mathcal{A}_{\zeta_1}\} \\ &\quad = \int_{Y_2} \cdots \int_{Y_{n-1}} \left(\int_{Y_n} f(\zeta_1, y_2, \ldots, y_n) q^{(n)}(dy_n, \zeta_1, y_2, \ldots, y_{n-1}) \right) \\ &\quad \times q^{(n-1)}(dy_{n-1}, \zeta_1, y_2, \ldots, y_{n-2}) \ldots q^{(2)}(dy_2, \zeta_1), \end{aligned} \tag{0.25}$$

$$\begin{aligned} &\mathbf{M}\{f(\zeta_1, \ldots, \zeta_n)\} \\ &\quad = \int_{Y_1} \cdots \int_{Y_{n-1}} \left(\int_{Y_n} f(y_1, \ldots, y_n) q^{(n)}(dy_n, y_1, \ldots, y_{n-1}) \right) \\ &\quad \times q^{(n-1)}(dy_{n-1}, y_1, \ldots, y_{n-2}) \ldots q^{(2)}(dy_2, y_1) q_1(dy_1). \end{aligned} \tag{0.26}$$

§0.4 Independence

Let $\{\Omega, \mathcal{F}, \mathbf{P}\}$ be the fixed probability space. $\mathcal{F}$-measurable subsets of Ω will be understood as events unless otherwise stipulated.

The events A and B are called ***independent***, if $\mathbf{P}(A \cap B) = \mathbf{P}(A)\mathbf{P}(B)$. In the case $\mathbf{P}(B) > 0$, this condition is equivalent to the following one: $\mathbf{P}(A|B) = \mathbf{P}(A)$. This definition immediately yields:

(a) Ω and A are independent for every A;

(b) if $\mathbf{P}(N) = 0$, then N and A are independent for an arbitrary A;

(c) if A and B_i $(i = 1, 2)$ are independent and $B_1 \supset B_2$, then A and $B_1 \backslash B_2$ are independent. In particular, A and $\Omega \backslash B_1$ are independent;

(d) if A and B_i $(i = 1, 2, \dots, n)$ are independent and, moreover, if $B_1, B_2, \dots, B_n$ are mutually disjoint, then A and $\bigcup_1^n B_i$ are also independent;

(e) A does not depend on A iff $\mathbf{P}(A) = 0$ or $\mathbf{P}(A) = 1$.

Let I be some set and $\{\mathcal{M}_i, i \in I\}$ be a set of classes of events.

Definition. Classes of events $\{\mathcal{M}_i, i \in I\}$ are called ***independent*** if for arbitrary pairwise nonequal $i_1, \dots, i_n$ $(i_k \in I)$ and arbitrary A_{i_k}, $A_{i_k} \in \mathcal{M}_{i_k}$, $k = 1, 2, \dots, n$,

$$\mathbf{P}(A_{i_1} \cap A_{i_2} \cap \dots \cap A_{i_n}) = \mathbf{P}(A_{i_1})\mathbf{P}(A_{i_2}) \dots \mathbf{P}(A_{i_n}).$$

Note that for an infinite set of classes of events, the definition of independence is equivalent to the following requirement: an arbitrary finite subset of classes of events must consist of independent classes of events.

A class of events will be called π-class if it is closed with respect to the operation of intersection of events (i.e., if $A \in \mathcal{M}$, $B \in \mathcal{M}$ implies $A \cap B \in \mathcal{M}$).

20°. Theorem. *Let $\{\mathcal{M}_i, i \in I\}$ be a collection of independent π-classes of events. Then the minimal σ-algebras $\sigma\{\mathcal{M}_i\}$, $i \in I$, are independent.*

Proof. We can restrict ourselves to the case of a finite number of classes $\mathcal{M}_1, \dots, \mathcal{M}_n$. It is sufficient to show that if one of the classes, e.g., $\mathcal{M}_1$, is replaced by $\sigma\{\mathcal{M}_1\}$, a new sequence of classes of events is also independent.

Let us denote by $\mathcal{A}$ a class of all events independent of $\mathcal{M}_2, \dots, \mathcal{M}_n$. By definition, $\mathcal{M}_1 \subset \mathcal{A}$, and $\mathcal{A}$ possesses the following properties: It is closed with respect to summation of a countable sequence of nonintersecting events and with respect to the operation of difference $B_2 \backslash B_1$ when $B_2 \supset B_1$. The next theorem now follows from the theorem on monotonic classes.

21°. Theorem. *Let $\{\mathcal{M}_i, i \in I\}$ be a set of independent π-classes of events, $I = I_1 \cup I_2$ $(I_i \cap I_2 = \varnothing)$, $\mathcal{B}_k = \sigma\{\mathcal{M}_i, i \in I_k\}$, $k = 1, 2$. Then $\mathcal{B}_1$ and $\mathcal{B}_2$ are independent.*

By virtue of Theorem 20°, we can restrict ourselves to the assumption that the $\mathcal{M}_i$ are σ-algebras. Consider the classes $\mathcal{A}_k$ $(k = 1, 2)$ consisting of all kinds of events like $A_{i_1} \cap A_{i_2} \cap \dots \cap A_{i_n}$, where n is an arbitrary integer, $i_r \in I_k$. They are closed with respect to intersections, $\mathcal{A}_k$ contains all $\mathcal{M}_i$, $i \in I_k$, and $\mathcal{A}_1$ and $\mathcal{A}_2$ are independent. By virtue of Theorem 20°, $\sigma\{\mathcal{A}_1\} = \sigma\{\mathcal{M}_i, i \in I_1\}$ and $\sigma\{\mathcal{A}_2\} = \sigma\{\mathcal{M}_i, i \in I_2\}$ are independent.

□

Corollary. *If I is split into an arbitrary collection of subsets such that $I = \bigcup_{k \in Q} I_k$, pairwise without common points, then the σ-algebras $\mathcal{B}_k = \sigma\{\mathcal{M}_i, i \in I_k\}$, $k \in Q$, are independent.*

Independent Random Variables

Let $\zeta_i = f_i(\omega)$ be a random element in $\{X_i, \mathcal{B}_i\}$, $i \in I$.

Definition. The random elements $\{\zeta_i, i \in I\}$ are called *independent* if for every n and arbitrary $B_k \in \mathcal{B}_{i_k}$, $i_k \in I$,

$$\mathbf{P}(\bigcap_{k=1}^{n}\{\zeta_{i_k} \in B_k\}) = \prod_{k=1}^{n} \mathbf{P}\{\zeta_{i_k} \in B_k\}.$$

More general is the definition of independence for the family of sets of random elements. Consider some family of sets $\{\zeta_i^\mu, i \in I_\mu\}$, $\mu \in M$, of random elements with values in $\{X_i^\mu, \mathcal{B}_i^\mu\}$.

Definition. The sets of random elements $\{\zeta_i^\mu, i \in I_\mu\}$, $\mu \in M$, are called *independent (mutually independent)* if the classes of events $\{\mathcal{M}_\mu, \mu \in M\}$ are independent, where $\mathcal{M}_\mu$ consists of all kinds of events like

$$\bigcap_{k=1}^{n}\{\zeta_{i_k}^\mu \in B_k^\mu\}, \quad n = 1, 2, \ldots, i_k \in I_\mu, \quad B_{i_k}^\mu \in \mathcal{B}_{i_k}^\mu.$$

Let $\sigma\{\zeta_i^\mu, i \in I_\mu\} = \mathcal{F}_\mu$ be a σ-algebra generated by the set of random elements ζ_i^μ, $i \in I_\mu$, i.e., the minimal σ-algebra with respect to which all random variables ζ_i^μ ($i \in I_\mu$, μ is fixed), are measurable.

22°. Theorem. *The sets of random elements $\{\zeta_i^\mu, i \in I_\mu\}$, $\mu \in M$, are independent iff the σ-algebras $\mathcal{F}_\mu$, $\mu \in M$, are independent.*

The proof follows from the fact that the classes of events introduced in the definition of independence of sets of random elements are π-classes, and from Theorem 20°.

Corollary. *Let $(\zeta_1^\mu, , \ldots, \zeta_{s_\mu}^\mu)$, $\mu \in M$, be a set of independent sequences of random elements, and let $g_\mu(x_1, \ldots, x_{s_\mu})$ be $\mathcal{B}_1 \otimes \cdots \otimes \mathcal{B}_{s_\mu}$-measurable functions ($\mu \in M$). Then the random variables*

$$\xi_\mu = g_\mu(\zeta_1^\mu, \ldots, \zeta_{s_\mu}^\mu), \quad \mu \in M,$$

are mutually independent.

Remark. The real random variables $\{\xi_\mu, \mu \in M\}$ are mutually independent iff for every n, arbitrary $\mu_1, \mu_2, \ldots, \mu_n$ from M and any real $a_1, \ldots, a_n$,

$$\mathbf{P}\{\xi_{\mu_1} < a_1, \ldots, \xi_{\mu_n} < a_n\} = \prod_{k=1}^{n} \mathbf{P}\{\xi_{\mu_k} < a_k\}.$$

Let ζ_k, $k = 1, 2, \ldots, n$ be a sequence of independent random elements (on $\{X_k, \mathcal{B}_k\}$, respectively), q_k be a distribution of ζ_k on $\mathcal{B}_k$, and $q^{(1,n)}$ be a joint distribution of the sequence $(\zeta_1, \ldots, \zeta_n)$ in

$$\left\{\prod_{k=1}^{n} X_k, \mathcal{B}_1 \otimes \cdots \otimes \mathcal{B}_n\}\right\}.$$

The definition of independence yields

$$q^{(1,n)}(B_1 \times B_2 \times \cdots \times B_n) = \prod_{k=1}^{n} q_k(B_k), \quad B_k \in \mathcal{B}_k. \tag{0.27}$$

The inverse statement is obvious: If (0.27) holds for all $B_k \in \mathcal{B}_k$, then the variables $\{\zeta_k, k = 1, \ldots, n\}$ are independent.

Let $g(x_1, x_2)$ be a $\sigma\{\mathcal{B}_i, i = 1, 2\}$-measurable function, ζ_1 and ζ_2 be independent random elements, and $\mathbf{M}g(\zeta_1, \zeta_2) < \infty$. Then the rule of change of variables and Fubini's theorem yields that $\mathbf{M}g(x_1, \zeta_2)$ is a $\mathcal{B}_i$-measurable function, finite for q_1-almost all x_1, and

$$\mathbf{M}g(\zeta_1, \zeta_2) = \int_{X_1} q_1(dx_1) \int_{X_2} g(x_1, x_2) q_2(dx_2). \tag{0.28}$$

As a corollary of (0.28), the following formula

$$\mathbf{M}g(\zeta_1)g(\zeta_2) = \mathbf{M}g(\zeta_1)\mathbf{M}g(\zeta_2) \tag{0.29}$$

is valid when $\mathbf{M}g(\zeta_k)$ are finite.

The next proposition can be regarded as strengthening the preceding one.

23°. Theorem. *If the random variable ξ with a finite mathematical expectation and the σ-algebra $\mathcal{A}$ are independent, then*

$$\mathbf{M}\{\xi | \mathcal{A}\} = \mathbf{M}\xi.$$

Proof. Independence of the random variable ξ and the σ-algebra $\mathcal{A}$ means that the σ-algebras $\mathcal{A}$ and $\mathcal{A}_\xi = \sigma\{\xi\}$ are independent. Thus, for every $F \in \mathcal{A}$, the random variables ξ and I_F are independent. Therefore,

$$\int_F \xi d\mathbf{P} = \mathbf{M}I_F\mathbf{M}\xi = \int_F (\mathbf{M}\xi) d\mathbf{P}.$$

Since $\mathbf{M}\xi$ is constant, it is $\mathcal{A}$-measurable; hence,

$$\mathbf{M}\xi = \mathbf{M}\{\xi | \mathcal{A}\}.$$

□

Law "0 or 1"

Let A_n, $n = 1, 2, \ldots$, be some sequence of events.

24°. Borel–Cantelli theorem. *If $\sum_{n=1}^{\infty} \mathbf{P}(A_k) < \infty$, then the event $\overline{\lim} A_n = \{A_n$ happening infinitely frequently (h.i.f.)$\}$ has probability 0. Moreover, if the events A_n, $n = 1, 2, \ldots$, are independent, the probability of the event $\overline{\lim} A_n$ is equal to 0 (if the series $\sum_{n=1}^{\infty} \mathbf{P}(A_n)$ is convergent) or to 1 (if this series is divergent).*

Proof. Since $\overline{\lim} A_n = \bigcap_{n=1}^{\infty} \bigcup_{k=n}^{\infty} A_k$, we have

$$\mathbf{P}\{A_n \text{ h.i.f. }\} = \lim_{n\to\infty} \mathbf{P}(\bigcup_{k=n}^{\infty} A_k) \le \lim_{n\to\infty} \sum_{k=n}^{\infty} \mathbf{P}(A_k) = 0,$$

proving the first part of the statement.

Suppose now that the events A_n are independent. We must only prove that if $\sum_{n=1}^{\infty} \mathbf{P}(A_n) = \infty$, then $\mathbf{P}\{\overline{\lim} A_n\} = 1$. Let $A^* = \overline{\lim} A_n$; then for $\Omega \backslash A^* = \bigcup_{n=1}^{\infty} \bigcap_{k=n}^{\infty} (\Omega \backslash A_k)$, we get

$$\mathbf{P}(\Omega \backslash A^*) = \lim_{n\to\infty} \mathbf{P}(\bigcap_{k=n}^{\infty} (\Omega \backslash A_k)) \lim_{n\to\infty} \prod_{k=n}^{\infty} \mathbf{P}(\Omega \backslash A_k)$$

$$= \lim_{n\to\infty} \prod_{k=n}^{\infty} (1 - \mathbf{P}(A_k)) = 0,$$

because the series $\sum_{n=1}^{\infty} \mathbf{P}(A_k)$ diverges.

□

Let $\mathcal{B}_k = \sigma(\mathcal{F}_i, j = k, k+1, \ldots)$, then $\mathcal{B}_k$ forms a monotonically decreasing sequence of σ-algebras. Their intersection $\mathcal{B} = \bigcap_{k=1}^{\infty} \mathcal{B}_k$ is also a σ-algebra.

By definition, we set

$$\mathcal{B} = \overline{\lim} \mathcal{F}_n = \bigcap_{k=1}^{\infty} \sigma\{\mathcal{F}_j, j = k, k+1, \ldots\}.$$

Obviously, the σ-algebra $\overline{\lim} \mathcal{F}_n$ will not be changed by replacement of any finite number of σ-algebras $\mathcal{F}_1, \mathcal{F}_2, \ldots, \mathcal{F}_n$ by the others.

25°. Theorem (general Kolmogorov law "0 or 1"). *If $\mathcal{F}_n$, $n = 1, 2, \ldots$, are mutually independent σ-algebras, then each event from $\overline{\lim} \mathcal{F}_n$ has probability 0 or 1.*

In fact, let $A \in \overline{\lim} \mathcal{F}_n$. Then $A \in \mathcal{B}_k$ for any k; hence, A and $\sigma\{\mathcal{F}_1, \ldots, \mathcal{F}_{k-1}\}$ are independent. Therefore, A and $\sigma\{\mathcal{F}_1, \ldots, \mathcal{F}_m, \ldots\}$ are independent. Since $A \in \sigma\{\mathcal{F}_k, k = 1, 2, \ldots\}$, A is independent of A. But this is possible only if $\mathbf{P}(A) = 0$ or $\mathbf{P}(A) = 1$.

□

CHAPTER 1

SUMS OF INDEPENDENT RANDOM VARIABLES

§1.1 Main Inequalities

In this section, we derive estimates for the probabilities

$$\mathbf{P}\{\sup_{k\in N} \zeta_k \geq a\}, \qquad \mathbf{P}\{\sup_{k\in N} |\zeta_k| \geq a\},$$

where N is either an interval $[0, n]$ or Z_+, and $\zeta_0 = 0$, $\zeta_k = \sum_{i=1}^{k} \xi_i$, ($\xi_k$, $k = 1, 2, \dots$) is a sequence of independent random variables. $\mathcal{F}_k$ is a σ-algebra generated by the variables $\xi_1, \dots, \xi_k$.

Sums of a Random Number of Random Variables

Let us consider a variable ζ_ν, where ν is a random variable with values in Z_+ for which the event $\{\nu = k\} \in \mathcal{F}_k$ ($\mathcal{F}_0$ is the trivial σ-algebra); ν is called a *stopping moment (s.m.) with respect to the sequence of σ-algebras* ($\mathcal{F}_k$, $k \geq 0$).

1°. *Let* $\mathbf{M}\xi_i = 0$, $\nu \leq n$; *then*

$$\mathbf{M}\zeta_\nu = 0.$$

In fact,

$$\begin{aligned}\mathbf{M}\zeta_\nu &= \mathbf{M}\sum_1^n \xi_k I_{\{\nu\geq k\}} = \mathbf{M}\sum_1^n \xi_k(1 - \sum_{i<k} I_{\{\nu=1\}})\\ &= \sum_{i<k\leq N} \mathbf{M}\xi_k I_{\{\nu=i\}} = \sum_{i<k\leq N} \mathbf{M}\xi_k \mathbf{M} I_{\{\nu=i\}} = 0\end{aligned}$$

(ξ_k does not depend on $\mathcal{F}_i$ for $i < k$).

□

2°. Let $\mathcal{F}_\nu$ denote the σ-algebra of the events A, $A \in \bigvee_k \mathcal{F}_k$, for which $\{\nu = k\} \cap A \in \mathcal{F}_k$.

If $B \in \mathcal{F}_\nu$, then under the conditions of 1°,

$$\mathbf{M}(\zeta_n - \zeta_\nu)I_B = 0. \tag{1.1}$$

In fact,

$$\mathbf{M}(\zeta_n - \zeta_\nu)I_B = \mathbf{M}\sum_{k=1}^{n}(\zeta_n - \zeta_k)I_{\{\nu=k\}\cap B}$$
$$= \sum_{k=1}^{n}\mathbf{M}(\zeta_n - \zeta_k)\mathbf{M}_{(\{\nu=k\}\cap B)} = 0.$$

□

Note that (1.1) means that

$$\mathbf{M}(\zeta_n - \zeta_\nu|\mathcal{F}_\nu) = 0. \tag{1.2}$$

3°. *Let* $\mathbf{M}\xi_i = 0$, $\mathbf{M}|\xi_i| \le c$, $\mathbf{M}\nu < \infty$. *Then* $\mathbf{M}\zeta_\nu = 0$.

It follows from 1° that $\mathbf{M}\zeta_{\nu\wedge n} = 0$, since $\nu \wedge n$ is the s.m., and $\nu \wedge n \le n$. Furthermore, $|\zeta_{\nu\wedge n}| \le \sum_1^\nu |\xi_k|$. We shall now show that $\mathbf{M}\sum_1^\nu |\xi_k| < \infty$, and therefore, we can pass to the limit as $n \to \infty$ in the equality $\mathbf{M}\zeta_{\nu\wedge n} = 0$. In fact,

$$\mathbf{M}\sum_1^\nu |\xi_k| = \lim_{n\to\infty}\mathbf{M}\sum_1^{\nu\wedge n}|\xi_k|$$
$$= \lim_{n\to\infty}\mathbf{M}\sum_{k=1}^{n}|\xi_k|\left(1 - \sum_{i<k} I_{\{\nu=i\}}\right)$$
$$= \lim_{n\to\infty}\sum_{k=1}^{n}\mathbf{M}|\xi_k|\left(1 - \sum_{i<k}\mathbf{M}I_{\{\nu=i\}}\right)$$
$$\le c\lim_{n\to\infty}\sum_{k=1}^{n}\mathbf{P}\{\nu \ge k\} = c\mathbf{M}\nu < \infty.$$

□

4°. *Let ξ_i be equally distributed and ν be the s.m.;*
(a) *if* $\mathbf{M}\xi_i = a$, $\mathbf{M}\nu < \infty$, *then*

$$\mathbf{M}\zeta_\nu = a\mathbf{M}\nu; \tag{1.3}$$

(b) *if for some λ we have* $\mathbf{M}e^{\lambda\xi_i} = f(\lambda) < \infty$, $\nu \le n$, *then*

$$\mathbf{M}e^{\lambda\zeta_\nu}f(\lambda)^{-\nu} = 1. \tag{1.4}$$

This is *Wald's formula.*

Statement (a) follows from 1° if we let $\xi_i' = \xi_i - a$, $\zeta_k' = \sum_1^k \xi_i'$. To prove (b), we introduce the variables $\eta_k = e^{\lambda\xi_k} f(\lambda)^{-1} - 1$. Then $\mathbf{M}\eta_k = 0$. It needs to be shown that $\mathbf{M}\prod_{k=1}^{\nu}(1+\eta_k) = 1$. We have

$$\prod_{k=1}^{\nu}(1+\eta_k) = \prod_{k=1}^{n}(1+\varepsilon_k),$$

where $\varepsilon_k = \eta_k I_{\{\nu \geq k\}} = \eta_k(1 - \sum_{i<k} I_{\{\nu=i\}})$. Whatever $k_1 < k_2 < \cdots < k_r$ are taken, we have

$$\mathbf{M}\varepsilon_{k_1}\varepsilon_{k_2}\ldots\varepsilon_{k_r} = \mathbf{M}\varepsilon_{k_1}\varepsilon_{k_2}\ldots\varepsilon_{k_{r-1}}\left(1 - \sum_{i<k_{r-1}} I_{\{\nu=i\}}\right)\eta_{k_r} = 0,$$

because η_{k_r} is independent of other factors under the sign of mathematical expectation. Thus,

$$\mathbf{M}\prod_{k=1}^{n}(1+\varepsilon_k) = 1 + \sum_{r\geq 1}\mathbf{M}\sum_{k_1<\cdots<k_r\leq n}\varepsilon_{k_1}\ldots\varepsilon_{k_r} = 1.$$

□

Remark. Suppose that ν is the s.m. for which the variable $e^{\lambda\zeta_{\nu\wedge n}} f(\lambda)^{-\nu\wedge n}$ is uniformly integrable. Then

$$\mathbf{M}e^{\lambda\zeta_\nu} f(\lambda)^{-\nu} = 1.$$

Kolmogorov's Inequalities and Their Generalizations

5°. *Let* $\mathbf{M}\xi_i = 0$, $\mathbf{D}\xi_i < \infty$. *Then*

$$\mathbf{P}\{\sup_{k\geq n}|\zeta_k| \geq a\} \leq \frac{1}{a^2}\mathbf{M}\zeta_n^2. \tag{1.5}$$

Let $\nu = k < n$ if $|\zeta_i| < a$, $i < k$, and $|\zeta_k| \geq a$; let $\nu = n$ if $|\zeta_i| < a$, $i < n$. Denote $B = \{\sup_{k\leq n}|\zeta_k| \geq a\}$, $B \in \mathcal{F}_\nu$. We have

$$\begin{aligned}\mathbf{M}\zeta_n^2 \geq \mathbf{M}\zeta_n^2 I_B &= \mathbf{M}(\zeta_n - \zeta_\nu)^2 I_B + \mathbf{M}(\zeta_n - \zeta_\nu)I_B\zeta_\nu + \mathbf{M}\zeta_\nu^2 I_B \\ &\geq \mathbf{M}\zeta_\nu^2 I_B \geq a^2\mathbf{P}(B),\end{aligned}$$

since

$$\mathbf{M}(\zeta_n - \zeta_\nu)I_B\zeta_\nu = \mathbf{M}[\mathbf{M}(\zeta_n - \zeta_\nu/\mathcal{F}_\nu)]I_B\zeta_\nu = 0, \qquad \zeta_\nu^2 I_B \geq a^2 I_B.$$

□

6°. *Let* $\mathbf{M}\xi_i = 0$, $\mathbf{P}\{|\xi_i| \le c\} = 1$. *Then*

$$\mathbf{M}\zeta_n^2 \le \frac{(a+c)^2}{1 - \mathbf{P}\{\sup_{1\le k\le n} |\zeta_k| \ge a\}}, \tag{1.6}$$

if a *is such that the variable on the r.h.s. is defined.*

In fact,

$$\zeta_n^2 \le a^2(1 - I_B) + I_B(\zeta_n - \zeta_\nu)^2 + I_B(\zeta_n - \zeta_\nu)\zeta_\nu + I_B\zeta_\nu^2,$$
$$D\zeta_n \le a^2(1 - \mathbf{M}I_B) + (a+c)^2\mathbf{M}I_B + \mathbf{M}I_B(\zeta_n - \zeta_\nu)^2$$
$$\le (a+c)^2 + \mathbf{M}I_B(\zeta_n - \zeta_\nu)^2.$$

But

$$\mathbf{M}I_B(\zeta_n - \zeta_\nu)^2 = \sum_{k=1}^{n} \mathbf{M}I_{B\cap\{\nu=k\}}(\zeta_n - \zeta_\nu)^2$$
$$= \sum_{k=1}^{n} \mathbf{M}I_{B\cap\{\nu=k\}}(\zeta_n - \zeta_k)^2 \le \mathbf{D}\zeta_n \cdot \mathbf{P}(B).$$

Hence,

$$\mathbf{D}\zeta_n \le (a+c)^2 + \mathbf{P}(B)\mathbf{D}\zeta_n.$$

□

Inequalities (1.5) and (1.6) are called *Kolmogorov's inequalities.* Other inequalities in which $\mathbf{P}\{\sup \zeta_n \ge a\}$ or $\mathbf{P}\{\sup_k |\zeta_k| \ge a\}$ are estimated are called *generalized Kolmogorov's inequalities.*

7°. *Suppose that for every* $k > n$, *some* α *and* $x \ge 0$,

$$\mathbf{P}\{\zeta_n - \zeta_k > x\} \le \alpha < 1.$$

Then

$$\mathbf{P}\{\sup_{k\le n} \zeta_k > a + x\} \le \frac{1}{1-\alpha}\mathbf{P}\{\zeta_n > a\}. \tag{1.7}$$

Proof. Let $\nu = k < n$ if $\zeta_1 < a + x, \dots, \zeta_{k-1} < a + x$, $\zeta_k > a + x$; and let $\nu = n$ in other cases, $B_+ = \{\sup_{k\le n} \zeta_k > a + x\}$. Then

$$\mathbf{P}\{B_+\} = \mathbf{P}\{B_+ \cap \{\zeta_n \le a\}\} + \mathbf{P}\{\{\zeta_n > a\} \cap B_+\}$$
$$\le \mathbf{P}\{\zeta_n > a\} + \sum_{k=1}^{n} \mathbf{P}\{B_+ \cap \{\nu = k\} \cap \{\zeta_n \le a\}\}$$
$$\le \mathbf{P}\{\zeta_n > a\} + \sum_{k=1}^{n} \mathbf{P}\{B_+ \cap \{\nu = k\} \cap \{\zeta_n - \zeta_k > x\}\}$$
$$= \mathbf{P}\{\zeta_n > a\} + \sum_{k=1}^{n} \mathbf{P}\{B_+ \cap \{\nu = k\}\}\mathbf{P}\{\zeta_n - \zeta_k > x\}$$
$$\le \mathbf{P}\{\zeta_n > a\} + \alpha\mathbf{P}(B_+).$$

This implies the inequality (1.7).

□

8°. *Let ξ_i be symmetric variables (i.e., ξ_i and $-\xi_i$ are equally distributed). Then*

$$\mathbf{P}\{\sup_{k\le n}\zeta_k > a\} \le 2\mathbf{P}\{\zeta_n > a\}. \tag{1.8}$$

Indeed, the quantity $\zeta_n-\zeta_k$ is also symmetric, and thus $\mathbf{P}\{\zeta_n-\zeta_k > 0\} \le \frac{1}{2}$. Therefore, the conditions of 7° are satisfied for $x = 0$ and $\alpha = \frac{1}{2}$, and (1.8) follows from (1.7).

□

9°. *Suppose that for every $k \le n$, some α and $x \ge 0$,*

$$\mathbf{P}\{|\zeta_n - \zeta_k| > x\} \le \alpha < 1.$$

Then

$$\mathbf{P}\{\sup_{k\le n}|\zeta_k| > a + x\} \le \frac{1}{1-\alpha}\mathbf{P}\{|\zeta_n| > a\}. \tag{1.9}$$

In particular, for symmetric ξ_i,

$$\mathbf{P}\{\sup_{k\le n}|\zeta_k| > a\} \le 2\mathbf{P}\{|\zeta_n| > a\}. \tag{1.10}$$

These inequalities can be obtained from (1.7) [or (1.8)] if one takes into account that

$$\mathbf{P}\{\sup_{k\le n}|\zeta_k| > a\} \le \mathbf{P}\{\sup_{k\le n}\zeta_k > a\} + \mathbf{P}\{\sup_{k\le n}(-\zeta_k) > a\}.$$

In the inequalities (1.9) and (1.10), the sign $>$ under the probability sign can be replaced by $\ge$. For example,

$$\mathbf{P}\{\sup_{k\le n}|\zeta_k| \ge a\} \le 2\mathbf{P}\{|\zeta_n| \ge a\}.$$

These inequalities are obtained by the limit transition on a.

Estimates for Moments of Sums of Bounded Variables

10°. *Let ξ_i be symmetric variables and* $\mathbf{P}\{|\xi_i| \le c\} = 1$ *for some* $c > 0$. *Then*

$$\mathbf{P}\{|\zeta_n| \ge rx + (r-1)c\} \le (2\mathbf{P}\{|\zeta_n| > x\})^r. \tag{1.11}$$

In fact, if $\nu = k < n$ when $|\zeta_k| \ge (r-1)x+(r-2)c$ and $|\zeta_i| < (r-1)x+(r-2)c$ for $i < k$, and if $\nu = n$ when $|\zeta_i| < (r-1)x + (r-2)c$ for $i < n$, then

$$\mathbf{P}\{|\zeta_n| \ge rx + (r-1)c\} = \sum_{k=1}^{n} \mathbf{P}\{\{|\zeta_n| \ge rx + (r-1)c\} \cap \{\nu = k\}\}.$$

Clearly, $|\zeta_\nu| \le (r-1)x + (r-1)c$ and $|\zeta_n - \zeta_\nu| > x$. Therefore,

$$\mathbf{P}\{|\zeta_n| \ge rx + (r-1)c\} \le \sum_{k=1}^{n-1} \mathbf{P}\{\nu = k\}\mathbf{P}\{|\zeta_n - \zeta_k| > x\}.$$

But

$$\begin{aligned}
&\mathbf{P}\{|\zeta_n - \zeta_k| > x\} \\
&\quad \le \mathbf{P}\{\zeta_n - \zeta_k > x\} \cdot 2\mathbf{P}\{\zeta_k \ge 0\} + \mathbf{P}\{\zeta_n - \zeta_k > -x\} \cdot 2\mathbf{P}\{\zeta_k \le 0\} \\
&\quad \le 2\mathbf{P}\{\zeta_n > x\} + 2\mathbf{P}\{\zeta_n < -x\} = 2\mathbf{P}\{|\zeta_n| > x\}.
\end{aligned}$$

Hence,

$$\begin{aligned}
\mathbf{P}\{|\zeta_n| \ge rx + (r-1)c\} &\le 2\mathbf{P}\{|\zeta_n| > x\}\mathbf{P}\{\nu < n\} \\
&\le 2\mathbf{P}\{|\zeta_n| \ge x\}\mathbf{P}\{|\zeta_n| \ge (r-1)x + (r-2)c\}.
\end{aligned}$$

This directly yields (1.11).

□

11°. *If α and x are such that*

$$2e^{\alpha(x+c)}\mathbf{P}\{|\zeta_n| \ge x\} < 1, \tag{1.12}$$

then under the conditions of 10°, the following inequality is valid

$$\mathbf{M}\exp\{\alpha|\zeta_n|\} \le \frac{e^{\alpha x}}{1 - 2e^{\alpha(x+c)}\mathbf{P}\{|\zeta_n| \ge x\}}. \tag{1.13}$$

It is a consequence of the estimate

$$\mathbf{M}\exp\{\alpha|\zeta_n|\} \le \sum_{r=1}^{\infty} e^{\alpha(rx+(r-1)c)}(2\mathbf{P}\{|\zeta_n| \ge x\})^{r-1}$$

[which follows from (1.11)].

□

12°. Theorem. *Let ξ_k be such that $\mathbf{P}\{|\xi_i| \le c\} = 1$ for some $c > 0$. If α and x are chosen so that*

$$4e^{2\alpha(x+c)}\mathbf{P}\{|\zeta_n| \ge x\} < 1, \tag{1.14}$$

then

$$\mathbf{M}\exp\{\alpha|\zeta_n|\} \le \frac{e^{3\alpha x}}{(1 - 4e^{2\alpha(x+c)}\mathbf{P}\{|\zeta_n| \ge x\})(1 - \mathbf{P}\{|\zeta_n| \ge x\})}. \tag{1.15}$$

Proof. We introduce the variables ξ_i' such that ξ_i and ξ_i' have the same distributions, and $\xi_1, \xi_1', \xi_2, \xi_2', \dots, \xi_n, \xi_n', \dots$ are independent. We also let $\tilde{\xi}_i = \xi_i - \xi_i'$. The variables ξ_i are symmetric and mutually independent, so $\mathbf{P}\{|\tilde{\xi}_i| \le 2c\} = 1$.

Let $\tilde{\zeta}_n = \sum_1^n \tilde{\xi}_i$; suppose α is such that

$$2e^{\alpha(2x+2c)}\mathbf{P}\{|\tilde{\zeta}_n| \ge 2x\} < 1. \tag{1.16}$$

Then

$$\mathbf{M}\exp\{\alpha|\tilde{\zeta}_n|\} \le \frac{e^{2\alpha x}}{1 - 2e^{2\alpha(x+c)}\mathbf{P}\{|\tilde{\zeta}_n| \ge 2x\}}. \tag{1.17}$$

However,

$$\begin{aligned}\mathbf{P}\{|\tilde{\zeta}_n| \ge 2x\} &\le \mathbf{P}\{|\zeta_n| + |\zeta_n'| \ge 2x\} \\ &\le \mathbf{P}\{|\zeta_n| \ge x\} + \mathbf{P}\{|\zeta_n'| \ge x\} = 2\mathbf{P}\{|\zeta_n| \ge x\}.\end{aligned}$$

Therefore, (1.16) follows from (1.14), and the inequality (1.17) is also satisfied. The r.h.s. of (1.17) can be replaced by the greater variable

$$\frac{e^{2\alpha x}}{1 - 4e^{2\alpha(x+c)}\mathbf{P}\{|\zeta_n| \ge x\}}$$

Note finally that $|\zeta_n| - |\zeta_n'| \le |\tilde{\zeta}_n|$, so

$$\mathbf{M}e^{\alpha|\zeta_n|}\mathbf{M}e^{-\alpha|\zeta_n'|} \le \frac{e^{2\alpha x}}{1 - 4e^{2\alpha(x+c)}\mathbf{P}\{|\zeta_n| \ge x\}};$$

and in addition,

$$\mathbf{M}e^{-\alpha|\zeta_n'|} \ge e^{-\alpha x}(1 - \mathbf{P}\{|\zeta_n| \ge x\}).$$

The two last inequalities imply (1.15).

□

13°. *Let* $\mathbf{P}\{|\zeta_n| \geq x\} \leq 1/(8e)$, $\mathbf{P}\{|\xi_i| \leq c\} = 1$. *Then for all* $m > 0$,

$$\mathbf{M}|\zeta_n|^m \leq m!(2x + 2c)^m H, \tag{1.18}$$

where H *is some absolute constant.*

Putting $\alpha = \frac{1}{2(x+c)}$ and making use of (1.15), we get

$$\mathbf{M}\exp\{\alpha|\zeta_n|\} \leq \frac{e^{\frac{3x}{2(x+c)}}}{(1-1/2)(1-1/(8e))} \leq \frac{2e^{3/2}}{1-1/(8e)} = H.$$

Therefore, $\mathbf{M}\frac{\alpha^m|\zeta_n|^m}{m!} \leq H$, and this yields (1.18).

□

14°. Theorem. *Let* $\mathbf{P}\{|\xi_i| \leq c\} = 1$, *and for all* $a \leq k < l \leq n$,

$$\mathbf{P}\{|\zeta_l - \zeta_m| \geq \frac{x}{2}\} \leq \frac{1}{1+2e}. \tag{1.19}$$

Then

$$\mathbf{M}\sup_{k \leq n}|\zeta_k|^m \leq 2cm!(2x+2c)^m. \tag{1.20}$$

Proof. Let $\nu = k < n$ if $|\zeta_i| < 2rx+(r-1)c$ for $i < k$ and $|\zeta_k| \geq 2rx+(r-1)c$, and let $\nu = n$ in other cases. The event $\{\sup_k |\zeta_k| \geq 2(r+1)x + rc\}$ implies the next one, $\{\nu < n\} \cap \{\sup_{\nu < l < n} |\zeta_l - \zeta_\nu| \geq 2x\}$ (since $|\zeta_\nu| \leq 2rx + rc$). Thus,

$$\begin{aligned}
\mathbf{P}\{\sup_k |\zeta_k| \geq 2(r+1)x + rc\} &\leq \sum_{k=1}^{n-1} \mathbf{P}\{\nu = k\}\mathbf{P}\{\sup_{k<l<n} |\zeta_l - \zeta_k| \geq 2x\} \\
&\leq \sum_{k=1}^{n-1} \mathbf{P}\{\nu = k\}\mathbf{P}\{\sup_{l<n} |\zeta_l| \geq x\} \\
&= \mathbf{P}\{\sup_{l<n} |\zeta_l| \geq x\}\mathbf{P}\{\sup_{k<n} |\zeta_k| \geq 2rx + (r-1)c\}.
\end{aligned}$$

Therefore,

$$\mathbf{P}\{\sup_{k\leq n} |\zeta_k| \geq 2(r+1)x + rc\} \leq \mathbf{P}\{\sup_{k\leq n} |\zeta_k| \geq x\}^{r+1},$$

$$\mathbf{M}e^{\alpha \sup_{k\leq n}|\zeta_k|} \leq \frac{e^{2\alpha x}}{1 - e^{2\alpha(x+c)}\mathbf{P}\{\sup_{k\leq n} |\zeta_k| \geq x\}},$$

if only the denominator on the r.h.s. is positive.

But by virtue of 9°,

$$\mathbf{P}\{\sup_{k\leq n} |\zeta_k| \geq x\} \leq \frac{1}{1 - \frac{1}{2e+1}} \cdot \mathbf{P}\{|\zeta_n| \geq \frac{x}{2}\} \leq \frac{1}{2e}.$$

Let $\alpha = \frac{1}{2(x+c)}$. Then

$$\mathbf{M} e^{\frac{1}{2(x+c)}\sup_{k\le n}|\zeta_k|} \le 2e.$$

This yields (1.20). □

Convergence of Series of Independent Random Variables

As an application of the above-mentioned inequalities, let us consider conditions of the existence (with probability 1) of the limit $\lim_{n\to\infty}\zeta_n$, i.e., the conditions of convergence (with probability 1) of the series $\sum_{i=1}^{\infty}\xi_i$. All these results belong to Kolmogorov.

15°. *Convergence of the series $\sum_{i=1}^{\infty}\mathbf{M}\xi_i$ and $\sum_{i=1}^{\infty}\mathbf{D}\xi_i$ is the sufficient condition for convergence of the series $\sum_{i=1}^{\infty}\xi_i$.*

Obviously, it is sufficient to prove the convergence of the series $\sum_{i=1}^{\infty}(\xi_i - \mathbf{M}\xi_i)$. Hence, one can assume that $\mathbf{M}\xi_i = 0$. Suppose that

$$\eta_n = \sup_{n\le k<l}|\zeta_l - \zeta_k|;$$

η_n is a decreasing sequence. The convergence of the series $\sum_{i=1}^{\infty}\xi_i$ is equivalent to the condition $\eta_n \to 0$ with probability 1, i.e., to the condition

$$\lim_{n\to\infty}\mathbf{P}\{\eta_n > \varepsilon\} = 0 \tag{1.21}$$

for all $\varepsilon > 0$. We have

$$\begin{aligned}\mathbf{P}\{\eta_n > 2\varepsilon\} &\le \mathbf{P}\{\sup_{k\le n}|\zeta_k - \zeta_n| > \varepsilon\}\\ &= \lim_{m\to\infty}\mathbf{P}\{\sup_{n<k\le m}|\zeta_k - \zeta_n| > \varepsilon\}\\ &\le \lim_{m\to\infty}\frac{1}{\varepsilon^2}\sum_{i=n+1}^{m}\mathbf{D}\xi_i = \frac{1}{\varepsilon^2}\sum_{i=n+1}^{\infty}\mathbf{D}\xi_i.\end{aligned}$$

Condition (1.21) is thus satisfied. □

16°. Lemma. *Let $\mathbf{P}\{|\xi_i| \le c\} = 1$, and let the sequence ζ_n be bounded in probability, i.e., $\lim_{x\to\infty}\sup_n \mathbf{P}\{|\zeta_n| > x\} = 0$. Then*

(a) *$\sum_{i=1}^{\infty}\mathbf{D}\xi_i < \infty$, and $\mathbf{M}\zeta_n$ is a bounded sequence;*

(b) *if the series* $\sum \xi_i$ *converges, the series* $\sum \mathbf{D}\xi_i$ *and* $\sum \mathbf{M}\xi_i$ *are also convergent.*

Proof. One can choose $x \geq 0$ such that for all n, $\mathbf{P}\{|\zeta_n| \geq x\} \leq 1/(8c)$. Then by virtue of (1.18), $\mathbf{M}\zeta_n \leq q$, where q is independent of n. Hence, $\sum_1^n \mathbf{D}\xi_i + (\mathbf{M}\zeta_n)^2 \leq q$, and this yields statement (a).

If the series $\sum \xi_i$ converges, then the convergence of the series $\sum \mathbf{D}\xi_i$ follows from (a), whereas the convergence of the series $\sum(\xi_i - \mathbf{M}\xi_i)$ arises from 15°. Consequently, the series $\sum \mathbf{M}\xi_i$ is also convergent.

□

17°. Theorem on three series (Kolmogorov). *Let* $\xi_i^c = \xi_i I_{\{|\xi_i| \leq c\}}$. *Then the convergence of the series*

$$\sum_{i=1}^{\infty} \mathbf{P}\{|\xi_i| > c\}, \qquad \sum_{i=1}^{\infty} \mathbf{M}\xi_i^c, \qquad \sum_{i=1}^{\infty} \mathbf{D}\xi_i^c, \tag{1.22}$$

for all c>0 is the necessary condition, and their convergence for some c>0 is the sufficient condition for the convergence of the series $\sum_{i=1}^{\infty} \xi_i$ *with probability 1.*

Proof. *Sufficiency.* Since the events $\{|\xi_i| > c\}$ are independent, the convergence of the first series (1.22) and the Borel–Cantelli theorem imply that $|\xi_i| \leq c$ with probability 1 for sufficiently large i. Thus, the series $\sum_{i=1}^{\infty}(\xi_i - \xi_i^c)$ converges with probability 1. The convergence of the series $\sum_{i=1}^{\infty} \xi_i^c$ follows from the convergence of the second and third series (1.22) and 15°. But in this case, the series $\sum_{i=1}^{\infty} \xi_i$ converges, too.

Necessity. If the series converges with probability 1, then $\xi_i \to 0$ with probability 1, and therefore for all $c > 0$, we have $\lim_{n\to\infty} \mathbf{P}\{\sup_{i\geq n} |\xi_i| \leq c\} = 1$. However,

$$\mathbf{P}\{\sup_{i\geq n} |\xi_i| \leq c\} = \prod_{i\geq n}(1 - \mathbf{P}\{|\xi_i| > c\}) \leq \exp\{-\sum_{i\geq n} \mathbf{P}\{|\xi_i| > c\}\}.$$

So, the first series (1.22) converges. But in this case, the series $\sum_{i=1}^{\infty}(\xi_i - \xi_i^c)$ converges with probability 1 as was shown in the proof of sufficiency. Thus, the series $\sum_{i=1}^{\infty} \xi_i^c$ converges, too. The convergence of the second and third series (1.22) follows from 16°.

□

18°. Theorem. *Suppose* ξ_k *to be a sequence of independent random variables such that the series* $\sum \xi_{n_k}$ *converges for whatever subsequence of different* $n_k \in Z_+$. *Then for every* $c > 0$,

$$\sum \mathbf{P}\{|\xi_k| > c\} < \infty, \qquad \sum |\mathbf{M}\xi_i^c| < \infty, \qquad \sum \mathbf{D}\xi_i^c < \infty. \tag{1.23}$$

If (1.23) holds for some $c > 0$, the series $\sum \xi_{n_k}$ converges for any subsequence of different $n_k \in Z_+$, and if n_k is a permutation of natural numbers, then $\sum \xi_{n_k} = \sum \xi_k$ with probability 1.

Proof. The necessity of convergence of the first and third series (1.23) follows from 17°. The convergence of the second series arises from the convergence of $\sum \mathbf{M}\xi_k^c$ and $\sum \mathbf{M}\xi_k^c I_{\{\mathbf{M}\xi_k^c > 0\}}$. If conditions (1.23) are satisfied, the convergence of $\sum \xi_{n_k}$ follows from 17°. Let n_i be a permutation of natural numbers. Then

$$\sum(\xi_k - \xi_k^c) = \sum(\xi_{n_i} - \xi_{n_i}^c),$$

because the series on the l.h.s. has a finite number of nonzero terms. Furthermore,

$$\sum \mathbf{M}\xi_k^c = \sum \mathbf{M}\xi_{n_i}^c,$$

by virtue of the absolute convergence of the series on the l.h.s. Finally,

$$\begin{aligned}
&\mathbf{M}(\sum[\xi_k^c - \mathbf{M}\xi_k^c] - \sum[\xi_{n_i}^c - \mathbf{M}\xi_{n_i}^c])^2 \\
&= \lim_{\substack{m\to\infty \\ l\to\infty}} \mathbf{M}(\sum_{k\le m}(\xi_k^c - \mathbf{M}\xi_k^c) - \sum_{i\le l}(\xi_{n_i}^c - \mathbf{M}\xi_{n_i}^c))^2 \\
&= \lim_{\substack{m\to\infty \\ l\to\infty}} \sum_{\substack{k\overline{\in}\{1,\dots,m\} \\ \cap\{n_1,\dots,n_l\}}} \mathbf{D}\xi_k^c = 0,
\end{aligned}$$

since $\bigcup_{m,l}\{1,\dots,m\} \cap \{n_1,\dots,n_l\}$ coincides with the natural series.

□

19°. Theorem. *Let ξ_i be symmetric random variables and the sequence ζ_n be bounded in probability. Then the series $\sum_{i=1}^{\infty} \xi_i$ converges with probability 1.*

Proof. Symmetricity of the variables ξ_k implies that the pairs of variables $\xi_k^c, \xi_k - \xi_k^c$ and $-\xi_k^c, \xi_k - \xi_k^c$ have the same distribution:

$$\begin{aligned}
\mathbf{P}\{\xi_k^c = 0, \xi_k - \xi_k^c \in A\} &= \mathbf{P}\{\xi_k^c = 0, -\xi_k + \xi_k^c \in A\}, \\
\mathbf{P}\{\xi_k^c \in B, \xi_k - \xi_k^c = 0\} &= \mathbf{P}\{\xi_k^c \in B, -\xi_k + \xi_k^c = 0\}.
\end{aligned}$$

Therefore,

$$\begin{aligned}
&\mathbf{P}\{\sum_{i=1}^{n}(\xi_i - \xi_i^c) \ge x\} \\
&\le \mathbf{P}\{\sum_{i=1}^{n}(\xi_i - \xi_i^c) \ge x, \sum_{i=1}^{n}\xi_i^c \ge 0\} + \mathbf{P}\{\sum_{i=1}^{n}(\xi_i - \xi_i^c) \ge x, \sum_{i=1}^{n}\xi_i^c \ge 0\} \\
&= 2\mathbf{P}\{\sum_{i=1}^{n}(\xi_i - \xi_i^c) \ge x, \sum_{i=1}^{n}\xi_i^c \ge 0\} \le 2\mathbf{P}\{\sum_{i=1}^{n}\xi_i \ge x\}.
\end{aligned}$$

Hence,

$$\mathbf{P}\{|\sum_{i=1}^{n}(\xi_i - \xi_i^c)| \geq c\} \leq 2\mathbf{P}\{|\sum_{i=1}^{n}\xi_i| \geq x\}.$$

It follows from the boundedness of ζ_n in probability that one can choose $c > 0$ such that $\mathbf{P}\{|\zeta_n| \geq c\} \leq 1/8$ for all n. Then

$$\mathbf{P}\{\sup_{k\leq n}|\sum_{i=1}^{k}(\xi_i - \xi_i^c)| \geq c\} \leq 2\mathbf{P}\{|\sum_{i=1}^{n}(\xi_i - \xi_i^c)| \geq c\}$$
$$\leq 4\mathbf{P}\{|\sum_{i=1}^{n}\xi_i| \geq c\} \leq \frac{1}{2};$$

and thus,

$$\mathbf{P}\{\sup_{k\leq n}|\xi_k - \xi_k^c| > 2c\} \leq \frac{1}{2}$$

for all n. This implies (just as in the proof of necessity in the theorem on three series) that $\sum \mathbf{P}\{|\xi_k| > 2c\} < \infty$ and that the series $\sum_{k=1}^{\infty}(\xi_k - \xi_k^{2c})$ converges.

Therefore, the variables $\sum_{k=1}^{n}\xi_k^{2c}$ are bounded in probability, the symmetry of ξ_k^{2c} yields $\sum \mathbf{P}\{|\xi_k| > 2c\} < \infty$, and from 16° it follows that $\sum_{k=1}^{\infty}\mathbf{D}\xi_k^{2c} = 0$. Using the theorem on three series, we complete the proof.

□

20°. Theorem. *The sequence ζ_n is bounded in probability iff there exists a real-valued sequence a_n such that the sequence $\sum_{k=1}^{n} a_k$ is bounded and the series $\sum_{k=1}^{\infty}(\xi_k - a_k)$ converges with probability 1.*

Let $\xi_k', \tilde{\xi}_k, \tilde{\zeta}_n$ be constructed as in 12°. Then $\tilde{\zeta}_n$ are also bounded in probability, and hence the series $\sum \tilde{\xi}_n$ converges with probability 1 (by virtue of 19°). Thus, for all $c > 0$,

$$\infty > \sum \mathbf{P}\{|\xi_k - \xi_k'|\} \geq \sum_k \mathbf{P}\{|\xi_k| > 2c, |\xi_k'| \leq c\}$$
$$\geq \inf_k \mathbf{P}\{|\xi_k| \leq c\} \sum_{k=1}^{\infty}\mathbf{P}\{|\xi_k| \geq 2c\}.$$

The variables $\xi_n = \zeta_n - \zeta_{n-1}$ are bounded in probability; this follows from the boundedness in probability of the variables ζ_n. Therefore, one can choose $c > 0$ such that $\mathbf{P}\{|\xi_k| \leq c\} \geq 1/2$ for all k. We have proved that $\sum_k \mathbf{P}\{|\xi_k| \geq 2c\} < \infty$. Thus, the series $\sum_{k=1}^{\infty}(\xi_k - \xi_k^{2c})$ converges with probability 1, and the variables $\sum_{k=1}^{n}\xi_k^{2c}$ are bounded in probability. From 16° it follows that $\sum_{k=1}^{n}\mathbf{D}\xi_k^{2c} < \infty$; and consequently, the series $\sum_{k=1}^{\infty}(\xi_k^{2c} - \mathbf{M}\xi_k^{2c})$ converges with probability 1. Moreover, the variables $\sum_{k=1}^{n}\mathbf{M}\xi_k^{2c}$ are bounded.

Statement 20° holds for $a_k = \mathbf{M}\xi_k^{2c}$.

□

Remark. If ζ_n are bounded in probability, then as is obvious from the proof of 20°, there exist a_k such that the series $\sum_k(\xi_{n_k} - a_{n_k})$ converges for any sequence of different natural numbers n_k.

Strong Law of Large Numbers

21°. Kolmogorov's theorem. *Suppose that $\chi_1, \chi_2, \dots$ is a sequence of independent random variables such that $\mathbf{M}\chi_i = 0$, $\mathbf{D}\chi_i < \infty$. If $\sum \frac{1}{k^2}\mathbf{D}\chi_k < \infty$, then*

$$\mathbf{P}\{\lim_{n\to\infty} \frac{1}{n}\sum_{k=1}^{n} \chi_k = 0\} = 1.$$

Proof. By virtue of 5°,

$$\mathbf{P}\{\sup_{n\le 2^m} |\sum_{k=1}^{m} \chi_k| > \varepsilon 2^m\} \le \frac{1}{\varepsilon^2 2^{2m}} \sum_{k=1}^{2^m} \mathbf{D}\chi_k,$$

and

$$\sum_{m=1}^{\infty} \frac{1}{2^{2m}} \sum_{k=1}^{2^m} \mathbf{D}\chi_k = \sum_{k=1}^{\infty} \mathbf{D}\chi_k \sum_{2^m\ge k} \frac{1}{2^{2m}} \le \sum_{k=1}^{\infty} \mathbf{D}\chi_k \sum_{m\ge\log_2 k} \frac{1}{4^m}$$
$$\le \frac{4}{3}\sum_{k=1}^{\infty} \mathbf{D}\chi_k \cdot \frac{1}{4}\log_2 k \le \frac{4}{3}\sum_{k=1}^{\infty} \frac{1}{k^2}\mathbf{D}\chi_k < \infty.$$

So, for every $\varepsilon > 0$, a finite number of the events $\{\sup_{n\le 2^m} |\sum_{k=1}^{n} \chi_k| > \varepsilon\cdot 2^m\}$ happen with probability 1, as follows from the Borel–Cantelli theorem. Hence, if

$$\psi_m = \frac{1}{2^m} \sup_{n\le 2^m} |\sum_{k=1}^{n} \chi_k|,$$

then $\psi_m \to 0$ with probability 1, but $\frac{1}{n}|\sum_{k=1}^{n} \chi_k| \le 2\psi_m$ for $2^{m-1} \le n \le 2^m$.

□

22°. Kolmogorov's theorem. *Suppose that $\xi_1, \xi_2, \dots$ is a sequence of independent equally distributed variables and that there exists $\mathbf{M}\xi_1 = a$. Then with probability 1,*

$$\lim_{n\to\infty} \frac{1}{n}\sum_{k=1}^{n} \xi_k = a.$$

Proof. It is sufficient to prove this for the case $a = 0$. Let $\xi_k' = \xi_k I_{\{|\xi_k| \le k\}}$, $\xi_k'' = \xi_k - \xi_k'$. Then

$$\mathbf{P}\{\xi_k'' \ne 0\} = \mathbf{P}\{|\xi_k| > k\},$$
$$\sum_{k=1}^{\infty} \mathbf{P}\{\xi_k'' \ne 0\} \le \sum_{k=1}^{\infty} \mathbf{P}\{|\xi_1| > k\} \le \mathbf{M}|\xi_1| < \infty.$$

Hence, $\xi_k'' = 0$ with probability 1 for all sufficiently large k. Therefore,

$$\mathbf{P}\{\sum_{k=1}^{\infty} |\xi_k''| < \infty\} = 1, \qquad \mathbf{P}\{\lim_{n\to\infty} \frac{1}{n}\sum_{k=1}^{n} \xi_k'' = 0\} = 1.$$

Let $F(x)$ be a distribution function of ξ_1. Then

$$\begin{aligned}\sum_{k=1}^{\infty} \frac{1}{k^2}\mathbf{D}\xi_k' &\le \sum_{k=1}^{\infty} \frac{1}{k^2}\int_{-k}^{k} |x|^2 dF(x) \\ &= \sum_{k=1}^{\infty} \frac{1}{k^2}\sum_{i=1}^{k} \int_{i-1<|x|\le i} |x|^2 dF(x) \\ &= \sum_{i=1}^{\infty} \int_{i-1<|x|\le i} |x|^2 dF(x) \sum_{k\ge i} k^{-2} \\ &\le c\sum_{i=1}^{\infty} \frac{1}{i}\int -i-1 < |x| \le i |x|^2 dF(x) \le c\int |x| dF(x) < \infty.\end{aligned}$$

By virtue of 21°,

$$\mathbf{P}\{\lim_{n\to\infty} \frac{1}{n}\sum_{k=1}^{n} (\xi_k' - \mathbf{M}\xi_k') = 0\} = 1.$$

Finally,

$$\lim_{n\to\infty} \sum_{k=1}^{n} \int_{-k}^{k} x dF(x) = \lim_{n\to\infty} \int_{-n}^{n} x dF(x) = \mathbf{M}\xi_1 = 0.$$

□

Remark. Clearly, the theorem is also valid in the case of $\mathbf{M}\xi_1 = +\infty\ (-\infty)$, because $\frac{1}{n}\sum_1^n \xi_k I_{\{\xi_k \le c\}} \le \frac{1}{n}\sum_1^n \xi_k$ and $\mathbf{M}\xi_1 I_{\{\xi_1 \le c\}}$ can be made arbitrarily large by the proper choice of c.

Generalizations for the Vector Case

We now state which of the inequalities obtained here can be transferred to the case of variables with values in finite-dimensional Euclidean, Hilbert, or Banach spaces.

23°. Kolmogorov's inequality in Hilbert space. Let H be a separable Hilbert space with a norm $|\cdot|_H$ and a scalar product $(\cdot,\cdot)_H$. If ξ_i are variables with values in a separable Hilbert space H, $\mathbf{M}\xi_i = 0$, $\mathbf{M}|\xi_i|_H^2 < \infty$, then for $a > 0$,

$$\mathbf{P}\{\sup_{k\le n} |\zeta_k|_H \ge a\} \le \frac{1}{a^2}\mathbf{M}|\zeta_n|_H^2.$$

This can be proved just as in 5°, by using the fact that $M(\zeta_n - \zeta_\nu, \eta)_H = 0$ for every $\mathcal{F}_\nu$-measurable element η for which $\mathbf{M}|\eta|_H^2 < \infty$. This follows from the equality

$$(\zeta_n - \zeta_\nu, \eta)_H = \sum_k [(\zeta_n, e_k)_H - (\zeta_\nu, e_k)_H](\eta, e_k),$$

where $\{e_k\}$ is an orthonormal basis in H, from 2°, and from the fact that (ξ_i, e_k) are independent real-valued variables.

Remarks. A. Inequalities (1.9) and (1.10) remain valid for the variables with values in the Banach space B. We prove (1.10) for example. If $\nu = k$ when $|\zeta_i|_B < a$ and $|\zeta_k|_B \ge a$ for some $k < n$, and $\nu = n$ in other cases, then the event $\{\nu < n\} \cap \{\zeta_n - \zeta_\nu \in \prod_\nu\}$, where $\prod_x$ is a half-space of those $y \in B$ for which $y + x$ and a sphere $\{z \in B : |z| \le x\}$ are situated on opposite sides of the support hyperplane Γ_x drawn to this sphere at the point x, implies the event $|\zeta_n|_B \ge |\zeta_\nu|_B$. Thus,

$$\mathbf{P}\{|\zeta_n|_B \ge a\} \ge \mathbf{P}\{\nu = n, |\zeta_n|_B \ge a\} + \sum_{k=1}^{n} \mathbf{P}\{\nu = k\}\mathbf{P}\{\zeta_n - \zeta_k \in \Pi\zeta_k\}.$$

The difference $\zeta_n - \zeta_k$ does not depend on ζ_k; and for every x, we have $\mathbf{P}\{\zeta_n - \zeta_k \in \prod_x\} \ge 1/2$, by virtue of the symmetry of $\zeta_n - \zeta_k$. Hence,

$$\mathbf{P}\{|\zeta_n|_B \ge a\} \ge \mathbf{P}\{\sup_{k\le n} |\zeta_k|_B \ge a\}\tfrac{1}{2}.$$

Statement (1.10) is proved.

B. Statement 14° is valid for every Banach space B. The proof remains the same; one should merely substitute the sign $|\cdot|_B$ for $|\cdot|$.

24°. The theorem on three series (17°) remains valid for the variables with values in a separable Hilbert space. The necessary and sufficient condition of convergence of the series $\sum \xi_i$ with probability 1 is the convergence of the series

$$\sum_{i=1}^{\infty} \mathbf{P}\{|\xi_i|_H > c\}, \qquad \sum_{i=1}^{\infty} \mathbf{M}\xi_i^c, \qquad \sum_{i=1}^{\infty} \mathbf{M}|\xi_i^c - \mathbf{M}\xi_i^c|_H^2.$$

The proof is completely analogous to the one-dimensional case, using remark B instead of statement 6°.

§1.2 Renewal Scheme

Let ξ_i be independent equally distributed positive random variables, $\zeta_0 = 0$, $\zeta_k = \sum_{s=1}^{k} \xi_s$. A process $\nu(t)$ defined for $t \in R_t^+$ by the equality $\nu(t) = n$ for $\zeta_{n-1} \le t < \zeta_n$ is called a *renewal process*, the variable ζ_n is called the *nth renewal moment*. This variable can be interpreted in the following way. Suppose that some device works during the random time ξ_1, then breaks and is replaced by an identical one which, in turn, works during the time ξ_2, and so on. It is natural to suppose that ξ_i are independent and equally distributed. If we assume that the time $t = 0$ coincides with the time when the first device starts working, then $\nu(t)$ coincides with the number of the device working at time t. Denote

$$N(t) = \mathbf{M}\nu(t);$$

$N(t)$ is called a *renewal function*. The main problem of the renewal theory is to determine the asymptotics of the renewal function $N(t)$.

25°. *The function $N(t)$ is bounded and satisfies the inequality*

$$N(t+h) \le N(t) + N(h). \tag{1.24}$$

Proof. One can find $a > 0$ such that $\mathbf{P}\{\xi_i > a\} = \beta < 1$ and thus

$$\mathbf{P}\{\nu(t) > n\} \le C_n^k (1-\beta)^{n-k},$$

where $k = [t/a]$ (the integral part of t/a). This statement follows from the fact that at most k variables among $\xi_1, \xi_2, \ldots, \xi_n$ will exceed a. Hence, $\nu(t)$ has all the moments.

Now we introduce the s.m. $\mu = \inf[k : \zeta_k \ge t]$ in order to deduce inequality (1.24). Then

$$\begin{aligned} \nu(t+h) - \nu(t) &= [\nu(t+h) - \nu(\zeta_\mu)] I_{\{\zeta_\mu < t+h\}} + 1 \\ &\le \nu(\zeta_\mu + h) - \nu(\zeta_\mu) + 1, \end{aligned} \tag{1.25}$$

$$\begin{aligned} &\mathbf{M}[\nu(\zeta_\mu + h) - \nu(\zeta_\mu) + 1] \\ &\quad = \sum \mathbf{P}\{\mu = k\} \mathbf{M}[\nu(\xi_k + h) - \nu(\xi_k) + 1]. \end{aligned}$$

But $\nu(\zeta_k + h) - \nu(\zeta_k)$ has the same distribution as $\nu(h) - 1$. Therefore,

$$\mathbf{M}[\nu(\zeta_\mu + h) - \nu(\zeta_\mu) + 1] = N(h).$$

Taking the mathematical expectation in (1.25), we get (1.24).

26°. *There exist a and b such that $N(t) \le a + bt$.*

This statement follows from (1.24):

$$\begin{aligned} N(t) &\le N([t] + 1) \le N(1) + [t]N(1) \\ &\le N(1) + (t+1)N(1) \le 2N(1) + tN(1). \end{aligned}$$

27°. Lemma. *For all $\lambda > 0$, the Laplace–Stieltjes transform*

$$n(\lambda) = \int_0^\infty e^{-\lambda t} dN(t) \tag{1.26}$$

is defined; it is given by the formula

$$n(\lambda) = \frac{1}{1 - f(\lambda)}, \tag{1.27}$$

where $f(\lambda) = \mathbf{M}e^{-\lambda\xi_1}$.

The existence of $n(\lambda)$ for $\lambda > 0$ follows from 26°. Further,

$$\int_0^\infty e^{-\lambda t} d\nu(t) = \sum_{k=0}^\infty e^{-\lambda\zeta_k} = 1 + \sum_{k=1}^\infty \prod_{i=1}^k e^{-\lambda\xi_i} \tag{1.28}$$

(we consider that $\nu(t) = 0$ for $t < 0$; the integral takes into account the unit jump of $\nu(t)$ at $t = 0$). Taking the mathematical expectation in (1.28) and allowing for $\mathbf{M}e^{-\lambda\xi_i} = f(\lambda)$, we obtain

$$\int_0^\infty e^{-\lambda t} dN(t) = \sum_{k=0}^\infty f^k(\lambda) \tag{1.29}$$

(one can easily prove that $\int_0^\infty e^{-\lambda t} dN(t) = \mathbf{M}\int_0^\infty e^{-\lambda t} d\nu(t)$ by integrating by parts:

$$\begin{aligned} \mathbf{M}\int_0^\infty e^{-\lambda t} d\nu(t) &= \mathbf{M}\lambda \int_0^\infty e^{-\lambda t}\nu(t)dt \\ &= \lambda \int_0^\infty e^{-\lambda t} N(t)dt = \lambda \int_0^\infty e^{-\lambda t} dN(t)). \end{aligned}$$

Equality (1.29) implies (1.27). □

28°. Theorem. *There exists*

$$\lim_{t\to\infty}\frac{N(t)}{t}=\frac{1}{\mathbf{M}\xi_1} \tag{1.30}$$

(*if* $\mathbf{M}\xi_1=+\infty$, *the quantity on the r.h.s. is considered to be zero*).

Proof. It follows from 26° that $\frac{N(t)}{t}$ is bounded for $t\geq 1$. We put $c=\inf_{t\geq 1}\frac{N(t)}{t}$. Let t_0 be such that $\frac{N(t_0)}{t_0}<c+\varepsilon$. Then

$$\frac{N(kt_0+\tau)}{kt_0+\tau}\leq\frac{kN(t_0)+N(\tau)}{kt_0}\leq c+\varepsilon+\frac{N(\tau)}{kt_0}.$$

Therefore, $\overline{\lim}_{t\to\infty}\frac{N(t)}{t}\leq c+\varepsilon$ for all $\varepsilon>0$, and thus

$$c\leq\underline{\lim}_{t\to\infty}\frac{N(t)}{t}\leq\overline{\lim}_{t\to\infty}\frac{N(t)}{t}\leq c.$$

The existence of the limit in (1.30) is proved.

From (1.27) we get

$$\int_0^\infty e^{-\lambda t}N(t)dt=\frac{1}{\lambda}\int_0^\infty e^{-\lambda t}dN(t)=\frac{1}{\lambda(1-f(\lambda))}.$$

Hence,

$$\frac{1}{\lambda(1-f(\lambda))}=\frac{1}{\lambda}\int_0^\infty e^{-u}N\left(\frac{u}{\lambda}\right)du,$$

$$\frac{1}{1-f(\lambda)}=\int_0^\infty e^{-u}\lambda N\left(\frac{u}{\lambda}\right)du.$$

The integrand has an integrable majorant $e^{-u}\lambda[a+b\frac{u}{\lambda}]=e^{-u}[a+bu]$ for $\lambda\leq 1$, and $\lim_{\lambda\to 0}\lambda N(\frac{u}{\lambda})=uc$. Therefore,

$$\lim_{\lambda\to 0}\frac{\lambda}{1-f(\lambda)}=c\int_0^\infty e^{-u}udu=c,$$

$$c=\lim_{\lambda\to 0}\frac{1}{\mathbf{M}\frac{1-e^{-\lambda\xi_1}}{\lambda}}=\begin{cases}0, & \text{if } \mathbf{M}\xi_1=+\infty,\\ \frac{1}{\mathbf{M}\xi_1}, & \text{if } \mathbf{M}\xi_1<\infty.\end{cases}$$

□

We introduce the next characteristics of the renewal process: (for the device working at the moment t) $\zeta_{\nu(t)}-t=\gamma_t^+$ is the time remaining until the breakage of this device; $t-\zeta_{\nu(t)-1}=\gamma_t^-$ is its working time from the beginning of the work until the moment t; $\gamma_t=\gamma_t^-+\gamma_t^+$ is the total working time of this device.

For more detailed investigation of the asymptotic behaviour of the renewal function, let us consider arithmetic (lattice) and nonlattice random variables separately.

Lattice Random Variables

The variable ξ has the *lattice distribution* if for some $h > 0$, the variable ξ/h is an integral one ($\mathbf{P}\{\xi/h \in Z\} = 1$). The greatest of the h is called a *step of distribution*. If h is the step of distribution and $\varphi(z) = \mathbf{M}e^{iz\xi}$ is the characteristic function of the variable ξ, then

$$\varphi\left(\frac{2\pi}{h}\right) = \mathbf{M}e^{i2\pi\frac{\xi}{h}} = 1.$$

On the other hand, if $1 = \varphi(z) = \sum e^{ikz_0h}\mathbf{P}\{\xi = kh\}$, then for all k for which $\mathbf{P}\{\xi = kh\} > 0$, we have $kz_0h = 2\pi n_k$, where n_k is an integer. If $0 < z_0 < 2\pi/h$, then $z_0 = \theta 2\pi/h$, where $0 < \theta < 1$. Hence, θ is rational. Suppose that $\theta = r/s$ is an irreducible fraction; then all k for which $\mathbf{P}\{\xi = kh\} > 0$ have the common factor s, and thus ξ/sh is also an integral variable, i.e., h is not a step of distribution.

29°. Lemma. *If ξ has the lattice distribution, then the step of distribution h is the least positive number for which $\varphi(2\pi/h) = 1$, and the inequality $\varphi(z) < 1$ is valid for $0 < |z| < 2\pi/h$.*

The validity of $\varphi(z) < 1$ for $-2\pi/h < z < 0$ follows from the fact that $-\xi$ also has the lattice distribution with the step h.

Further, we shall consider integral variables with the step 1. We define the variables

$$q_n = \sum_{k=0}^{\infty} \mathbf{P}\{\zeta_k = n\} = N(n) - N(n-1).$$

30°. Lemma. *The next representation of q_n is valid:*

$$q_n = \lim_{\lambda\uparrow 1} \frac{1}{\pi i} \int_{-\pi}^{\pi} \frac{\sin nz}{1 - \lambda\varphi(z)} dz, \tag{1.30'}$$

where $\varphi(z) = \mathbf{M}e^{iz\xi_1}$ is the characteristic function of the variable ξ_1.

In fact,

$$\begin{aligned} q_n &= \lim_{\lambda\uparrow 1} \sum_{k=0}^{\infty} \lambda^k \mathbf{P}\{\zeta_k = n\} \\ &= \lim_{\lambda\uparrow 1} \sum_{k=0}^{\infty} \lambda^k \frac{1}{2\pi} \int_{-\pi}^{\pi} e^{-izn} \sum_{l} \mathbf{P}\{\zeta_k = l\} e^{ilz} dz \\ &= \lim_{\lambda\uparrow 1} \sum_{k=0}^{\infty} \lambda^k \frac{1}{2\pi} \int_{-\pi}^{\pi} e^{-izn} \mathbf{M}e^{iz\zeta_k} dz \\ &= \lim_{\lambda\uparrow 1} \sum_{k=0}^{\infty} \lambda^k \frac{1}{2\pi} \int_{-\pi}^{\pi} e^{-izn} \varphi_k(z) dz = \lim_{\lambda\uparrow 1} \frac{1}{2\pi} \int_{-\pi}^{\pi} \frac{e^{-izn}}{1 - \lambda\varphi(z)} dz. \end{aligned}$$

Taking into account that the variables ξ_k are nonnegative, we similarly obtain

$$0 = \lim_{\lambda\uparrow 1} \frac{1}{2\pi}\int_{-\pi}^{\pi} \frac{e^{izn}}{1-\lambda\varphi(z)}dz.$$

Subtracting this equality from the previous one, we get (1.30′).

□

31°. Lemma.

$$\lim_{n\to\infty}[q_{n+1}+q_{n-1}-2q_n] = 0. \tag{1.31}$$

We have

$$\begin{aligned} q_{n+1}+q_{n-1}-2q_n &= \lim_{\lambda\uparrow 1}\frac{1}{2\pi}\int_{-\pi}^{\pi}\frac{e^{-iz(n+1)}+e^{-iz(n-1)}-2e^{-izn}}{1-\lambda\varphi(z)}dz \\ &= \lim_{\lambda\uparrow 1}\frac{1}{2\pi}\int_{-\pi}^{\pi} e^{-izn}\frac{e^{iz}+e^{-iz}-2}{1-\lambda\varphi(z)}dz. \end{aligned} \tag{1.32}$$

The variable $\frac{z^2}{1-\varphi(z)}$ is bounded for $|z|\le\pi$, because the denominator does not vanish for $z\neq 0$, and

$$1-\varphi(z) = \sum \mathbf{P}\{\xi_1 = k\}(1-e^{izk}),$$

$|1-\varphi(z)| \ge \operatorname{Re}(1-\varphi(z)) \ge \alpha(1-\cos lz) \ge \beta z^2$ for $|z| < \frac{\pi}{2l}$, where l is such that $\mathbf{P}\{\xi_1 = l\} = \alpha > 0$, $\beta > 0$. Hence, we can take the limit under the integral sign in (1.32) and get

$$q_{n+1}+q_{n-1}-2q_n = \frac{1}{\pi}\int_{-\pi}^{\pi} e^{-izn}\frac{\cos z - 1}{1-\varphi(z)}dz.$$

Equation (1.31) follows from Riemann's theorem on the Fourier coefficients (the function $\frac{\cos z-1}{1-\varphi(z)}$ is bounded).

□

32°. Lemma.

$$\lim_{n\to\infty}(q_n - q_{n-1}) = 0. \tag{1.33}$$

If for some sequence n_k we have $\underline{\lim}_{k\to\infty}(q_{n_k}-q_{n_k-1}) = c \neq 0$, then $q_{n_k+1}-q_{n_k} \to c$ since $q_{n_k+1}+q_{n_k-1}-2q_{n_k}\to 0$. Therefore, $q_{n_k+m}-q_{n_k+m-1}\to c$, $q_{n_k+m}+q_{n_k}\to mc$ for every m, which is impossible by virtue of the boundedness of q_n (this follows from the inequalities $q_n \le N(n)-N(n-1)\le N(1)$). Hence, the bounded sequence $q_n - q_{n-1}$ has the single limit point 0, i.e., equation (1.33) holds.

□

33°. Blackwell's theorem. *If ξ_n are integral variables with the step 1, then*

$$\lim_{n\to\infty} q_n = \frac{1}{\mathbf{M}\xi_1} \quad \left(\frac{1}{+\infty} = 0\right). \tag{1.34}$$

Proof. Let $r_k = \sum_{l>k} \mathbf{P}\{\xi_1 = l\}$. Then for all n,

$$\sum r_k q_{n-k} = 1. \tag{1.35}$$

In fact, let $Q(\lambda) = \sum_{n=0}^{\infty} q_n \lambda^n$, $R(\lambda) = \sum_{k=0}^{\infty} r_k \lambda^k$. Then for $|\lambda| < 1$,

$$\begin{aligned} Q(\lambda) = \sum_{n=0}^{\infty}\sum_{k=0}^{\infty} \mathbf{P}\{\zeta_k = n\}\lambda^n &= \sum_{k=0}^{\infty} \mathbf{M}\lambda^{\zeta_k} \\ &= \sum_{k=0}^{\infty} (\mathbf{M}\lambda^{\zeta_1})^k = \frac{1}{1-\psi(\lambda)}, \end{aligned}$$

where

$$\psi(\lambda) = \mathbf{M}\lambda^{\xi_1} = \sum \mathbf{P}\{\xi_1 = n\}\lambda^n,$$

$$\begin{aligned} R(\lambda) &= \sum_{n=0}^{\infty} \lambda^n \sum_{k>n} \mathbf{P}\{\xi_1 = k\} = \sum_{k=1}^{\infty} \mathbf{P}\{\xi_1 = k\} \sum_{n=1}^{k-1} \lambda^n \\ &= \sum_{k=1}^{\infty} \mathbf{P}\{\xi_1 = k\} \frac{\lambda^k - 1}{\lambda - 1} = \frac{1-\psi(\lambda)}{1-\lambda}. \end{aligned}$$

Hence,

$$R(\lambda)Q(\lambda) = \frac{1}{1-\lambda}. \tag{1.36}$$

Expanding both sides of (1.36) in powers of λ and equating the coefficients of the same powers of λ, we obtain (1.35).

Let $q = \lim_{k\to\infty} q_{n_k}$. If $\mathbf{M}\xi_1 = \sum_{k=0}^{\infty} r_k = +\infty$, we have

$$1 \geq \lim \sum_{i=0}^{N} r_i q_{n_k - i} = q \sum_{i=0}^{N} r_i, \quad q \leq 1 \Big/ \sum_{i=0}^{N} r_i.$$

Passing to the limit as $N \to \infty$, we get $q = 0$. Hence, the single limit point of the sequence q_n is the point 0. The relation (1.34) is thus proved for the case of $\mathbf{M}\xi = +\infty$. Moreover, it follows from the aforesaid that we always have

$$q \leq 1/\mathbf{M}\xi_1, \qquad \text{i.e., } \overline{\lim}\, q_n \leq 1/\mathbf{M}\xi_1.$$

Suppose that $\sum r_k < \infty$ and choose N such that $(\sum_{i>N} r_i) \sup q_n < \varepsilon$. Then

$$1 \le \sum_{i=0}^{N} r_i q_{n_k - i} + \varepsilon, \quad 1 \le q \sum_{i=0}^{N} r_i + \varepsilon, \quad q \ge (1-\varepsilon) / \sum_{i=0}^{N} r_i.$$

Passing to the limit as $N \to \infty$ and then as $\varepsilon \to \infty$, we get

$$\varliminf_{n\to\infty} q_n \ge \frac{1}{\mathbf{M}\xi_1}.$$

□

Let us find the limit distribution of the variables γ_t^+, γ_t^-, γ_t^*. They are obtained similarly by using Blackwell's theorem.

34°. Theorem. *The following limit relations ($u \in Z_+$) hold:*

$$\begin{aligned}
\lim_{t\to\infty} \mathbf{P}\{\gamma_t^+ = u\} &= \frac{1}{\mathbf{M}\xi_1}\mathbf{P}\{\xi_1 \ge u\}, \\
\lim_{t\to\infty} \mathbf{P}\{\gamma_t^- = u\} &= \frac{1}{\mathbf{M}\xi_1}\mathbf{P}\{\xi_1 \ge u\}, \\
\lim_{t\to\infty} \mathbf{P}\{\gamma_t^* = u\} &= \frac{u}{\mathbf{M}\xi_1}\mathbf{P}\{\xi_1 = u\}.
\end{aligned} \tag{1.37}$$

Let us obtain the first formula

$$\begin{aligned}
\mathbf{P}\{\gamma_t^+ = u\} &= \sum_{k=0}^{\infty} \sum_{v} \mathbf{P}\{\zeta_k = t - v\} \mathbf{P}\{\xi_{k+1} = v + u\} \\
&= \sum_{v} q_{t-v} \mathbf{P}\{\xi_{k+1} = u + v\}.
\end{aligned}$$

Passing to the limit as $t \to \infty$ under the summation sign (it is possible since q_{t-v} are bounded and $\sum_v \mathbf{P}\{\xi_{k+1} = u+v\}$ converges), we get the first formula (1.37). Other formulas are obtained similarly from the equalities

$$\mathbf{P}\{\gamma_t^- = u\} = \sum_{k=0}^{\infty} \mathbf{P}\{\zeta_k = t - u\} \mathbf{P}\{\xi_{k+1} > u\},$$

$$\mathbf{P}\{\gamma_t^* = u\} = \sum_{k=0}^{\infty} \sum_{r=0}^{t} \mathbf{P}\{\zeta_k = t - v\} \mathbf{P}\{\xi_{k+1} = u + v\}.$$

□

Nonlattice Variables

35°. Blackwell's theorem. *Suppose that the ξ_k have the nonlattice distribution. Then for all $h > 0$,*

$$\lim_{t\to\infty}[N(t+h) - N(t)] = h/\mathbf{M}\xi_1.$$

Proof. We shall prove this theorem in several steps. Let us first show that for every twice continuously differentiable function $\theta(u)$ for which $\int \theta(u)du = 0$, $\int u\theta(u)du = 0$, $\theta(u) = O(u^{-2})$ when $u \to \infty$, the following relation holds:

$$\lim_{t\to\infty} \int \theta(u-t)dN(u) = 0. \tag{1.38}$$

Let $\theta(z) = \int e^{izu}\hat{\theta}(z)dz$. Then $\hat{\theta}(0) = 0$, $\hat{\theta}'(0) = 0$, $\hat{\theta}(z^2) = O(z^2)$. Suppose first that $\hat{\theta}(z)$ is finite. We put $N_\lambda(t) = \sum_{k=0}^{\infty} \lambda^k \mathbf{P}\{\zeta_k \le t\}$. Then

$$N(t) = \lim_{\lambda\uparrow 1} N_\lambda(t),$$

$$\begin{aligned}\int \theta(u+t)dN(u) &= \lim_{\lambda\uparrow 1}\int \theta(u-t)dN_\lambda(u)\\ &= \lim_{\lambda\uparrow 1}\iint e^{iz(u-t)}\hat{\theta}(z)dz\, dN_\lambda(u)\\ &= \lim_{\lambda\uparrow 1}\int e^{izt}\hat{\theta}(z)\int e^{izu}du\sum \lambda^k\mathbf{P}\{\zeta_k \le u\}\\ &= \lim_{\lambda\uparrow 1}\int \frac{e^{-izt}\hat{\theta}(z)}{1-\lambda\varphi(z)}dz,\end{aligned}$$

where $\varphi(z) = \int e^{izu}d\mathbf{P}\{\xi_1 \le u\}$. Since $\hat{\theta}(z)$ is finite, $\varphi(z) \ne 1$ when $z \ne 0$, and

$$\operatorname{Re}(1-\varphi(z)) = \int (1-\cos zx)dF(x) \ge \alpha z^2$$

for sufficiently small z, where $\alpha > 0$, then $\frac{\hat{\theta}(z)}{1-\varphi(z)}$ is a finite bounded function. Passing to the limit in the equality

$$\int \theta(u-t)dN(u) = \int e^{-izt}\frac{\hat{\theta}(z)}{1-\varphi(z)}dz,$$

we get (1.38) (by virtue of Riemann's theorem), assuming finiteness of $\hat{\theta}(z)$.

In the general case, let us introduce a function

$$\chi_n(z) = \begin{cases} 1 & \text{for } |z| \le n,\\ 0 & \text{for } |z| > n+1,\end{cases}$$

which is twice continuously differentiable with bounded derivatives

$$\theta_n(u) = \frac{1}{2\pi}\int e^{-izu}\hat{\theta}(z)\chi_n(z)dz.$$

Clearly, $\theta_n(u)$ converges uniformly to $\theta(u)$, and $\theta_n(u) = O(u^{-2})$ uniformly in n. Using estimate 26°, one can ascertain that

$$\lim_{t\to\infty}\int \theta(u-t)dN(u) = \lim_{n\to\infty}\lim_{t\to\infty}\int \theta_n(u+t)dN(u) = 0.$$

Suppose now that $\theta(u)$ is an arbitrary continuous finite function for which $\int\theta(u)du = 0$, $\int u\theta(u)du = 0$. Then (1.38) holds. This can be established by using the uniform approximation of θ by twice continuously differentiable finite functions.

Let us take $\theta_1(u) = \theta(u+h) - \theta(u)$, where $\theta(u)$ is continuous and finite and $\int u\theta(u)du = 0$. Then $\int\theta_1(u)du = 0$, $\int u\theta_1(u)du = \int h\theta(u)du = 0$; and applying (1.38) for θ_1, we get

$$\lim_{t\to\infty}\int[\theta(u+h-t) - \theta(u-t)]dN(u) = 0. \qquad (1.39)$$

Let $t_k \to \infty$ be such that the functions $N_{t_k}(u) = N(t_k+u) - N(t_k)$ have the limit $\overline{N}(u)$ for all rational u. Now we shall show that $\overline{N}(u) = \lambda u$ for some λ. In fact, it follows from (1.39) that for all c and $\varepsilon > 0$, t_0 and h,

$$\int_{t_0}^{t_0+\varepsilon}[\overline{N}(s+c+h) + \overline{N}(s-c+h) - 2\overline{N}(s+h)]ds$$

$$= \int_{t_0}^{t_0+\varepsilon}[\overline{N}(s+c) + \overline{N}(s-c) - 2\overline{N}(s)]ds \qquad (1.40)$$

(we have applied formula (1.39) to the function $\theta(u)$ for which $\theta(t_0 - c) = 0$ and $\theta'(u) = I_{(t_0-c,t_0-c+\varepsilon)}(u) + I_{(t_0+c,t_0+c+\varepsilon)}(u) - 2I_{(t_0,t_0+\varepsilon)}(u)$).

It follows from (1.40) that

$$\overline{N}(s+c) + \overline{N}(s-c) - 2\overline{N}(s) = \alpha,$$

where α is some constant. Hence, ($\overline{N}(0) = 0$) $\overline{N}(t) = \lambda t + \alpha t^2/2$. $\overline{N}(t)$ is a nondecreasing function, and thus $\alpha \geq 0$; but since $\overline{N}(t) \leq a + bt$ for some a and b, we get $\alpha = 0$. Let us show that $\lambda = 1/\mathbf{M}\xi_1$ ($\lambda = 0$ for $\mathbf{M}\xi_1 = 0$). To make this, we use the equality

$$N(t) = \sum_{k=0}^{\infty}\mathbf{P}\{\zeta_k \leq t\}$$

$$= 1 + \int_0^t \mathbf{P}\{\xi_1 \in ds\}\sum_{k=1}^{\infty}\mathbf{P}\{\sum_{i=2}^{K}\xi_i \leq t-s\}$$

$$= 1 + \int_0^t N(t-s)\mathbf{P}\{\xi_1 \in ds\}.$$

Denoting the distribution function of the variable ξ_1 by $F(t)$, we obtain from the last relation

$$1 = \int_0^t [1 - F(t-s)]dN(s).$$

Suppose that $\mathbf{M}\xi_1 = +\infty$. Then

$$1 \geq \int_{t_k}^{t_k+t} [1 - F(t + t_k - s)]dN(s) = \int_0^t [1 - F(t-s)]dN_{t_k}(s);$$

and therefore,

$$1 \geq \int_0^t [1 - F(t-s)]d\overline{N}(s) = \lambda \int_0^t [1 - F(t-s)]ds$$
$$= \lambda \int_0^t [1 - F(u)]du.$$

Since $\int_0^\infty [1 - F(u)]du = \mathbf{M}\xi_1 = +\infty$, we have $\lambda = 0$. In addition, $\lambda \mathbf{M}\xi_1 \leq 1$ if $\mathbf{M}\xi_1 < \infty$. Let $\int_0^\infty [1 - F(u)]du < \infty$. Since

$$\int_0^n [1 - F(n + t - u)]dN(u) = \sum_{0 \leq k < n} \int_k^{k+1} [1 - F(n + t - u)]dN(u)$$
$$\leq N(1) \sum_{0 \leq k < n} [1 - F(n + t - k)]$$
$$\leq N(1) \sum_{0 \leq k \leq n} \int_{k-1}^k [1 - F(n + t - u)]du$$
$$\leq N(1) \int_{t-1}^\infty [1 - F(u)]du,$$

we have for all t_k for sufficiently large t,

$$\int_0^{t_k} [1 - F(t + t_k - s)]dN(s) < \varepsilon;$$

and consequently,

$$1 - \varepsilon \leq \int_{t_k}^{t_k+t} [1 - F(t + t_k - s)]dN(s),$$
$$1 - \varepsilon \leq \int_0^t [1 - F(t-u)]dN_{t_k}(u),$$
$$1 - \varepsilon \leq \lambda \int_0^t [1 - F(t-u)]du = \lambda \int_0^t [1 - F(u)]du.$$

Passing to the limit as $t \to \infty$ and $\varepsilon \to 0$, we get $1 \le \lambda \mathbf{M}\xi_1$. Thus,

$$\bar{N}(t) = \frac{1}{\mathbf{M}\xi_1} t.$$

Since $N_s(t)$ is a compact family of nondecreasing functions with a single limit point, we have $N_s(t) \to \frac{1}{\mathbf{M}\xi_1} t$. This implies the proof of the theorem. □

We now give the formulas for the limit distributions of the variables γ_t^+, γ_t^-, γ_t^*; they are established using Blackwell's theorem, quite similarly to the case of arithmetic variables.

$$\lim_{t\to\infty} \mathbf{P}\{\gamma_t^+ > u\} = \lim_{t\to\infty} \mathbf{P}\{\gamma_t^- > u\} = \frac{1}{\mathbf{M}\xi_1} \int_u^\infty [1 - F(v)]dv, \quad (1.41)$$

$$\lim_{t\to\infty} \mathbf{P}\{\gamma_t^* > u\} = \frac{1}{\mathbf{M}\xi_1} \int_u^\infty v dF(v). \quad (1.42)$$

§1.3 Random Walks. Recurrence

A sequence of random variables $\{\zeta_n\}_{n\ge 0}$ in R^d, where $\zeta_0 = x \in R^d$ is an initial position of the "walking" particle and $\zeta_n = x + \sum_{k=1}^n \xi_k$ (here $\{\xi_k\}_{k\ge 1}$ are independent equally distributed random variables in R^d), is called a *random walk in R^d*. The variable ξ_n is called the *nth step* of the random walk, and ζ_n is called the position at the moment n. The random walk described is denoted as *starting from the point x*. Walks starting from the point 0 are usually considered.

The multidimensional variable ξ is called *latticed* if one can find linearly independent vectors $b_1, \dots, b_r \in R^d$ such that

$$\sum_{k_1,\dots,k_r \in Z} \mathbf{P}\{\xi = \sum_{i=1}^r k_i b_i\} = 1.$$

By means of a linear transformation, every latticed variable can be transformed into an arithmetical one, i.e., into a variable ξ' for which $\mathbf{P}\{\xi' \in Z^d\} = 1$. Walks with arithmetically distributed steps will also be called *arithmetical.* A variable that is not latticed is called *nonlatticed*; a walk with a nonlatticed step is also called *nonlatticed.*

A walk in R^d is called *nondegenerate* if there is no proper subspace $L \in R^d$ such that $\mathbf{P}\{\xi_k \in L\} = 1$.

Henceforth, $\varphi(z)$ will denote the characteristic function of the step of a random walk:

$$\varphi(z) = \int e^{i(z,x)} \mathbf{P}\{\xi_1 \in dx\}. \quad (1.43)$$

36°. Lemma. *If a random walk is nondegenerate, then there exists $\delta > 0$ and $\varepsilon > 0$ such that*

$$\mathrm{Re}(1 - \varphi(z)) \geq \varepsilon |z|^2 \quad \textit{for } |z| \leq \delta. \tag{1.44}$$

Proof. Denote by $S \in R^d$ a set of those $x \in R^d$ for which $\mathbf{P}\{|\xi_1 - x| \leq \rho\} > 0$ for all $\rho > 0$; S is the least closed set on which the measure $\mathbf{P}\{\xi_1 \in dx\}$ is concentrated. For a nondegenerate random walk, a linear span of S coincides with R^d. Therefore, d linearly independent vectors $b_1, \dots, b_d$ can be found in S. Hence,

$$\begin{aligned}\mathrm{Re}(1 - \varphi(z)) &\geq \sum_{k=1}^{d} \int_{|b_k - x| \leq p} (1 - \cos(z, x))\mathbf{P}\{\xi_1 \in dx\} \\ &\geq q \sum_{k=1}^{d} \int_{|b_k - x| \leq p} (z, x)^2 \mathbf{P}\{\xi_1 \in dx\},\end{aligned} \tag{1.45}$$

where $q = \inf_{0<x<1} \dfrac{1 - \cos x}{x^2}$, if $|z|(\max_k |b_k| + \rho) \leq 1$. Since

$$\lim_{\rho \to 0} \frac{1}{\mathbf{P}\{|\xi_1 - b_k| \leq \rho\}} \int_{|b_k - x| \leq \rho} (z, x)^2 \mathbf{P}\{\xi_1 \in dx\} = (b_k, z)^2$$

and $\sum_{k=1}^{d} (b_k, z)^2 \geq c|z|^2$ for some $c > 0$ (by virtue of the linear independence of $b_1, \dots, b_k$), the quadratic form (with respect to z) on the r.h.s. of (1.45) is positive.

□

Arithmetic Random Walks

Any subset $S \subset Z^d$ that forms an additive group is called an *integral lattice* in Z^d (d-dimensional integral lattice).

An arithmetic random walk is called *irreducible* if the facts that S is an integral lattice in Z^d and that $\mathbf{P}\{\xi_1 \in S\} = 1$ imply $S = Z^d$.

A minimal lattice S for which $\mathbf{P}\{\xi_1 \in S\} = 1$ is called a *support* of a random walk.

We shall need some auxiliary algebraic statements.

37°. Lemma. *Suppose that S is an integral lattice in Z^d, the linear span of which coincides with R^d. Then there exist vectors $b_1, \dots, b_d$ in Z^d such that S coincides with the set of vectors x representable in the form $\sum_{k=1}^{d} n_k b_k$, where $n_k \in Z$ (such vectors are called a system of generators for S); moreover, the vectors b_k can be chosen so that*

$$b_k = \alpha_{k1} e_1 + \cdots + \alpha_{kk} e_k, \qquad e_k = (\delta_{k1}, \dots, \delta_{kd}), \qquad \alpha_{ki} \in Z, \qquad k = 1, \dots, d. \tag{1.46}$$

Proof. Let us denote a set of integral linear combinations of the vectors $e_1, \ldots, e_r$ by $L(e_1, \ldots, e_r)$. Also write $S_k = S \cap L(e_1, \ldots, e_k)$. Let V be a set of integers t for each of which the vector x can be found such that $x = te_d + y$, where $y \in L(e_1, \ldots, e_{d-1})$. V is clearly an additive group in Z; and therefore, V consists of the points of the form nt_0, where $n \in Z$ and t_0 is a fixed positive integer. If $b_d \in S$, $b_d = y_0 + t_0 e_d$, then for all $x \in S$ for which $x - nt_0 e_d \in L(e_1, \ldots, e_{d-1})$, we have $x - nb_d \in S_{d-1}$.

Similarly, we find that for all k, $b_k \in S_k$ can be constructed such that for all $x \in S_k$, there exist $n \in Z$ such that $x - nb_k \in S_{k-1}$.

□

Corollary. *The necessary and sufficient condition of $S = Z$ is $\alpha_{kk} = 1$ in the representation* (1.46). *If $\alpha_{kk} = 1$ for all k, then $b_1 = e_1$, $e_2 = b_2 - \alpha_{21} b_1$, etc.; i.e., $e_k \in S_k$ for all k. Now suppose that $\alpha_{ll} > 1$ for some l. Then* $\det(b_1, \ldots, b_d) > 1$, *where* $\det(b_1, \ldots, b_d)$ *is the determinant of the matrix* (b_j^i) *and b_j^i are the coordinates of the vector b_j. Whatever integers β_{ij} we take,*

$$\det\Big(\sum \beta_{1j} b_j, \ldots, \sum \beta_{dj} b_j\Big) = \det(\beta_{ij})_{i,j=1,\ldots,d} \cdot \det(b_1, \ldots, b_d);$$

the first factor on the r.h.s. is an integer. Consequently, $\det(x_1, \ldots, x_d)$ *is a multiple of* $\det(b_1, \ldots, b_d)$ *for every $x_1, \ldots, x_d \in S$. Hence, the situation when all $e_1, \ldots, e_d$ belong to S is impossible, since* $\det(e_1, \ldots, e_d) = 1$. *Therefore, $S \neq Z$.*

38°. *Let z_0 be such that $\varphi(z_0) = \mathbf{M} e^{i(z_0, \xi_0)} = 1$. Then $\varphi(z + z_0) = \varphi(z)$.*

In fact, $(1/2\pi)(z_0, x) \in Z$ for all $x \in S$, where S is a support of a random walk. Thus,

$$\begin{aligned}\varphi(z + z_0) &= \sum_{x \in S} \mathbf{P}\{\xi_1 = x\} e^{i(z+z_0, x)} \\ &= \sum_{x \in S} \mathbf{P}\{\xi_1 = x\} e^{i(z,x)} e^{i(z_0,x)} = \varphi(z),\end{aligned}$$

because $e^{i(z_0,x)} = 1$ for $x \in S$.

□

39°. Theorem. *The nondegenerate random walk is irreducible iff $\varphi(z) < 1$ for $z \in K_0 = \{z = (z^1, \ldots, z^d) : |z^j| \leq \pi, j = 1, \ldots, d\}$.*

Proof. Let $z_0 \in K_0$, $\varphi(z_0) = 1$. Then

$$1 = \sum_{x \in Z^d} \mathbf{P}\{\xi_1 = x\} \exp\{i \sum_{j=1}^{d} x^j z_0^j\};$$

and therefore, for all $x \in Z^d$ for which $\mathbf{P}\{\xi_1 = x\} > 0$, we have $\sum_{j=1}^d x^j z_0^j = 2\pi m$, where $m \in Z$. Clearly, the set of those x for which $(1/2\pi)(x, z_0) \in Z$ forms a group; and therefore, this relation also holds for $x \in S$, where S is a support of a random walk. So for all $x \in S$, we have $\cos(x, z_0) = 1$. Nondegeneracy of a random walk implies the existence of d linearly independent vectors $x_1, \dots, x_d$ in Z^d for which $\mathbf{P}\{\xi_1 = x_k\} > 0$, $k = 1, \dots, d$. It follows from the system of equations $\sum_{j=1}^d x_k^j z_0^j = 2\pi m_k$, $k = 1, \dots, d$, $x_k = (x_k^1, \dots, x_k^d)$, $m_k \in Z$, that $z_0^j = 2\pi s_j / r_j$, where s_j and r_j are relatively prime integers (for $s_j \neq 0$), $r_j > 0$, and $|s_j| < r_j$. Thus, if S is the least integral lattice in Z^d for which $\mathbf{P}\{\xi_1 \in S\} = 1$, then $\sum_{j=1}^d x^j s_j / r_j \in Z$, $x \in S$. If $S = Z^d$, then $x^1, \dots, x^d$ can take arbitrary integer values and $s_j / r_j \in Z$ for all j, which is possible only for $s_1 = \dots = s_d = 0$, $z_0 = 0$.

Now suppose that $S \neq Z^d$. Then, as follows from 36°, in S we can find a system of generators $b_k \in Z^d$, $k = 1, \dots, d$ of the form $b_k = \sum_{i=1}^k \alpha_{ki} e_i$; moreover, $\alpha_{ki} \in Z$, $\alpha_{kk} > 0$, $\prod \alpha_{kk} > 1$. Hence, we can find l such that $\alpha_{kk} = 1$ for $k > l$, and $\alpha_{ll} = m > 1$, and the vectors

$$\hat{b}_{l+1} = c_1 + m_1 e_l + e_{l+1}, \dots, \qquad \hat{b}_d = c_{d-l} + m_{d-l} e_l + e_d,$$

where $c_1, \dots, c_{d-1} \in L(e_1, \dots, e_{l-1})$, $0 \le m_i \le m$, such that the vectors $b_1, \dots, b_l, \hat{b}_{l+1}, \dots, \hat{b}_d$ will also be a system of generators in S. In fact,

$$\begin{aligned} b_{l+1} &= c_1' + \alpha_{l+1} e_{l+1}, \\ b_{l+2} &= c_2' + \alpha_{l+2\, l+1} e_{l+1} + e_{l+2}, \\ &\dots \\ b_d &= c_d' + \alpha_{d\, l+1} e_{l+1} + \dots + e_d, \end{aligned}$$

where $c_i' \in L(e_1, \dots, e_l)$. Assume (for $l < j \le d$) that

$$\hat{b}_j = b_j - \sum_{s=l+1}^{j-1} \alpha_{js}^* b_s - \beta_j e_l,$$

where $\begin{pmatrix} 1 & 0 & \dots & 0 \\ \alpha_{l+2\, l+1}^* & 1 & \dots & 0 \\ \dots & \dots & \dots & \dots \\ \alpha_{d\, l+1}^* & \dots & \dots & 1 \end{pmatrix}$ is the integral matrix reciprocal of $\begin{pmatrix} 1 & 0 & \dots & 0 \\ \alpha_{l+2\, l+1} & 1 & \dots & 0 \\ \dots & \dots & \dots & \dots \\ \alpha_{d\, l+1} & \dots & \dots & 1 \end{pmatrix}$, and β_j is an integer such that

$$0 \le \alpha_{lj} - \sum_{s=l+1}^{j-1} \alpha_{js}^* \alpha_{sm} - \beta_j < m.$$

Let $z^i = 0$ for $i < l$; $z^l = 1/m$, and $z^j = -m_{j-l}/m$, $j > l$. Then $(b_k, z) = 0$ for $k < l$; $(b_l, z) = 1$ and $(\hat{b}_j, z) = m_{j-l}/m - m_{j-l}/m = 0$ for $j > l$. Hence, (z, x) is an integer for all $x \in S$; and therefore, $2\pi z \in K_0$ and $\varphi(2\pi z) = 1$. □

Let $(\zeta_n, n \geq 0)$ be a random walk. Introduce the variable $\tau_a = k$ if $\zeta_1 \neq a, \dots, \zeta_{k-1} \neq a$, $\zeta_k = a$; and $\tau_a = +\infty$ if $\zeta_k \neq a$ for all $k > 0$.

40°. Lemma. *Let $\{\zeta_n\}$ be an arithmetic random walk ($\zeta_0 = 0$). For all $|\lambda| < 1$,*

$$\mathbf{M}\lambda^{\tau_a} = \sum_{n=1}^{\infty} \mathbf{P}\{\zeta_n = a\}\lambda^n \Big/ \sum_{n=0}^{\infty} \mathbf{P}\{\zeta_n = 0\}\lambda^n \tag{1.47}$$

(we assume that $\lambda^{+\infty} = 0$).

Proof. The event $\{\zeta_n = a\}$ involves the event $\{\tau_a \leq n\}$; and thus,

$$\begin{aligned}
\mathbf{P}\{\zeta_n = a\} &= \sum_{k=1}^{n} \mathbf{P}\{\zeta_n = a, \tau_a = k\} \\
&= \sum_{k=1}^{n} \mathbf{P}\{\tau_a = k, \xi_{k+1} + \cdots + \xi_n = 0\} \\
&= \sum_{k=1}^{n} \mathbf{P}\{\tau_a = k\}\mathbf{P}\{\xi_{k+1} + \cdots + \xi_n = 0\}
\end{aligned}$$

(the event $\{\tau_a = k\}$ is independent of $\xi_{k+1}, \dots, \xi_n$). Hence,

$$\mathbf{P}\{\zeta_n = a\} = \sum_{k=1}^{n} \mathbf{P}\{\tau_a = k\}\mathbf{P}\{\zeta_{n-k} = 0\}.$$

Multiplying both parts of this equality by λ^n and summing over n from 1 to infinity, we get

$$\sum_{n=1}^{\infty} \mathbf{P}\{\zeta_n = a\}\lambda^n = \sum_{k=1}^{\infty} \mathbf{P}\{\tau_a = k\}\lambda^k \sum_{n=0}^{\infty} \mathbf{P}\{\zeta_n = 0\}\lambda^n,$$

which implies (1.47). □

A walk is called *recurrent* if $\mathbf{P}\{\tau_0 < +\infty\} = 1$.

41°. Theorem. *A walk is recurrent iff*

$$\sum_{n=1}^{\infty} \mathbf{P}\{\zeta_n = 0\} = +\infty. \tag{1.48}$$

Proof. Formula (1.47) yields

$$\sum_{n=1}^{\infty} \mathbf{P}\{\tau_0 = n\}\lambda^n = \frac{\sum_{n=1}^{\infty} \mathbf{P}\{\zeta_n = 0\}\lambda^n}{1 + \sum_{n=1}^{\infty} \mathbf{P}\{\zeta_n = 0\}\lambda^n}.$$

Clearly,

$$\begin{aligned} \mathbf{P}\{\tau_0 < +\infty\} &= \sum_{n=1}^{\infty} \mathbf{P}\{\tau_0 = n\} = \lim_{\lambda \uparrow 1} \sum_{n=1}^{\infty} \mathbf{P}\{\tau_0 = n\}\lambda^n \\ &= \lim_{\lambda \uparrow 1} \left(1 - \frac{1}{\sum_{n=1}^{\infty} \mathbf{P}\{\zeta_n = 0\}\lambda^n}\right)^{-1}. \end{aligned} \tag{1.49}$$

If the series in (1.48) converges, then

$$\mathbf{P}\{\tau_0 < +\infty\} = \left(1 - \frac{1}{\sum_{n=1}^{\infty} \mathbf{P}\{\zeta_n = 0\}}\right)^{-1} < 1;$$

and if (1.48) holds, then the limit on the r.h.s. of (1.49) is equal to 1.

□

42°. Corollary. *A walk is recurrent iff*

$$\begin{aligned} &\lim_{\lambda \uparrow 1} \int_{C_\pi} \mathrm{Re}\, \frac{1}{1 - \varphi(z)\lambda} dz = +\infty, \\ &C_\pi = \{z : |z^j| \leq \pi, j = 1, \ldots, d\}. \end{aligned} \tag{1.50}$$

In fact,

$$\sum_{x \in Z^d} \mathbf{P}\{\zeta_n = x\} e^{i(z,x)} = \mathbf{M} e^{i(z,\zeta_n)} = \mathbf{M} e^{i\left(z, \sum_{k=1}^{n} \xi_k\right)} = \varphi^n(z),$$

$$\mathbf{P}\{\zeta_n = 0\} = \left(\frac{1}{2\pi}\right)^d \int_{C_\pi} \varphi^n(z) dz,$$

$$\begin{aligned} \sum_{n=0}^{\infty} \lambda^n \mathbf{P}\{\zeta_n = 0\} &= \left(\frac{1}{2\pi}\right)^d \int_{C_\pi} \sum_{n=0}^{\infty} \lambda^n \varphi^n(z) dz \\ &= \left(\frac{1}{2\pi}\right)^d \int_{C_\pi} \frac{1}{1 - \lambda\varphi(z)} dz = \left(\frac{1}{2\pi}\right)^d \int_{C_\pi} \mathrm{Re}\, \frac{1}{1 - \lambda\varphi(z)} dz. \end{aligned}$$

This equality together with (1.48) implies (1.50).

□

43°. Lemma. *Let $x \neq 0$, $|\lambda| < 1$. Then*

$$\sum_{n=0}^{\infty} \lambda^n \mathbf{P}\{\zeta_n = 0, \tau_x > n\} \leq \left(\frac{1}{2\pi}\right)^d \int_{C_\pi} (2 - e^{i(z,x)} - e^{-i(z,x)}) \frac{1}{1 - \lambda\varphi(z)} dz. \tag{1.51}$$

Proof. We have

$$\begin{aligned}
&\sum_{n=0}^{\infty} \lambda^n \mathbf{P}\{\zeta_n = 0, \tau_x > n\} \\
&\quad = \sum_{n=0}^{\infty} \lambda^n \mathbf{P}\{\zeta_n = 0\} - \sum_{n=0}^{\infty} \lambda^n \mathbf{P}\{\zeta_n = 0, \tau_x \leq n\} \\
&\quad = \sum_{n=0}^{\infty} \lambda^n \mathbf{P}\{\zeta_n = 0\} - \sum_{n=0}^{\infty} \lambda^n \sum_{k=1}^{n} \mathbf{P}\{\tau_x = k, \zeta_n = 0\} \\
&\quad = \sum_{n=0}^{\infty} \lambda^n \mathbf{P}\{\zeta_n = 0\} - \sum_{n=0}^{\infty} \lambda^n \sum_{k=1}^{n} \mathbf{P}\{\tau_x = k\}\mathbf{P}\{\zeta_{n-k} = -x\} \\
&\quad = \sum_{n=0}^{\infty} \lambda^n \mathbf{P}\{\zeta_n = 0\} - \sum_{n=1}^{\infty} \lambda^n \mathbf{P}\{\tau_x = n\} \sum_{k=1}^{\infty} \mathbf{P}\{\zeta_k = -x\}\lambda^k.
\end{aligned}$$

But it has been proved in 40° that

$$\sum_{k=1}^{\infty} \mathbf{P}\{\zeta_k = -x\}\lambda^k = \sum_{k=1}^{\infty} \lambda^k \mathbf{P}\{\tau_{-x} = k\} \sum_{n=0}^{\infty} \lambda^n \mathbf{P}\{\zeta_n = 0\}. \tag{1.52}$$

Hence, for $0 < \lambda < 1$,

$$\begin{aligned}
&\sum_{n=0}^{\infty} \lambda^n \mathbf{P}\{\zeta_n = 0, \tau_x > n\} \\
&\quad = \sum_{n=0}^{\infty} \mathbf{P}\{\zeta_n = 0\}(1 - \mathbf{M}\lambda^{\tau_x}\mathbf{M}\lambda^{\tau_{-x}}) \\
&\quad \leq \sum_{n=0}^{\infty} \lambda^n \mathbf{P}\{\zeta_n = 0\}(2 - \mathbf{M}\lambda^{\tau_x} - \mathbf{M}\lambda^{\tau_{-x}}) \\
&\quad = \sum_{n=0}^{\infty} (2\mathbf{P}\{\zeta_n = 0\} - \mathbf{P}\{\zeta_n = x\} - \mathbf{P}\{\zeta_n = -x\})\lambda^n
\end{aligned}$$

(we have used the inequality $2 - a - b > 1 - ab$ (for $0 < a < 1$, $0 < b < 1$) and (1.52)).

Since

$$\mathbf{P}\{\zeta_n = x\} = \left(\frac{1}{2\pi}\right)^d \int_{C_\pi} \varphi^n e^{-i(z,x)} dz,$$

we have

$$\sum_{n=0}^{\infty} \lambda^n \mathbf{P}\{\zeta_n = 0, \tau_x > n\}$$
$$\leq \left(\frac{1}{2\pi}\right)^d \int_{C_\pi} (2 - e^{i(z,x)} - e^{-i(z,x)}) \sum_{n=0}^{\infty} \lambda^n \varphi^n(z) dz.$$

The inequality (1.51) is thus proved.

□

44°. Theorem. *A random walk is recurrent iff*

$$\int_{C_\pi} \operatorname{Re} \frac{1}{1 - \varphi(z)} dz = +\infty. \tag{1.53}$$

Proof. If (1.53) is valid, then by virtue of Fatou's theorem, we also have

$$\lim_{\lambda \uparrow 1} \int_{C_\pi} \operatorname{Re} \frac{1}{1 - \lambda\varphi(z)} dz = +\infty;$$

i.e., (1.50) holds and the walk is recurrent. Now suppose that

$$\int_{C_\pi} \operatorname{Re} \frac{1}{1 - \varphi(z)} dz < \infty.$$

The variable $\frac{1-\varphi(z)}{1-\lambda\varphi(z)} = \frac{1-\varphi(z)}{(1-\lambda)+\lambda(1-\varphi(z))}$ is bounded for $1/2 < \lambda < 1$, and

$$\sum_{n=0}^{\infty} \lambda^n \mathbf{P}\{\zeta_n = 0, \tau_x > n\}$$
$$\leq 2\left(\frac{1}{2\pi}\right)^d \int_{C_\pi} (1 - \cos(z,x)) \frac{1}{1 - \lambda\varphi(z)} dz \tag{1.54}$$
$$= 2\left(\frac{1}{2\pi}\right)^d \int_{C_\pi} \frac{1 - \cos(z,x)}{1 - \varphi(z)} \cdot \frac{1 - \varphi(z)}{1 - \lambda\varphi(z)} dz.$$

Let S be a lattice support of a random walk. If $z_0 \in C_\pi$ and $\varphi(z_0) = 1$, then $(x, z_0) = 2\pi m$, $m \in Z$ for all $x \in S$ (as was shown in 38°). Therefore,

$$\frac{1 - \cos(z,x)}{1 - \varphi(z)} = \frac{1 - \cos(z - z_0, x)}{1 - \varphi(z - z_0)} \leq \frac{1}{2\varepsilon} \cdot \frac{(z - z_0, x)^2}{|z - z_0|^2}$$

for sufficiently small $|z - z_0|$. The function $\frac{1-\cos(z,x)}{1-\varphi(z)}$ is bounded for $z \in C_\pi$.

The integrand on the r.h.s. of (1.54) is bounded for $z \in C_\pi$, $\lambda \in [1/2, 1)$. Passing to the limit as $\lambda \uparrow 1$, we get for $x \in S$,

$$\begin{aligned}\sum_{n=0}^{\infty} \mathbf{P}\{\zeta_n = 0, \tau_x > n\} &\\ &\le 2\left(\frac{1}{2\pi}\right)^d \int_{C_\pi} \frac{1-\cos(z,x)}{1-\varphi(z)} dz \\ &= 2\left(\frac{1}{2\pi}\right)^d \int_{C_\pi} (1-\cos(z,x)) \operatorname{Re} \frac{1}{1-\varphi(z)} dz.\end{aligned}$$

For every m,

$$\begin{aligned}\sum_{n=0}^{m} \mathbf{P}\{\zeta_n = 0, \tau_x > n\} &\\ &\le 2\left(\frac{1}{2\pi}\right)^d \int_{C_\pi} (1-\cos(z,x)) \operatorname{Re} \frac{1}{1-\varphi(z)} dz.\end{aligned}$$

If $|x| \to \infty$, then $\tau_x \to +\infty$; and consequently,

$$\begin{aligned}\sum_{n=0}^{m} \mathbf{P}\{\zeta_n = 0\} &= \lim_{|x|\to\infty} \sum_{n=0}^{m} \mathbf{P}\{\zeta{=}0, \tau_x > n\} \\ &\le 2 \cdot \left(\frac{1}{2\pi}\right)^d \int_{C_\pi} \operatorname{Re} \frac{1}{1-\varphi(z)} dz\end{aligned}$$

(by virtue of the Riemann–Lebesgue theorem, $\lim_{|x|\to\infty} \int_{C_\pi} \cos(z,x) \times \operatorname{Re} \frac{1}{1-\varphi(z)} dz = 0$). Therefore,

$$\sum_{n=0}^{\infty} \mathbf{P}\{\zeta_n = 0\} \le 2(\frac{1}{2\pi})^d \int_{C_\pi} \operatorname{Re} \frac{1}{1-\varphi(z)} dz < \infty;$$

i.e., a walk is transient. □

Corollary. *For $d \ge 3$, each nondegenerate walk is transient.*

If $\varphi(z_0) = 1$, $z_0 \in C_\pi$, then $|1/(1-\varphi(z))| \le 1/\varepsilon|z - z_0|^2$ for sufficiently small $|z - z_0|$. The function $1/|z - z_0|^2$ is integrable in the neighbourhood of z_0 for $d \ge 3$. Thus,

$$\int_{C_\pi} \operatorname{Re} \frac{1}{1-\varphi(z)} dz < \infty.$$

□

Nonlattice Distribution

Let $\{\zeta_n\}_{n\geq 0}$ be a walk in R^d, and let B be some Borel set in R^d. Let $\tau(B)$ denote a random variable for which $\tau(B) = k \geq 1$ if $\zeta_1 \overline{\in} B, \ldots, \zeta_{k-1} \overline{\in} B$, $\zeta_k \in B$; and $\tau(B) = +\infty$ if $\zeta_k \overline{\in} B$ for all $k \geq 1$.

A random walk is called *recurrent* (we consider $\zeta_0 = 0$) if $\mathbf{P}\{\tau(U) < \infty\} = 1$ for every neighbourhood U of the point 0.

Let us introduce a *resolvent* of a random walk: For each Borel set B and $0 < \lambda < 1$, we put

$$R(\lambda, B) = \sum_{n=0}^{\infty} \lambda^n \mathbf{P}\{\zeta_n \in B\}. \tag{1.55}$$

The *Fourier transform* of the resolvent [$R(\lambda, \cdot)$ is a measure (as a function of the set)] is defined by the formula

$$r(\lambda, z) = \int e^{i(z,x)} R(\lambda, dx) = \frac{1}{1 - \lambda\varphi(z)}. \tag{1.56}$$

The resolvent satisfies the equation

$$R(\lambda, B) = I_B(0) + \lambda \int \mathbf{P}\{\xi_1 \in dy\} R(\lambda, B - y) \tag{1.57}$$

(here $B - y = \{x : x + y \in B\}$). The formula (1.56) is obtained by applying the Fourier transform to (1.55). The Fourier transform of the r.h.s. of (1.57) coincides with

$$1 + \lambda\varphi(z) \frac{1}{1 - \lambda\varphi(z)} = \frac{1}{1 - \lambda\varphi(z)},$$

i.e., with the Fourier transform of the l.h.s.

Introduce a function defining the joint distribution of $\tau(B)$ and $\zeta_{\tau(B)}$ under the assumption $\tau(B) < \infty$:

$$N_B(k, C) = \mathbf{P}\{\tau(B) = k, \zeta_{\tau(B)} \in C\} \tag{1.58}$$

(C is a Borel set), and let $\widehat{N}_B(\lambda, C)$ be its generating function

$$\widehat{N}_B(\lambda, C) = \sum_{k=1}^{\infty} \lambda^k N_B(k, C). \tag{1.59}$$

45°. Lemma. *The function $\widehat{N}_B(\lambda, C)$ satisfies the equation*

$$R(\lambda, B) = I_B(0) + \int \widehat{N}_B(\lambda, dy) R(\lambda, B - y). \tag{1.60}$$

Proof. For $n > 0$, we have

$$\begin{aligned}
\mathbf{P}\{\zeta_n \in B\} &= \mathbf{P}\{\zeta_n \in B, \tau(B) \le n\} \\
&= \sum_{k=1}^{n} \mathbf{P}\{\zeta_n \in B, \tau(B) = k\} \\
&= \sum_{k=1}^{n} \int \mathbf{P}\{\zeta_n \in B, \tau(B) = k, \zeta_k \in dy\} \\
&= \sum_{k=1}^{n} \int \mathbf{P}\{\zeta_n - \zeta_k \in B - y, \tau(B) = k, \zeta_k \in dy\} \\
&= \sum_{k=1}^{n} \mathbf{P}\{\zeta_{n-k} \in B - y\} N_B(k, dy).
\end{aligned}$$

Hence,

$$R(\lambda, B) = \mathbf{P}\{\zeta_0 \in B\} + \sum_{n=1}^{\infty}\sum_{k=1}^{n} \lambda^n \int \mathbf{P}\{\zeta_{n-k} \in B - y\} N_B(k, dy),$$

This immediately yields (1.60). □

Let U_ε denote the closed sphere of radius ε centered at the origin.

46°. Lemma. *If for all sufficiently small* $\varepsilon > 0$,

$$\lim_{\lambda \uparrow 1} R(\lambda, U_\varepsilon) = R(1, U_\varepsilon) < \infty,$$

then the random walk is transient.

Proof. Let $0 < \lambda < 1$. Then it follows from (1.60) that

$$\begin{aligned}
R(\lambda, U_{\delta+\varepsilon}) &\ge 1 + \int_{|y| \le \varepsilon} \widehat{N}_{U_{\delta+\varepsilon}}(\lambda, dy) R(\lambda, U_\delta) \\
&= 1 + R(\lambda, U_\delta) \widehat{N}_{U_{\delta+\varepsilon}}(\lambda, U_\varepsilon).
\end{aligned}$$

Since $\tau(U_\varepsilon) \ge \tau(U_{\delta+\varepsilon})$, then for $0 < \lambda < 1$,

$$\widehat{N}_{U_{\delta+\varepsilon}}(\lambda, U_\varepsilon) \ge \widehat{N}_{U_\varepsilon}(\lambda, U_\varepsilon) = \mathbf{M}\lambda^{\tau(U_\varepsilon)}.$$

Thus,

$$\begin{aligned}
\mathbf{M}\lambda^{\tau(U_\varepsilon)} &\le \frac{R(\lambda, U_{\delta+\varepsilon}) - 1}{R(\lambda, U_\delta)}, \\
\mathbf{P}\{\tau(U_\varepsilon) < +\infty\} &\le \frac{R(1, U_{\delta+\varepsilon}) - 1}{R(1, U_\delta)}.
\end{aligned} \tag{1.61}$$

Since $R(1, U_{\delta+\varepsilon}) < \infty$, and $\lim_{\varepsilon\downarrow 0} R(1, U_{\delta+\varepsilon}) = R(1, U_\delta)$, the l.h.s. of (1.61) is less than 1 for sufficiently small ε.

□

47°. Lemma. *If* $\lim_{\lambda\uparrow 1} R(x, U) = +\infty$ *for every open set* $U \ni 0$, *then there exist functions* $\alpha(\lambda) > 0$, $\lambda \in (0, 1)$ *such that* $\alpha(\lambda) \to 0$ *as* $\lambda \uparrow 1$, *and the measures* $\alpha(\lambda)R(\lambda, U)$ *converge weakly on finite continuous functions to the Lebesgue measure in* R^d.

Proof. Let us choose some $c > 0$, and let L_0^c denote the set of functions $g(x)$ for which

$$\int (1 + |x|^2)|g(x)|dx < \infty, \tag{1.62}$$

and if $\tilde{g}(z) = \int e^{i(z,x)} g(x)dx$, then $\tilde{g}(z) = 0$ for $|z| \geq c$. The set L_0^c contains a strictly positive function: Let $\psi_1(z)$ be an even positive function for $|z| < c/2$, $\psi_1(z) = 0$ when $|z| \geq c/2$, and $g_1(x) = (2\pi)^{-d} \int e^{-i(z,x)} \psi(z)dz$. Then for the function $g_c(x) = \int g_1^2(x - y) g_1^2(y)dy$, we have

$$\tilde{g}_c(z) = \left(\int \psi_1(z - u)\psi_1(u)du\right)^2$$

and $\tilde{g}_c(z) = 0$ for $|z| \geq c$. The condition (1.62) holds if $\psi_1(z)$ is twice continuously differentiable.

Strict positiveness of $g_c(z)$ follows from the fact that $g_1^2(x) \geq 0$ and cannot vanish in a certain neighbourhood as an entire analytic function.

We put

$$a_\lambda(c) = (2\pi)^{-d} \int_{|z|\leq c} \frac{1}{1 - \lambda\varphi(z)} dz,$$
$$b_\lambda(c) = (2\pi)^{-d} \int_{|z|\leq c} \frac{z}{1 - \lambda\varphi(z)} dz,$$
$$(b_\lambda(c) \in R^d).$$

Let $g(x) \in L_0^c$. Then

$$\int g(x)R(\lambda, dx) = (2\pi)^{-d} \int_{|z|\leq c} \frac{\tilde{g}(z)}{1 - \lambda\varphi(z)} dz$$
$$= (2\pi)^{-d}(a_\lambda(c)\tilde{g}(0) + (b_\lambda(c), \tilde{g}'(0)) + O(1),$$

since $\overline{\lim}_{\lambda\uparrow 1} \int_{|z|\leq c} \frac{|z|^2}{|1-\varphi(z)|^2} dz < \infty$. If g is even, then

$$(2\pi)^d \int g(x)R(\lambda, dx) = a_\lambda(c)\tilde{g}(0) + O(1) = a_\lambda(c) \int g(x)dx + O(1);$$

and therefore, $a_\lambda(c) \to +\infty$ as $\lambda \uparrow 1$. If g is odd, then

$$(2\pi)^d \int g(x)R(\lambda, dx) = \left(b_\lambda(c), \int xg(x)dx\right) + O(1),$$
$$\left|\left(b_\lambda(c), \int xg(x)dx\right)\right| \le a_\lambda(c)\int |g(x)|dx + O(1).$$

Since $\int |g(x)|dx(|\int xg(x)dx|)^{-1}$ can be made arbitrary small, $|b_\lambda(c)|$=$0(a_\lambda(c))$.
Hence, assuming that $\alpha(\lambda) = a_\lambda^{-1}(c)(2\pi)^d$, we obtain

$$\lim_{\lambda\uparrow 1} g(x)\alpha(\lambda)R(\lambda, dx) = \tilde{g}(0) = \int g(x)dx. \tag{1.63}$$

Moreover, $\overline{\lim}_{\lambda\uparrow 1}\int g_c(x)\alpha(\lambda)R(\lambda, dx) < \infty$. It follows from the last relation that $\overline{\lim}\int \alpha(\lambda)R(\lambda, U_r) < \infty$ for every sphere U_r; and therefore, the measures $\alpha(\lambda)R(\lambda, dx)$ are compact with respect to the weak convergence on finite continuous functions. For functions from L_0^c, (1.63) is valid (c can be chosen arbitrarily in this case). Hence, if $m(dx)$ is some limit measure for $\alpha(\lambda)R(\lambda, dx)$, then $\int g(x)dx = \int g(x)m(dx)$ for $g \in L_0^c$, such that $|g(x)| \le g_c(x)$ for some $c > 0$; but such functions clearly form a total set. Thus, the single limit point of $\alpha(\lambda)R(\lambda, dx)$, in the sense of the weak convergence on continuous finite functions, is the Lebesgue measure.

□

48°. Theorem. *For a nonlattice random walk, one of two possibilities is valid: either* (1) $R(1, V) < \infty$ *for all bounded open sets* V*; or* (2) $R(1, V) = +\infty$ *for all bounded open sets* V.

Proof. If $R(1, U_\delta) < \infty$ for some $\delta > 0$, then $R(1, U_\rho) \le R(1, U_\delta)$ for all $\rho < \delta$. Suppose that S is a sphere of radius $\delta/2$ centered at the point x.
Then $R(1, S) \le R(1, U_\delta)$ for $|x| \le \delta/2$. By virtue of (1.60), we have for $|x| > \delta/2$,

$$R(\lambda, S) = \int_S \widehat{N}(\lambda, dy)R(\lambda, S - y) \le R(\lambda, U_\delta)\widehat{N}(\lambda, S) \le R(\lambda, U_\delta).$$

Hence, $R(1, S) \le R(1, U_\delta)$. Any bounded set V can be covered by a finite number of spheres of radius $\delta/2$, so $R(1, V) < \infty$. And if $R(1, U_\delta) = +\infty$ for all δ, then the validity of statement (2) follows from Lemma 47°.

□

49°. Theorem. *The nonlattice random walk is recurrent iff for some* $c > 0$,

$$\lim_{\lambda\uparrow 1}\int_{|z|\le c}\frac{1}{1 - \lambda\varphi(z)}dz = +\infty.$$

Proof. Using the proof of Lemma 47°, one can see that the condition

$$\lim_{\lambda\uparrow 1}\int_{|z|\le c}\frac{1}{1-\lambda\varphi(z)}dz<\infty$$

implies

$$\lim_{\lambda\uparrow 1}\int g_c(x)R_\lambda(dx)<\infty.$$

This inequality, in turn, yields the validity of $R(1,V)<\infty$ for every bounded open set V. By virtue of Lemma 46°, the walk is transient. The necessity of the condition of the theorem is thus proved. If the condition of the theorem is satisfied, then $\alpha(\lambda)$ (introduced in Lemma 47°) approaches zero. By virtue of Eq. (1.60), we get

$$\alpha(\lambda)R(\lambda,U_\delta)=\int\widehat{N}_{U_\delta}(\lambda,dy)\alpha(\lambda)R(\lambda,U_\delta-y)+\alpha(\lambda).$$

The variable $\alpha(\lambda)R(\lambda,U_\delta-y)\le\alpha(\lambda)R(\lambda,U_{2\delta})$ is bounded and tends to $m(U_\delta)$, where m is the Lebesgue measure. Passing to the limit as $\lambda\uparrow 1$, we obtain $m(U_\delta)=m(U_\delta)\widehat{N}_{U_\delta}(1,U_\delta)$, $\widehat{N}_{U_\delta}(1,U_\delta)=\mathbf{P}\{\tau_{U_\delta}<\infty\}=1$.

□

50°. Corollary. *If for some $c>0$,*

$$\int_{|z|\le c}\operatorname{Re}\frac{1}{1-\varphi(z)}dz=+\infty,$$

then a walk is recurrent.

In fact, since $\operatorname{Re}\frac{1}{1-\lambda\varphi(z)}\ge 0$, then by virtue of Fatou's theorem,

$$\lim_{\lambda\uparrow 1}\int\operatorname{Re}\frac{1}{1-\lambda\varphi(z)}dz\ge\int\operatorname{Re}\frac{1}{1-\varphi(z)}dz=+\infty.$$

□

Consider the recurrence of walks for concrete values of d. The results are formulated for lattice and nonlattice walks simultaneously.

51°. Theorem. **1.** *If $d\ge 3$, the walk is transient.*

2. *Suppose that $\mathbf{M}\xi_1$ exists and $\mathbf{M}\xi_1\ne 0$. Then the walk is transient.*

3. *Suppose that $\mathbf{M}\xi_1=0$ and $\mathbf{M}|\xi_1|^2<\infty$, $d\le 2$. Then the walk is recurrent.*

4. *If $d = 1$, $\mathbf{M}\xi_1$ exists, and $\mathbf{M}\xi_1 = 0$, then the walk is transient.*

Proof. 1. We have for sufficiently small $c > 0$ and some $\beta > 0$,

$$\int_{|z|\le c} \frac{1}{|1-\lambda\varphi(z)|}dz \le \frac{1}{\lambda}\int_{|z|\le c} \frac{1}{|1-\varphi(z)|}dz \le \beta \int_{|z|\le c} \frac{dz}{|z|^2} < \infty.$$

2. The strong law of large numbers implies that $\zeta_n \sim na$ for $\mathbf{M}\xi_1 = a \neq 0$; and thus, whatever r we take, only a finite number of the events $\{\zeta_n \in U_r\}$ happen. But a recurrent walk visits any open set infinitely many times, so the walk is transient.

3. Suppose $\mathbf{M}\xi_1 = 0$, $\mathbf{M}\xi_1^2 < \infty$. Then there exists a function $\alpha(z)$ continuous for $z \neq 0$, $|z| \le c$, such that $|\alpha(z)| > \delta$ and $\varphi(z) = 1 - |z|^2\alpha(z)$. Therefore,

$$\int_{|z|\le c} \frac{dz}{1-\lambda\varphi(z)} = \int_{|z|\le c} \frac{dz}{(1-\lambda)+\lambda\alpha(z)z^2} \to +\infty$$

as $\lambda \uparrow 1$, because $\int_{|z|\le c} \frac{dz}{|z|^2} = +\infty$ for $d \le 2$.

4. Let $d = 1$, $\mathbf{M}\xi_1 = 0$. By virtue of the law of large numbers, for every $\varepsilon > 0$, one can find n_ε such that the inequality $\mathbf{P}\{|\zeta_n| < n\varepsilon\} \ge 1/2$ holds for $n > n_\varepsilon$. Thus, $\mathbf{P}\{|\zeta_n| \le a\} \ge 1/2$ for an integer $a > 0$ if $n_\varepsilon < n < a/\varepsilon$. Therefore,

$$R(1,[-a,a]) \ge \frac{1}{2}\left(\frac{a}{\varepsilon} - n_\varepsilon\right).$$

Since

$$R(1,[k-1,k]) \le R(1,[-1,1]),$$

we have

$$\frac{1}{2}\left(\frac{a}{\varepsilon} - n_\varepsilon\right) \le R(1,[-a,a]) \le \sum_{k=-a+1}^{a} R(1,[k-1,k]) \le 2aR(1,[-1,1]).$$

Hence, $R(1,[-1,1]) \ge \frac{1}{4}(\frac{1}{\varepsilon} - \frac{n_\varepsilon}{a})$. Passing to the limit as $a \to \infty$, $\varepsilon \to \infty$, we obtain $R(1,[-1,1]) = +\infty$. The walk is recurrent.

□

§1.4 Distribution of Ladder Functions

Consider a random walk $\{\zeta_n, n \ge 0\}$, $\zeta_0 = 0$ in R. We put $\tau_1 = \tau(R_+\backslash\{0\})$, $\eta_1 = \zeta_{\tau_1}$, $\tau_2 = \inf[k > \tau_1 : \zeta_k > \eta_1]$, $\eta_2 = \zeta_{\tau_2}, \ldots, \tau_r = \inf[k > \tau_{r-1} : \zeta_k > \eta_{r-1}]$, $\eta_r = \zeta_{\tau_r}$, etc. The variable τ_1 can take the value $+\infty$; in this case, we assume that all other τ_r also take the value $+\infty$. The variable η_r is defined provided that $\tau_r < +\infty$; furthermore, $\eta_r > \eta_{r-1}$. Let us consider that η_r is defined for $\tau_r = +\infty$ as well, and in this case $\eta_r = \eta_{r-1}$. The moments τ_k are called *ladder moments*, and η_k are called *ladder values*.

Let $\mathcal{F} = \sigma(\xi_1, \xi_2, \ldots)$.

Denote by $\mathcal{F}_{\tau_r}$ a σ-algebra generated by the events $A \in \mathcal{F}$ for which $A \cap \{\tau_r = k\} \in \mathcal{F}_k$, where $\mathcal{F}_k$ is a σ-algebra generated by the random variables $\xi_1, \xi_2, \ldots, \xi_k$ (steps of a random walk).

52°. *For $\tau_r < \infty$, we have*

$$\mathbf{P}\{\tau_{r+1} - \tau_r = k, \eta_{r+1} - \eta_r \in C/\mathcal{F}_{\tau_r}\} = \mathbf{P}\{\tau_1 = k, \eta_1 \in C\} \tag{1.64}$$

for all $k \in Z_+$ and Borel sets $C \in R_+$.

Proof. If $A \in \mathcal{F}_{\tau_r}$, then $A \cap \{\tau_r = l\} = \{(\xi_1, \ldots, \xi_l) \in D\}$, where D is a Borel set in R^l. To prove formula (1.64), we are to show that

$$\begin{aligned}\mathbf{P}\{\tau_{r+1} - \tau_r = k, \eta_{r+1} - \eta_r \in C, \tau_r = l, (\xi_1, \ldots, \xi_l) \in D\} \\ = \mathbf{P}\{\tau_r = l, (\xi_1, \ldots, \xi_l) \in D\}\mathbf{P}\{\tau_1 = k, \eta_1 \in C\}.\end{aligned}$$

Here the probability on the left coincides with the probability

$$\begin{aligned}\mathbf{P}\{(\xi_1, \ldots, \xi_l) \in D, \tau_r = l, \zeta_{l+1} - \zeta_l \leq 0, \ldots, \\ \zeta_{l+k-1} - \zeta_l \leq 0, \zeta_{l+k} - \zeta_l > 0, \zeta_{l+k} - \zeta_l \in C\} \\ = \mathbf{P}\{(\xi_1, \ldots, \xi_l) \in D, \tau_r = l\}\mathbf{P}\{\zeta_1 \leq 0, \ldots, \zeta_{k-1} \leq 0, \zeta_k > 0, \zeta_k \in C\};\end{aligned}$$

The second factor on the l.h.s. is actually $\mathbf{P}\{\tau_1 = k, \eta_1 \in C\}$.

□

Formula (1.64) allows us to find the joint distribution of $\tau_1, \eta_1, \ldots, \tau_r, \eta_r$ for every r

$$\begin{aligned}\mathbf{P}\{\tau_1 = k_1, \eta_1 \in C_1, \tau_2 - \tau_1 = k_2, \eta_2 - \eta_1 \in C_2, \ldots, \\ \tau_r - \tau_{r-1} = k_r, \eta_r - \eta_{r-1} \in C_r\} \\ = \prod_{i=1}^{r} \mathbf{P}\{\tau_1 = k_i, \eta_1 \in C_i\}.\end{aligned} \tag{1.65}$$

Let us denote $\tau(x) = \inf[k > 1, \zeta_k > x]$, $\gamma(x) = \zeta_{\tau(x)} - x$ for $x \geq 0$ ($\tau(x) = +\infty$ if $\sup_{k>0} \zeta_k \leq x$; $\gamma(x)$ is not defined in this case). $\tau(x)$ is called a *time of the first jump* (random walk) *over the value x*, and $\gamma(x)$ is called a *value of the jump*.

$$\begin{aligned}\tau_1 &= \tau(0), & \eta_1 &= \gamma(0), \\ \tau_2 &= \tau(\gamma(0)), & \eta_2 &= \eta_1 + \gamma(\eta_1), \ldots, \\ \tau_r &= \tau(\eta_{r-1}), & \eta_r &= \eta_{r-1} + \gamma(\eta_{r-1}).\end{aligned}$$

53°. Introduce the functions

$$Q_k(x, y) = \mathbf{P}\{\tau(x) = k, \gamma(x) < y\}.$$

Then

$$\begin{aligned}Q_1(x, y) &= \mathbf{P}\{x < \xi_1 < x + y\}, \\ Q_r(x, y) &= \int_{(-\infty, x]} \mathbf{P}\{\xi_1 \in dz\} Q_{r-1}(x - z, y), \quad r > 1\end{aligned} \tag{1.66}$$

(for $r > 1$, the event $\tau(x) = r$ implies the events $\xi_1 < x$, $\sum_{i=2}^{k} \xi_i < x - \xi_1$ for $k = 2, \ldots, r-1$, and $\sum_{i=2}^{r} \xi_i \in (x-\xi_1, x-\xi_1+y)$). Multiplying the first relation by λ, the second one by λ^2, ..., the rth one by λ^r ($|\lambda| < 1$) and summing them, we get the equation for the function $\widehat{Q}(\lambda, x, y) = \sum_{k=1}^{\infty} \lambda^k Q_k(\lambda, x, y)$:

$$\widehat{Q}(\lambda, x, y) = \lambda \mathbf{P}\{x < \xi_1 < x + y\} + \lambda \int_{(-\infty, x]} \widehat{Q}(\lambda, x - z, y) dF(z).$$

Here $F(x) = \mathbf{P}\{\xi_1 < x\}$.

We extend the definition of $\widehat{Q}(\lambda, x, y)$ for negative x by putting $\widehat{Q}(\lambda, x, y) = 0$ if $x < 0$. Then the previous equation can be rewritten as follows:

$$\begin{aligned} \widehat{Q}(\lambda, x, y) &= \lambda \mathbf{P}\{x < \xi_1 < x + y\} \\ &\quad + \lambda \int \widehat{Q}(\lambda, x - z, y) dF(z), \quad x \geq 0; \\ \widehat{Q}(\lambda, x, y) &= 0, \quad x < 0. \end{aligned} \tag{1.67}$$

Such an equation is called a *convolution equation on the half-line* (or *Wiener–Hopf equation*).

In Eq. (1.67), y can be considered as a fixed parameter. Let us show how to solve equations of this type:

$$\begin{aligned} \widehat{Q}(\lambda, x) &= \Phi(x) + \lambda \int Q(\lambda, x - z) dF(z), \quad x \geq 0, \\ \widehat{Q}(\lambda, x) &= 0, \quad x < 0. \end{aligned} \tag{1.68}$$

Let $\varepsilon(t) = 1$ for $t \geq 0$ and $\varepsilon(t) = 0$ for $t < 0$. Then assuming that $\phi(x) = \varepsilon(x)\phi_1(x)$, we obtain from (1.68)

$$\begin{aligned} \widehat{Q}(\lambda, x) &= \Phi_1(x) + \lambda \varepsilon(x) \int Q(\lambda, x - z) dF(z), \\ \Phi_1(x) &= \varepsilon(x) \int Q(\lambda, x - z) d[\varepsilon(z) - \lambda F(z)]. \end{aligned} \tag{1.69}$$

If $v_1(t)$ is some function with bounded variation for which $v_1(t) = v_1(0)$ when $t > 0$, then

$$\int \Phi_1(x - t) dv_1(t) = \int \varepsilon(x - t) \int Q(\lambda, x - z - t) d[\varepsilon(z) - \lambda F(z)] dv_1(t).$$

Since $dv_1(t) = 0$ for $t > 0$, we have $\varepsilon(x) = \varepsilon(x)\varepsilon(x - t)$ for $x \geq 0$; and thus,

$$\begin{aligned} \varepsilon(x) \int \Phi_1(x - t) dv_1(t) &= \varepsilon(x) \iint Q(\lambda, x - z - t) d[\varepsilon(z) - \lambda F(z)] dv_1(t) \\ &= \varepsilon(x) \int Q(\lambda, x - u) dv_2(u), \end{aligned}$$

where

$$v_2(u) = \int_{-\infty}^{\infty} [\varepsilon(u-t) - \lambda F(u-t)]dv_1(u). \tag{1.70}$$

Suppose that $v_2(u) = 0$ for $u \le 0$. Then $\int Q(\lambda, x-u)dv_2(u) = 0$ for $x < 0$; and therefore,

$$\varepsilon(x) \int \Phi_1(x-t)dv_1(t) = \int Q(\lambda, x-u)dv_2(u). \tag{1.71}$$

Equation (1.71) is the usual convolution equation, which can be solved using the Fourier transform (as will be shown later).

54°. Lemma. *Denote by $F_n(t)$ the distribution function of the variable ζ_n : $F_n(t) = \mathbf{P}\{\zeta_n < t\}$. Then the functions*

$$\tilde{v}_2(z) = \exp\left\{-\sum_{n=1}^{\infty} \frac{\lambda^n}{n} \int_{[0,\infty)} e^{izt} dF_n(t)\right\}, \tag{1.72}$$

$$\tilde{v}_1(z) = \exp\left\{\sum_{n=1}^{\infty} \frac{\lambda^n}{n} \int_{(-\infty,0]} e^{izt} dF_n(t)\right\} \tag{1.73}$$

are Fourier–Stieltjes transforms of the functions with bounded variation $v_1(t)$ and $v_2(t)$; $v_2(t) = 0$ for $t \le 0$ and $v_2(t) = v_2(0)$ for $t > 0$; moreover, $v_1(t)$ and $v_2(t)$ satisfy relation (1.70).

Proof. We put

$$w_1^+(\lambda, t) = \begin{cases} \sum_{n=1}^{\infty} \frac{\lambda^n}{n}[F_n(t) - F_n(0)], & t \ge 0, \\ 0, & t < 0, \end{cases}$$

$$w_k^+(\lambda, t) = \int w_{k-1}^+(\lambda, t-s)dw_1^+(\lambda, s), \quad k > 1.$$

Then

$$\tilde{v}_2(z) = 1 + \sum_{k=1}^{\infty} \frac{(-1)^k}{k!} \int_{[0,\infty)} e^{izt} dw_k^+(\lambda, t),$$

$$v_2(z) = \varepsilon(t-) + \sum_{k=1}^{\infty} \frac{(-1)^k}{k!} w_k^+(\lambda, t),$$

$$\operatorname{var} v_2(t) \le 1 + \sum_{k=1}^{\infty} \operatorname{var} w_k^+(\lambda, t) \le 1 + \sum_{k=1}^{\infty} \frac{1}{k!}(\operatorname{var} w_1^+(\lambda, t))^k,$$

and

$$\operatorname{var} w_1^+(t) \le \sum_{n=1}^{\infty} \frac{\lambda^n}{n} = \ln \frac{1}{1-\lambda}.$$

Similarly, we obtain that

$$v_1(t) = \varepsilon(t) + \sum_{k=1}^{\infty} \frac{\lambda^n}{n} \frac{1}{k!} w_k^-(\lambda, t),$$

where $w_1^-(\lambda,t) = \sum_{n=1}^{\infty} \frac{\lambda^n}{n} F_n(t)$, $t \le 0$; for $t > 0$, we have $w_1^-(\lambda,t) = w_1^-(\lambda,0)$; $w_k^-(\lambda,t) = \int w_{k-1}^-(\lambda,t-s)dw_1^-(s)$; $\operatorname{var} w_1^-(\lambda,t) \le \ln\frac{1}{1-\lambda}$, $\operatorname{var} v_1(t) \le 1+\sum_{k=1}^{\infty} \frac{1}{k!}(\ln\frac{1}{1-\lambda})^k$. On the r.h.s. of (1.70), there is a convolution of the functions $\varepsilon(t)-\lambda F(t)$ and $v_1(t)$; its Fourier–Stieltjes transform is $(1-\lambda\varphi(z))v_1(z)$, where $\varphi(z) = \int e^{izt}dF(t)$. The functions v_1 and v_2 satisfy (1.70) if $v_2(z) = (1-\lambda\varphi(z))v_1(z)$, i.e.,

$$\begin{aligned}\frac{1}{1-\lambda\varphi(z)} &= \frac{v_1(z)}{v_2(z)} \\ &= \exp\left\{\sum_{n=1}^{\infty} \frac{\lambda^n}{n} \int e^{izt}dF_n(t)\right\} = \exp\left\{\sum_{n=1}^{\infty} \frac{\lambda^n}{n}\varphi^n(z)\right\},\end{aligned}$$

and this is valid for $|\lambda\varphi| < 1$. □

55°. Lemma. *Let*

$$v_+(\lambda,t) = \varepsilon(t-) + \sum_{k=1}^{\infty} \frac{1}{k!} w_k^+(\lambda,t).$$

Then the equation

$$b(x) = \int q(x-t)dv_2(t), \tag{1.74}$$

where $b(x)$ and $q(x)$ are functions with bounded variation, $b(x) = 0$ for $x > 0$, and $b(x)$ is a given function, whereas $q(x)$ is a desired one, has a solution given by the formula

$$q(x) = \int b(x-t)dv_+(t). \tag{1.75}$$

In fact, applying the Fourier–Stieltjes transform to (1.74), we get

$$\int e^{izt}db(t) = \int e^{izt}dq(t) \cdot \tilde{v}_2(z).$$

Hence,

$$\int e^{izt}dq(t) = \frac{1}{\tilde{v}_2(z)} \int e^{izt}db(t). \tag{1.76}$$

It remains to note that

$$\int e^{izt}dv_+(t) = \exp\left\{\sum_{n=1}^{\infty} \frac{\lambda^n}{n} \int_{(-\infty,0)} e^{izt}dF_n(t)\right\}, \tag{1.77}$$

which coincides with $\frac{1}{\tilde{v}_2(z)}$; thus, (1.75) follows from (1.76). □

56°. Lemma. *The function $Q(\lambda, x)$ satisfying (1.71) is given for $x \geq 0$ by the equality*

$$Q(\lambda, x) = \iint \varepsilon(x - z)\Phi_1(x - z - t)dv_-(\lambda, t)dv_+(\lambda, z), \tag{1.78}$$

where $v_-(\lambda, t) = v_1(t)$.

This formula is a consequence of (1.75).

57°. *For $x \geq 0$ and $y > 0$, the function $\widehat{Q}(\lambda, x, y)$ is defined by the equality*

$$\widehat{Q}(\lambda, x, y) = \lambda \int_{-\infty}^{-0} dv_-(\lambda, t) \int_{-0}^{x} dv_+(\lambda, z)[F(x + y - z - t) - F(x - z - t+)]. \tag{1.79}$$

58°. Theorem. *Let*

$$S_n(x) = \mathbf{P}\{\sup_{0 \leq k \leq n} \zeta_k \leq x\}, \qquad s(\lambda, x) = \sum_{n=0}^{\infty} \lambda^n S_n(x). \tag{1.80}$$

Then

$$s(\lambda, x) = \exp\left\{\sum_{n=1}^{\infty} \frac{\lambda^n}{n} F_n(0)\right\} v_+(\lambda, x).$$

Proof. It is easy to obtain (similarly to (1.67))

$$\begin{aligned} s(\lambda, x) &= \varepsilon(x) + \lambda \int s(\lambda, x - t)dF(t), & x \geq 0, \\ s(\lambda, x) &= 0, & x < 0. \end{aligned}$$

Rewriting it in the form of (1.71) and using (1.78), we get

$$s(\lambda, x) = \int_{-\infty}^{-0} dv_-(\lambda, t) \int_{-0}^{x} dv_+(\lambda, t) = v_-(\lambda, 0)v_+(\lambda, x).$$

Formula (1.73) yields

$$v_-(\lambda, 0) = v_1(0) = \tilde{v}_1(0) = \exp\left\{\sum_{n=1}^{\infty} \frac{\lambda^n}{n} F_n(0)\right\}.$$

59°. Lemma. *For every $a > 0$,*

$$\overline{\lim_{n \to \infty}} \sqrt{n}\mathbf{P}\{|\zeta_n| < a\} < \infty. \tag{1.81}$$

Proof. Suppose that $g(x)$ is a positive continuous function for which the function $\tilde{g}(z) = \int g(x)e^{izx}dx$ is finite and $\tilde{g}(z) = 0$ for $|z| \geq c$, and c is chosen so that $|\varphi(z)| < 1 - \delta|z|^2$ for $|z| \neq 0$ and some $\delta > 0$. The function $g(x)$ can be chosen as follows:

$$g(x) = \int \frac{\sin^2 \varepsilon(x-y)}{\varepsilon^2(x-y)^2} \cdot \frac{\sin^2 \varepsilon y}{\varepsilon^2 y^2} dy.$$

In this case,

$$\tilde{g}(z) = \begin{cases} 2\pi\frac{1}{\varepsilon}(2\varepsilon - |z|)^2, & |z| \leq 2\varepsilon; \\ 0, & |z| > 2\varepsilon. \end{cases}$$

It suffices to show that

$$\begin{aligned} \overline{\lim_{n\to\infty}} \sqrt{n} \int g(x)dF_n(x) &= \overline{\lim_{n\to\infty}} \frac{1}{2\pi}\sqrt{n} \int \tilde{g}(z)\varphi^n(z)dz \\ &\leq \overline{\lim_{n\to\infty}} \frac{1}{2\pi}\sqrt{n} \int_{-c}^{c} |\tilde{g}(z)|e^{-\delta|z|^2 n}dz \\ &\leq c_1 \overline{\lim_{n\to\infty}} \int_{-c\sqrt{n}}^{c\sqrt{n}} e^{d|z|^2} dz < \infty \end{aligned}$$

(here $c_1 = \sup_z |\tilde{g}(z)|/(2\pi)$). □

60°. Lemma. *For all $t > 0$, there exists the limit*

$$\lim_{\lambda \uparrow 1} v_+(\lambda, t) = v_+(1, t),$$

where $v_+(1,t)$ is a nondecreasing function.

Indeed, 59° implies that the series

$$w_1^+(1,t) = \sum_{n=1}^{\infty} \frac{1}{n}[F_n(t) - F_n(0)]$$

converges and that the function $w_1^+(1,t)$ is nondecreasing, $w_1^+(1,0) = 0$. Hence,

$$w_k^+(1,t) \leq (w_1^+(1,t))^k, \qquad v_+(\lambda,t) \leq e^{w_1^+(1,t)}$$

for all $t > 0$. In addition, $v_+(\lambda, t)$ is a monotonically increasing function of λ.

61°. Theorem. *The necessary and sufficient condition of* $\mathbf{P}\{\sup_{k>0} \zeta_k < \infty\} = 1$ *is*

$$\sum_{n=1}^{\infty} \frac{1}{n} \mathbf{P}\{\zeta_n \geq 0\} < \infty. \tag{1.82}$$

The function

$$q_+(x) = \mathbf{P}\{\sup_{k\geq 0} \zeta_k < x\}$$

is defined in this case by its Fourier transform:

$$\int_0^\infty e^{izx} dq_+(x) = \exp\left\{\sum_{n=1}^\infty \frac{1}{n}\int_0^\infty (e^{-izx} - 1)dF_n(x)\right\}. \tag{1.83}$$

Proof. The relation $q_+(x) = \lim_{n\to\infty} S_n(x)$ yields

$$\begin{aligned} q_+(x) &= \lim_{\lambda\uparrow 1}(1-\lambda)s(\lambda, x) \\ &= \lim_{\lambda\uparrow 1}\exp\left\{-\ln\frac{1}{1-\lambda} + \sum_{n=1}^\infty \frac{\lambda^n}{n}F_n(0)\right\} v_+(\lambda, x) \\ &= v_+(1, x)\lim_{\lambda\uparrow 1}\exp\left\{-\sum_{n=1}^\infty \frac{\lambda^n}{n} + \sum_{n=1}^\infty \frac{\lambda^n}{n}F_n(0)\right\} \\ &= v_+(1, x)\lim_{\lambda\uparrow 1}\exp\left\{-\sum_{n=1}^\infty \frac{\lambda^n}{n}\mathbf{P}\{\zeta_n \geq 0\}\right\}. \end{aligned}$$

Suppose that $\sum_{n=1}^\infty \frac{1}{n}\mathbf{P}\{\zeta_n \geq 0\} = +\infty$. Then

$$\lim_{\lambda\uparrow 1}\exp\left\{-\sum_{n=1}^\infty \frac{\lambda^n}{n}\mathbf{P}\{\zeta_n \geq 0\}\right\} = 0;$$

and consequently, $q_+(x) = 0$ for all $\lambda > 0$, and $\mathbf{P}\{\sup_{k\geq 0}\zeta_k < \infty\} = 0$. Assume that (1.82) holds. Then $w_1^+(1, t)$ is a function with bounded variation,

$$w_1^+(1, +\infty) = \sum_{n=1}^\infty \frac{1}{n}\mathbf{P}\{\zeta_n \geq 0\},$$

and for all k,

$$w_k^+(1, +\infty) = \left(\sum_{n=1}^\infty \frac{1}{n}\mathbf{P}\{\zeta_n \geq 0\}\right)^k.$$

Hence,

$$\begin{aligned} v_+(1, +\infty) &= 1 + \sum_{k=1}^\infty \frac{1}{k!}\left(\sum_{n=1}^\infty \frac{1}{n}\mathbf{P}\{\zeta_n \geq 0\}\right)^k = \exp\left\{\sum_{n=1}^\infty \frac{1}{n}\mathbf{P}\{\zeta_n \geq 0\}\right\}, \\ q_+(+\infty) &= v_+(1, +\infty)\exp\left\{-\sum_{n=1}^\infty \frac{1}{n}\mathbf{P}\{\zeta_n \geq 0\}\right\} = 1. \end{aligned}$$

(Note that the inequality $q_+(+\infty) > 0$ implies that $q_+(+\infty) = 1$ by virtue of Kolmogorov's "0 or 1" law.) To prove formula (1.83), we mention that

$$\begin{aligned}\int e^{izx} dq_+(x) &= \lim_{\lambda\uparrow 1} \exp\left\{-\sum_{n=1}^{\infty} \frac{\lambda^n}{n} \mathbf{P}\{\zeta_n \geq 0\}\right\} \\ &\quad \times \exp\left\{\sum_{n=1}^{\infty} \frac{\lambda^n}{n} \int_{[0,\infty)} e^{izx} dF_n(x)\right\} \\ &= \lim_{\lambda\uparrow 1} \exp\left\{\sum_{n=1}^{\infty} \frac{\lambda^n}{n} \int_{[0,\infty)} (e^{izx} - 1) dF_n(x)\right\}.\end{aligned}$$

The last limit coincides with the r.h.s. of (1.83).

□

63°. The following equalities hold:

$$\mathbf{P}\{\sup_{k>0} \zeta_k < 0\} = q_+(+0) = \exp\left\{-\sum_{n=1}^{\infty} \frac{1}{n} \mathbf{P}\{\zeta_n > 0\}\right\}. \tag{1.84}$$

In fact,

$$\begin{aligned} w_1^+(1, 0+) &= \sum_{n=1}^{\infty} \frac{1}{n} [F_n(0+) - F_n(0)] = \sum_{n=1}^{\infty} \frac{1}{n} \mathbf{P}\{\zeta_n > 0\}, \\ w_k^+(1, 0+) &= (w_1^+(1, 0+))^k, \\ v_+(1, 0+) &= \exp\left\{\frac{1}{n} \sum_{n=1}^{\infty} \frac{1}{n} \mathbf{P}\{\zeta_n = 0\}\right\}, \\ q_+(0+) &= \exp\left\{-\sum_{n=1}^{\infty} \frac{1}{n} \mathbf{P}\{\zeta_n > 0\} + \sum_{n=1}^{\infty} \frac{1}{n} \mathbf{P}\{\zeta_n = 0\}\right\}. \end{aligned}$$

□

The next result is obtained by analogy with Theorem 61°.

63°. Theorem. *The relation* $\mathbf{P}\{\inf_k \zeta_k > -\infty\} = 1$ *is valid iff*

$$\sum_{n=1}^{\infty} \frac{1}{n} \mathbf{P}\{\zeta_n \leq 0\} < \infty. \tag{1.85}$$

The function

$$q_-(x) = \mathbf{P}\{\inf_{k\geq 0} \zeta_k < x\}$$

is defined in this case by its Fourier transform:

$$\int e^{izx}dq_-(x) = \exp\left\{\sum_{n=1}^{\infty}\frac{1}{n}\int_{-\infty}^{-0}(e^{izx}-1)dF_n(x)\right\}, \tag{1.86}$$

$$\mathbf{P}\{\inf_{k>0}\zeta_k > 0\} = 1 - q_-(0) = \exp\left\{-\sum_{n=1}^{\infty}\frac{1}{n}\mathbf{P}\{\zeta_n < 0\}\right\}. \tag{1.87}$$

We now mention conditions of the unilateral boundedness or unboundedness of random walks that have mathematical expectations.

64°. Theorem. *Suppose that* $\mathbf{M}\xi_1$ *exists (it is possible that* $\mathbf{M}\xi_1 = +\infty$ *or* $\mathbf{M}\xi_1 = -\infty$*).*

(a) *If* $\mathbf{M}\xi_1 > 0$, *then*

$$\mathbf{P}\{\sup_n \zeta_n = +\infty\} = \mathbf{P}\{\inf_n \zeta_n > -\infty\} = 1;$$

(b) *if* $\mathbf{M}\xi_1 < 0$, *then*

$$\mathbf{P}\{\sup_n \zeta_n < +\infty\} = \mathbf{P}\{\inf_n \zeta_n = -\infty\} = 1;$$

(c) *if* $\mathbf{M}\xi_1 = 0$, *and* $\mathbf{P}\{\zeta_1 \neq 0\} < 1$, *then*

$$\mathbf{P}\{\sup_n \zeta_n = +\infty\} = \mathbf{P}\{\inf_n \zeta_n = -\infty\} = 1.$$

Proof. Statements (a) and (b) follow from the strong law of large numbers (Theorem 22°). We prove (c). By virtue of 51°, the walk is recurrent. Thus, the event $\bigcup_{n=1}^{\infty}\{0 < \zeta_n < \delta\}$ has probability 1. Let ν_1 be the first time for which $0 < \zeta_{\nu_1} < \delta$ ($\mathbf{P}\{\nu_1 < +\infty\} = 1$). Also let ν_2 denote the first time $k > \nu_1$ for which $0 < \zeta_k - \zeta_{\nu_1} < \delta$, etc. The event $\{\sup_{0<i<\nu_1}\zeta_i > c\}$ has a positive probability for all $c > 0$. Let us show that the event $\{\sup_{\nu_1<i<\nu_2}\zeta_i - \zeta_1 > c\}$ is independent of the first one and has the same probability. Indeed,

$$\begin{aligned}
&\{\sup_{0<i<\nu_1}\zeta_i > c\} \\
&\quad = \bigcup_k\{\zeta_1\overline{\in}(0,\delta),\ldots,\zeta_{k-1}\overline{\in}(0,\delta),\zeta_k\in(0,\delta),\sup_{0<i<k}\zeta_i > c\},\\
&\{\sup_{1<i<\nu_1}\zeta_i > c\}\cap\{\sup_{\nu_1<i<\nu_2}\zeta_i - \zeta_1 > c\}\\
&\quad = \bigcup_{k,j}\{\zeta_1\overline{\in}(0,\delta),\ldots,\zeta_{k-1}\overline{\in}(0,\delta),\zeta_k\in(0,\delta),\sup_{0<i<k}\zeta_i > c\}\\
&\quad\cap\{\zeta_{k+1}-\zeta_k\overline{\in}(0,\delta),\ldots,\zeta_{k+j-1}-\zeta_k\overline{\in}(0,\delta),\zeta_{k+j}-\zeta_k\in(0,\delta),\\
&\qquad \sup_{k<i<j+k}\zeta_i - \zeta_k > c\}.
\end{aligned}$$

Hence,

$$\begin{aligned}
&\mathbf{P}(\{\sup_{0<i<\nu_1} \zeta_i > c\} \cap \{\sup_{\nu_1<i<\nu_2} (\zeta_i - \zeta_{\nu_1}) > c\}) \\
&\quad = \sum_{k,j} \mathbf{P}\{\zeta_1 \overline{\in}(0,\delta), \dots, \zeta_{k-1}\overline{\in}(0,\delta), \zeta_k \in (0,\delta), \sup_{i<k} \zeta_i > c\} \\
&\qquad \times \mathbf{P}\{\zeta_{k+1} - \zeta_1 \overline{\in}(0,\delta), \dots, \zeta_{k+j-1} - \zeta_k \overline{\in}(0,\delta), \zeta_{k+j} - \zeta_k \overline{\in}(0,\delta), \\
&\qquad\quad \sup_{0<i<j} (\zeta_{k+i} - \zeta_k) > c\} \\
&\quad = \sum_{k} \mathbf{P}\{\zeta_1 \overline{\in}(0,\delta), \dots, \zeta_{k-1}\overline{\in}(0,\delta), \zeta_k \overline{\in}(0,\delta), \sup_{i<k} \zeta_i > c\} \\
&\qquad \times \sum_{j} \mathbf{P}\{\zeta_1 \overline{\in}(0,\delta), \dots, \zeta_{j-1}\overline{\in}(0,\delta), \zeta_j \overline{\in}(0,\delta), \sup_{i<j} \zeta_i > c\}.
\end{aligned}$$

(We have used the fact that $\zeta_{k+j} - \zeta_k$ are independent of $\zeta_1, \dots, \zeta_k$ and have the same joint distribution as ζ_j, $j > 0$.)

Similarly, we conclude that all events

$$A_k = \{\sup_{\nu_k<i<\nu_{k+1}} (\zeta_i - \zeta_{\nu_k}) > c\}, \quad k = 0, 1, \dots \quad (\nu_0 = 0)$$

are independent and have the same probability. Since $\mathbf{P}\{\xi_1 > 0\} > 0$, we have $\mathbf{P}\{\xi_1 > \delta\} \geq \delta$ for some δ. Therefore,

$$\mathbf{P}\{\sup_{i<\nu_1} \zeta_i > c\} \geq \mathbf{P}\{\xi_1 > \delta, \dots, \xi_m > \delta\} \geq \delta^m > 0 \quad (m\delta > c).$$

Thus, at least one event of A_k takes place with probability 1. Consequently, for all $c > 0$,

$$\mathbf{P}\{\sup_n \zeta_n \geq c\} = 1.$$

Whence $\mathbf{P}\{\sup_n \zeta_n = +\infty\} = 1$.

□

CHAPTER 2

General Processes with Independent Increments (Random Measures)

§2.1 Nonnegative Random Measures with Independent Values (r.m.i.v.)

Let $(X, \mathcal{B})$ be a measurable space. The σ-algebra $\mathcal{B}$ is generated by some countable algebra $\mathcal{B}_0$. We shall consider the random functions given on the elements of $\mathcal{B}$ (or $\mathcal{B}_0$).

Definition. A random function $\mu(A)$ given on $\mathcal{B}_0$ is called a *nonnegative random measure with independent values (r.m.i.v.)* if

(1) for every collection of mutually disjoint sets $A_1, \ldots, A_k \in \mathcal{B}_0$, the random variables $\mu(A_1), \ldots, \mu(A_k)$ are independent (jointly);

(2) $\mu(A_1 \cup A_2) = \mu(A_1) + \mu(A_2)$ for $A_1, A_2 \in \mathcal{B}_0$, $A_1 \cap A_2 \neq \emptyset$;

(3) $\mu(A) \geq 0$ for all $A \in \mathcal{B}_0$.

Let $\mathcal{B}_0 = \bigcup_n \mathcal{A}_n$, where $\mathcal{A}_n$ are finite algebras of sets and $\mathcal{A}_n \subset \mathcal{A}_{n+1}$. We say that $\mu(A)$ is *stochastically continuous with respect to* $\{\mathcal{A}_n\}_{n\geq 1}$ if for any sequence $A_n \in \mathcal{B}_0$ such that A_n is an atom of $\mathcal{A}_n$ and $A_n \supset A_{n+1}$, we have $\mu(A_n) \to 0$ with probability 1.

Sequences (A_n), where A_n is an atom of $\mathcal{A}_n$ and $A_n \supset A_{n+1}$, are called *fundamental with respect to* $\{\mathcal{A}_n\}$.

If $(\tilde{A}_n)$ is a certain sequence fundamental with respect to $\{\mathcal{A}_n\}$, the limit $\lim \mu(\tilde{A}_n)$ exists for every nonnegative r.m.i.v. We shall call this limit a *value of a measure* μ *on the fundamental sequence* $(\tilde{A}_n)$ and denote it by $\mu((\tilde{A}_n))$.

A measure μ is called *purely discrete* if there exists a countable collection of fundamental sequences (with respect to $\{\mathcal{A}_n\}$) (A_n^k), $k = 1, 2, \ldots$, such that for any $C \in \mathcal{A}_m$,

$$\mu(C) = \sum_{A_m^k \subset C} \mu((A_n^k)). \tag{2.1}$$

Suppose that $V \in \mathcal{B}_0$. Denote by $\mathcal{F}_V$ a σ-algebra generated by the variables $\{\mu(A), A \in \mathcal{B}_0, A \subset V\}$. One can easily check that the σ-algebras $\mathcal{F}_V$ and $\mathcal{F}_U$ are independent if $U \cap V \neq \emptyset$.

1°. Theorem. *Every nonnegative r.m.i.v. admits the representation*

$$\mu(A) = \mu_1(A) + \mu_2(A),$$

where $\mu_1(A)$ *is stochastically continuous with respect to* $\{A_n\}_{n\geq 1}$, *and* $\mu_2(A)$ *is purely discrete. The measures* $\mu_k(A)$, $k = 1, 2$, *are nonnegative mutually independent r.m.i.v.*

Proof. Let us split all sequences (A_n) that are fundamental with respect to $\{A_n\}$ into two sets. All those sequences (A_n) for which $\mu((A_n)) > 0$ with positive probability are included in the first set S_1, while all the others are in the second one. The first set is at most countable. In fact, let us estimate the number of sequences (A_n) for which $\mathbf{P}\{\mu((A_n)) > 1/k\} > 1/k$. If one can find l such sequences, then there exist l mutually disjoint sets $C_1, \ldots, C_l \in \mathcal{B}_0$ such that $\mathbf{P}\{\mu(C_i) > 1/k\} > 1/k$.

Then

$$\mathbf{P}\{\mu(X) > r\} \geq \sum_{s > kr} C_l^s (1/k)^s (1 - 1/k)^{l-s},$$

and the last expression tends to 1 as $l \to \infty$ for any r. Thus, the set of those fundamental sequences (A_n) for which $\mathbf{P}\{\mu((A_n)) > 1/k\} > 1/k$ is finite. Let (A_n^k), $k = 1, 2, \ldots$, be all fundamental sequences from S_1. We put

$$\mu_{nr}^1(C) = \mu(C \setminus \bigcup_{k\leq r} A_n^k),$$

$$\mu_{nr}^2(C) = \mu(C \cap (\bigcup_{k\leq r} A_n^k)).$$

Whatever $C_1, C_2, \ldots, C_l$ we take, the random vectors $(\mu_{nr}^1(C_1), \mu_{nr}^1(C_2), \ldots, \mu_{nr}^1(C_l))$ and $(\mu_{nr}^2(C_1), \mu_{nr}^2(C_2), \ldots, \mu_{nr}^2(C_l))$ are independent, because the first one is measurable with respect to $\mathcal{F}_{X\setminus\cup_{k\leq r}A_n^k}$, and the second one, with respect to $\mathcal{F}_{\cup_{k\leq r}A_n^k}$.

Clearly, $\mu_{nr}^1(C)$ and $\mu_{nr}^2(C)$ are nonnegative mutually independent r.m.i.v. The measure μ_{nr}^1 increases with n, while μ_{nr}^2 decreases. Hence, the following limits exist:

$$\mu_i^r(C) = \lim_n \mu_{nr}^i(C), \quad i = 1, 2; \qquad \mu_1^r(C) + \mu_2^r(C) = \mu(C).$$

They are mutually independent nonnegative r.m.i.v. measurable with respect to the σ-algebras $\mathcal{F}_1^r$ and $\mathcal{F}_2^r$, respectively, where

$$\mathcal{F}_1^r = \bigvee_n \mathcal{F}_{X\setminus\cup_{k\leq r}A_n^k}, \qquad \mathcal{F}_2^r = \bigcap_n \mathcal{F}_{\cup_{k\leq r}A_n^k},$$

the σ-algebras $\mathcal{F}_1^r$ and $\mathcal{F}_2^r$ are independent, since $\mathcal{F}_{X\setminus\cup_{k\leq r}A_n^k}$ is independent of $\mathcal{F}_{\cup_{k\leq r}A_n^k}$. Further, the measure $\mu_1^r(C)$ decreases with r, and the measure $\mu_2^r(C)$ increases. We put

$$\mu_i(C) = \lim_{r\to\infty} \mu_i^r(C).$$

Then $\mu_1(C)+\mu_2(C)=\mu(C)$, the measure μ_1 is measurable with respect to the σ-algebra $\bigcap \mathcal{F}_1^r=\mathcal{F}_1$, and μ_2, with respect to $\mathcal{F}_2=\vee_r\mathcal{F}_2^r$; $\mathcal{F}_1$ and $\mathcal{F}_2$ are independent by virtue of the independence of $\mathcal{F}_1^r$ and $\mathcal{F}_2^r$. Clearly, $\mu_i(C)$ are nonnegative r.m.i.v.

Show that μ_1 is stochastically continuous. If $(A_n)\overline{\in}S_1$, then $\mu_1(A_n)\le\mu(A_n)=0$. Consider $\mu_2^r((A_n^k))$ for $r>k$,

$$\mu_2^r((A_n^k))=\lim_{n\to\infty}\lim_{m\to\infty}\mu(A_n^k\cap\bigcup_{i\le r}A_m^i)\ge\lim_{n\to\infty}\lim_{m\to\infty}\mu(A_m^k)=\mu((A_n^k)).$$

But the inequality $\mu_2^r(C)\le\mu(C)$ implies $\mu_2^r((A_n^k))\le\mu((A_n^k))$. Thus,

$$\mu_2^r((A_n^k))=\mu((A_n^k)),\qquad \mu_2((A_n^k))=\mu((A_n^k)),$$
$$\mu_1((A_n^k))=\mu((A_n^k))-\mu_2((A_n^k))=0.$$

So μ_1 is stochastically continuous. Let A_m be an atom of $\mathcal{A}_m$. To prove that μ_2 is purely discrete, it suffices to show that

$$\mu_2(A_m)=\sum_{A_m^k=A_m}\mu_2((A_n^k))=\sum_{A_m^k=A_m}\mu((A_n^k)). \tag{2.2}$$

We have

$$\mu_2^r(A_m)=\lim_{n\to\infty}\mu(A_m\cap\bigcup_{k<r}A_n^k)=\sum_{k\le r}\lim_{n\to\infty}\mu(A_m\cap A_n^k)$$

($A_n^1,\dots,A_n^r$ are mutually disjoint for sufficiently large n). Therefore,

$$\mu_2(A_m)=\lim_{r\to\infty}\sum_{\substack{k\le r\\ A_m^k=A_m}}\mu((A_n^k))=\sum_{A_m^k=A_m}\mu((A_n^k));$$

(2.2) is proved.

□

Consider a purely discrete nonnegative r.m.i.v. $\mu(A)$ on $\mathcal{B}_0$. $\mathcal{B}_0$ is said to be complete with respect to μ if for every fundamental sequence (A_n) for which $\mu((A_n^k))\ne0$ with probability 1, we have $\bigcap A_n\ne\varnothing$.

2°. Theorem. *Suppose that the algebra $\mathcal{B}_0$ is complete with respect to a purely discrete nonnegative r.m.i.v. μ. Then μ can be extended onto the σ-algebra $\mathcal{B}$, generated by $\mathcal{B}_0$, as a countably additive random measure: $\mu(\bigcup_k B_k)=\sum\mu(B_k)$ with probability 1 for every sequence of mutually disjoint sets $B_k\in\mathcal{B}$, and if $B\in\mathcal{B}_0$, the value of the measure coincides with the initial one $\mu(B)$.*

Proof. Let (A_n^k), $k = 1, 2, \ldots$, be those fundamental sequences on which the measure μ is concentrated, $A^k = \bigcap A_n^k$. A^k is nonempty, since $\mathcal{B}_0$ is complete. Note that A^k are atoms of the σ-algebra $\mathcal{B}$: For all $B \in \mathcal{B}$, either $B \cap A^k = \varnothing$ or $B \cap A^k = A^k$. This follows from the fact that a class of sets B satisfying these conditions is monotonic and, moreover, it contains $\mathcal{B}_0$. Let us define μ for all $B \in \mathcal{B}$ by

$$\mu(B) = \sum_{A^k \subset B} \mu((A_n^k)).$$

Defined in this way, the measure is countably additive, since it is atomistic. Its coincidence with the initial one for $B \in \mathcal{B}_0$ follows from formula (2.1) and from the following property: For $B \subset \mathcal{A}_m$, we have $A^k \subset B$ iff $A_m^k \subset B$.

□

Remark. Let us carry out the following procedure of completion of X. For every fundamental sequence (A_n^k) such that $\bigcap_n A_n^k = \varnothing$ and $\mathbf{P}\{\mu((A_n^k)) = 0\} < 1$, we define a point $u_k \overline{\in} X$. Consider that u_k belongs to each of the sets A_n^k. Then $\bigcap A_n^k = \{u_k\}$; we denote $X \cup \{u_1, \ldots, u_k, \ldots\}$ by $\widetilde{X}$. An algebra $\widetilde{\mathcal{A}}_m$ on $\widetilde{X}$ consists of sets that are unions of atoms $A_m^k \cup \{u_k\}$. The measure $\tilde{\mu}$ is defined on $\widetilde{\mathcal{B}}_0 = \bigcup \widetilde{\mathcal{A}}_m$ by the equality $\tilde{\mu}(\widetilde{A}_m^k) = \mu(A_m^k)$. $\widetilde{\mathcal{B}}_0$ is now complete with respect to $\tilde{\mu}$; and hence, Theorem 2° can be applied.

Integral Random Measures

Consider nonnegative r.m.i.v. $\mu(A)$ on the algebra $\mathcal{B}$ such that $\mu(A)$ is an integral random variable for all $A \in \mathcal{B}_0$. Decomposition of such measures into stochastically continuous and purely discrete components yields integral stochastically continuous measures and integral purely discrete ones, respectively. Construction of purely discrete measures has been described above. Consider integral stochastically continuous measures.

4°. Lemma. *Let $\mu(A)$ be an integral stochastically continuous nonnegative r.m.i.v. There exists a finitely additive nonnegative measure $m(A)$ on $\mathcal{B}_0$ such that*

$$\mathbf{P}\{\mu(A) = 0\} = e^{-m(A)}. \tag{2.3}$$

Proof. Note that $\{\mu(A) > 0\} = \{\mu(A) > 1/2\}$ and by virtue of stochastic continuity,

$$\lim_{m \to \infty} \max_{i \le r_m} \mathbf{P}\{\mu(A_m^i) > \frac{1}{2}\} = 0, \tag{2.4}$$

where $A_m^1, \ldots, A_m^{r_m}$ are all the atoms of the algebra $\mathcal{A}_m$. If it is not true, then for some sequence $A_m^k \supset A_{m+1}^k$, we find $\underline{\lim}\, \mathbf{P}(\mu(A_m^k) > \frac{1}{2}) > 0$, which contradicts the stochastic continuity of μ. Hence, $\min_{i\le r_m} \mathbf{P}\{\mu(A^i_m)=0\}\to 1$. This implies, in particular, that

$$\mathbf{P}\{\mu(X)=0\} = \mathbf{P}\{\mu(\bigcup_{k=1}^{r_m} A^k_m)=0\}$$
$$= \mathbf{P}\{\sum_{k=1}^{r_m}\mu(A^k_m)=0\} = \prod_{k=1}^{r_m}\mathbf{P}\{\mu(A^k_m)=0\} > 0.$$

If $A\in\mathcal{B}_0$, then $\mathbf{P}\{\mu(A)=0\} = \frac{\mathbf{P}\{\mu(X)=0\}}{\mathbf{P}\{\mu(X\setminus A)=0\}} > \mathbf{P}\{\mu(X)=0\}$. Suppose that

$$m(A) = -\ln\mathbf{P}\{\mu(A)=0\},\quad A\in\mathcal{B}_0.$$

Then $m(A)\ge 0$ and for $A\cap B=\varnothing$,

$$m(A\cup B) = -\ln\mathbf{P}\{\mu(A)+\mu(B)=0\} = -\ln\mathbf{P}\{\mu(A)=0\}\mathbf{P}\{\mu(B)=0\}$$
$$= -\ln\mathbf{P}\{\mu(A)=0\} - \ln\mathbf{P}\{\mu(B)=0\} = m(A)+m(B).$$

□

We call an integral nonnegative r.m.i.v. *counting* if for all sufficiently large n, $\max_{i\le r_n}\mu(A^i_n)\le 1$ with probability 1.

5°. Theorem. *Every stochastically continuous counting measure is a Poisson measure, i.e., $\mu(A)$ has the Poisson distribution for all $A\in\mathcal{B}_0$.*

Proof. We have

$$\mathbf{P}\{\max_{i\le r_n}\mu(A^i_n)\le 1\} = \prod_{i=1}^{r_n}\mathbf{P}\{\mu(A^i_n)\le 1\}$$
$$= \prod_{i=1}^{r_n}(1-\mathbf{P}\{\mu(A^i_n)>1\}) \le \exp\left\{-\sum_{i=1}^{r_n}\mathbf{P}\{\mu(A^i_n)>1\}\right\}.$$

The l.h.s. approaches 1 as $n\to\infty$, so

$$\lim_{n\to\infty}\sum_{i=1}^{r_n}\mathbf{P}\{\mu(A^i_n)>1\} = 0. \tag{2.5}$$

Let $B\in\mathcal{A}_m$. Then

$$\mathbf{P}\{\mu(B)=k\}$$
$$= \lim_{n\to\infty}\mathbf{P}\{\sum_{A^i_n\in B}\mu(A^i_n)=k,\ \max_{\substack{i\le r_n\\ A^i_n\subset B}}\mu(A^i_n)\le 1\}$$
$$= \lim_{n\to\infty}\sum_{\substack{i_1<\cdots<i_k\\ A^{i_1}_n\subset B,\dots,A^{i_k}_n\subset B}}\frac{\mathbf{P}\{\mu(A^{i_1}_n)=1\}}{\mathbf{P}\{\mu(A^{i_1}_n)=0\}}\cdots\frac{\mathbf{P}\{\mu(A^{i_k}_n)=1\}}{\mathbf{P}\{\mu(A^{i_k}_n)=0\}}\prod_{A^j_n\subset B}\mathbf{P}\{\mu(A^j_n)=0\}$$
$$= \mathbf{P}\{\mu(B)=0\}\lim_{n\to\infty}\sum_{\substack{i_1<\cdots<i_k\\ A^{i_1}_n\subset B,\dots,A^{i_k}_n\subset B}}\mathbf{P}\{\mu(A^{i_1}_n)=1\}\dots\mathbf{P}\{\mu(A^{i_k}_n)=1\},$$

since $\inf_k \mathbf{P}\{\mu(A_n^k) = 0\} \to 1$ by virtue of (2.4). The last equality implies the existence of the limit

$$\lim_{n\to\infty} \sum_{A_n^i \subset B} \mathbf{P}\{\mu(A_n^i) = 1\} = \frac{\mathbf{P}\{\mu(B) = 1\}}{\mathbf{P}\{\mu(B) = 0\}}$$

(the denominator is not zero, because of (2.3)). But (2.5) yields

$$\begin{aligned}\lim_{n\to\infty} \sum_{A_n^i \subset B} \mathbf{P}\{\mu(A_n^i) = 1\} &= \lim_{A_n^i \subset B} \mathbf{P}\{\mu(A_n^i) > 0\} \\ &= \lim_{n\to\infty} \sum_{A_n^i \subset B} (1 - e^{-m(A_n^i)}) \\ &= \sum_{A_n^i \subset B} [m(A_n^i) + O(m^2(A_n^i))] = m(B),\end{aligned}$$

since

$$\sum m^2(A_n^i) \le \max_i m(A_n^i) m(B) \to 0.$$

To go on with the proof, we need one simple lemma.

6°. Lemma. *Let* $\alpha_n^i \ge 0$, $i = 1, \dots, r_n$, $\sum_{i=1}^{r_n} \alpha_n^i \to a < \infty$, $\max_{i \le r_n} \alpha_n^i \to 0$. *Then*

$$\sum_{1 < i_1 < \dots < i_k \le r_n} \alpha_n^{i_1} \dots \alpha_n^{i_k} \to \frac{a^k}{k!}.$$

Proof. We have

$$\begin{aligned}\left(\sum_{i=1}^{r_n} \alpha_n^i\right)^k &= k! \sum_{i_1 < \dots < i_k \le r_n} \alpha_n^{i_1} \dots \alpha_n^{i_k} \\ &+ \sum_{\substack{\max m_i \ge 2,\\ m_1 + \dots + m_{r_n} = k}} \frac{k!}{m_1! \dots m_{r_n}!} (\alpha_n^1)^{m_1} \dots (\alpha_n^{r_n})^{m_{r_n}}.\end{aligned}$$

Hence,

$$\begin{aligned}\frac{1}{k!}\left(\sum_{i=1}^{r_n} \alpha_n^i\right)^k &> \sum_{i_1 < \dots < i_k \le r_n} \alpha_n^{i_1} \dots \alpha_n^{i_k} \\ &> \frac{1}{k!}\left(\sum_{i=1}^{r_n} \alpha_n^i\right)^k - \max_j \alpha_n^j \sum_{\substack{m_1 + \dots + m_{r_n} = k-1\\ m_1 \ge 0, \dots, m_{r_n} \ge 0}} \frac{(\alpha_n^1)^{m_1} \dots (\alpha_n^{r_n})^{m_{r_n}}}{m_1! \dots m_{r_n}!} \\ &> \frac{1}{k!}\left(\sum_{i=1}^{r_n} \alpha_n^i\right)^k - \max_j \alpha_n^j \frac{1}{(k-1)!}\left(\sum_{i=1}^{r_n} \alpha_n^i\right)^{k-1}.\end{aligned}$$

Passing to the limit in this inequality, we obtain the required result.

□

Returning to the proof of the theorem, we note that

$$\lim_{n\to\infty} \sum_{\substack{i_1<\cdots<i_k \\ A_n^{i_1}\subset B,\ldots,A_n^{i_k}\subset B}} \mathbf{P}\{\mu(A_n^{i_1})\}\ldots\mathbf{P}\{\mu(A_n^{i_k})=1\}$$

$$= \lim_{n\to\infty}\frac{1}{k!}\left(\sum_{A_n^i\subset B}\mathbf{P}\{\mu(A_n^i)>0\}\right)^k = \frac{1}{k!}m^k(B).$$

Therefore,

$$\mathbf{P}\{\mu(B)=k\} = \mathbf{P}\{\mu(B)=0\}\cdot\frac{1}{k!}m^k(B) = \frac{m^k(B)}{k!}e^{-m(B)}.$$

□

For further analysis of integral random measures, the following lemma will be very useful.

7°. Lemma. *Suppose that a σ-algebra $\mathcal{F}_A$ corresponds to each $A \subset \mathcal{B}_0$, and moreover, that $\mathcal{F}_A \subset \mathcal{F}_B$ for $A \subset B$, and $\mathcal{F}_A$ and $\mathcal{F}_B$ are independent if $A \cap B = \varnothing$. Suppose also that μ_1 and μ_2 are two nonnegative stochastically continuous integral r.m.i.v. such that: (1) $\mu_1(A)$ and $\mu_2(A)$ are $\mathcal{F}_A$-measurable and (2) $\max_{i\le r_n}\mu_1(A_n^i)\mu_2(A_n^i) = 0$ for sufficiently large n. Then the measures $\mu_1(A)$ and $\mu_2(A)$ are independent.*

Proof. It suffices to show that the variables $\mu_1(A)$ and $\mu_2(A)$ are independent for every $A \in \mathcal{B}_0$. In fact, in this case, the variables $\mu_1(A_m^i)$ and $\mu_2(A_m^i)$ are independent for any m; the pairs of variables $\mu_1(A_m^i)$, $\mu_2(A_m^i)$, $i = 1,\ldots,r_n$ are also independent (different pairs are measurable with respect to independent σ-algebras). Therefore, the sets $(\mu_1(A_m^1),\ldots,\mu_1(A_m^{r_m}))$ and $(\mu_2(A_m^1),\ldots,\mu_2(A_m^{r_m}))$ are independent; and thus, the measures μ_1 and μ_2 are independent on each of the algebras $\mathcal{A}_m$. This implies independence of μ_1 and μ_2 on $\mathcal{B}_0$.

Now let $A \subset \mathcal{B}_0$. We put

$$\xi_{mi} = \mu_1(A\cap A_m^i), \qquad \eta_{mi} = \mu_2(A\cap A_m^i).$$

We are to prove that

$$\mathbf{M}z^{\mu_1(A)}u^{\mu_2(A)} = \mathbf{M}z^{\mu_1(A)}\mathbf{M}u^{\mu_2(A)} \tag{2.6}$$

for $|z| \le 1$, $|u| \le 1$. Since $\mu_1(A) = \Sigma\xi_{mi}$, and $\mu_2(A) = \Sigma\eta_{mi}$, the proof of equality (2.6) is reduced to the proof of

$$\lim_{m\to\infty}(\mathbf{M}z^{\Sigma\xi_{mi}}u^{\Sigma\eta_{mi}} - \mathbf{M}z^{\Sigma\xi_{mi}}\mathbf{M}u^{\Sigma\eta_{mi}}) = 0. \tag{2.7}$$

Using the independence of the pairs (ξ_{mi}, η_{mi}) for different i, we find

$$\mathbf{M}z^{\Sigma\xi_{mi}}u^{\Sigma\eta_{mi}} = \prod_i \mathbf{M}z^{\xi_{mi}}u^{\eta_{mi}},$$

$$\mathbf{M}z^{\Sigma\xi_{mi}}\mathbf{M}u^{\Sigma\eta_{mi}} = \prod_i \mathbf{M}z^{\xi_{mi}}\mathbf{M}u^{\eta_{mi}}.$$

Having used the fact that $|\prod a_k - \prod b_k| \le \Sigma|a_k - b_k|$ for $|a_k| \le 1$, $|b_k| \le 1$, we get

$$\begin{aligned}&|\prod_i z^{\xi_{mi}}u^{\eta_{mi}} - \prod_i \mathbf{M}z^{\xi_{mi}}\mathbf{M}u^{\eta_{mi}}|\\ &\qquad \le \sum_i |\mathbf{M}z^{\xi_{mi}}u^{\eta_{mi}} - \mathbf{M}z^{\xi_{mi}}\mathbf{M}u^{\eta_{mi}}|.\end{aligned}$$

However,

$$\mathbf{M}z^{\xi_{mi}}u^{\eta_{mi}} = \mathbf{M}z^{\Sigma\xi_{mi}} + \mathbf{M}u^{\Sigma\eta_{mi}} + \mathbf{M}(1 - z^{\xi_{mi}})(1 - u^{\Sigma\eta_{mi}}) - 1.$$

Hence,

$$\begin{aligned}&\sum_i |\mathbf{M}z^{\xi_{mi}}u^{\eta_{mi}} - \mathbf{M}z^{\xi_{mi}}\mathbf{M}u^{\eta_{mi}}|\\ &\qquad \le \sum_i |-\mathbf{M}(z^{\xi_{mi}} - 1)\mathbf{M}(u^{\eta_{mi}} - 1) + \mathbf{M}(z^{\xi_{mi}} - 1)(u^{\eta_{mi}} - 1)| \qquad (2.8)\\ &\qquad \le \sum_i |\mathbf{M}z^{\xi_{mi}} - 1| \cdot |\mathbf{M}u^{\eta_{mi}} - 1| + \sum_i |\mathbf{M}(z^{\xi_{mi}} - 1)(u^{\eta_{mi}} - 1)|.\end{aligned}$$

We can estimate the first sum as follows:

$$\begin{aligned}&\sum_i |\mathbf{M}z^{\xi_{mi}} - 1|\,|\mathbf{M}u^{\eta_{mi}} - 1|\\ &\qquad \le \sup_{i\le r_m} |\mathbf{M}u^{\eta_{mi}} - 1| \sum_{k=1}^{r_m} 2\mathbf{P}\{\xi_{mk} > 0\}\\ &\qquad \le \sup_{i\le r_m} (2\mathbf{P}\{\eta_{mi} > 0\}) \sum_{k=1}^{r_m} 2\mathbf{P}\{\xi_{mk} > 0\}.\end{aligned}$$

The factor before the sum tends to zero, because μ_2 is stochastically continuous and

$$\sum_{k=1}^{r_m} \mathbf{P}\{\xi_{mk} > 0\} \le \sum_{k=1}^{r_m} \mathbf{P}\{\mu_1(A_m^i) > 0\}$$
$$\le \sum_{k=1}^{r_m} (1 - e^{-m_1(A_m^i)}) \le m_1(X),$$

where $m_1(A) = -\ln \mathbf{P}\{\mu_1(A) = 0\}$ (it is a finitely additive function on $\mathcal{B}_0$ (by virtue of 4°)). Therefore,

$$\lim_{m\to\infty} \sum_i |\mathbf{M}z^{\xi_{mi}} - 1|\,|\mathbf{M}u^{\eta_{mi}} - 1| = 0.$$

Further,

$$\mathbf{M}(z^{\xi_{mi}} - 1)(u^{\eta_{mi}} - 1) \le 4\mathbf{P}\{\xi_{mi}\eta_{mi} > 0\}. \tag{2.9}$$

Since

$$\mathbf{P}\{\max_{i\le r_m} \xi_{mi}\eta_{mi} < 1\} = \prod_{i\le r_m} \mathbf{P}\{\xi_{mi}\eta_{mi} < 1\}$$
$$\le \exp\left\{-\sum_{i\le r_m} \mathbf{P}\{\xi_{mi}\eta_{mi} > 0\}\right\}$$

and the l.h.s. approaches 1 as $n \to \infty$, we have

$$\lim \sum_{i=1}^{r_m} \mathbf{P}\{\xi_{mi}\eta_{mi} > 0\} = 0.$$

Hence, the second sum on the r.h.s. of (2.8) tends to zero as $n \to \infty$. This proves (2.7).

□

8°. Lemma. *Let $\mu(A)$ be an integral nonnegative r.m.i.v. on $\mathcal{B}_0$. Then*

$$\lim_{m\to\infty} \sup_{i\le r_m} \inf_{A_m^i \supset A_{m+1}^j} [\mu(A_m^i) - \mu(A_{m+1}^j)] = 0.$$

In fact, let us denote by ν_m a number of those $i \le r_m$ for which $\mu(A_m^i) > 0$. Obviously, $\nu_m \le \nu_{m+1} \le \mu(X)$. Thus, $\nu_m = \nu_{m+1} = \nu_{m+2} = \cdots = \nu$, beginning with a certain number m (ν and m depend on ω). Hence, for all sufficiently large m, one can find numbers $i_1^{(m)}, \ldots, i_\nu^{(m)}$ such that

$$\mu(A_m^{i_1^{(m)}}) > 0, \ldots, \mu(A_m^{i_\nu^{(m)}}) > 0.$$

Since $A^j_{m+1} \subset A^{i_1^{(m)}}_m$ can be found such that $\mu(A^j_{m+1}) > 0$ and since m (the number of sets A^j_{m+1} for which $\mu(A^j_{m+1}) > 0$) is equal to ν, these sets can be enumerated so that $A^{i_k^{(m)}}_m \supset A^{i_k^{(m+1)}}_{m+1}$, $k = 1, 2, \ldots, \nu$. We have $\mu(A^j_{m+1}) = 0$ for $A^j_{m+1} \subset A^{i_k^{(m)}}_m$, $j \neq i_k^{(m+1)}$; and thus, $\mu(A^{i_k^{(m)}}_m) = \mu(A^{i_k^{(m+1)}}_{m+1})$. Therefore,

$$\inf_{A^j_{m+1} \subset A^i_m} [\mu(A^i_m) - \mu(A^j_{m+1})] = 0.$$

□

Let $A^{i_k^{(l)}}_l$ be the atom of the σ-algebra $\mathcal{A}_l$ that contains $A^{i_k^{(m)}}_m$ for all sufficiently large m (it is unique). We have constructed a finite number (ν) of random fundamental sequences $(A^{i_k^{(m)}}_m)$, $k = 1, 2, \ldots, \nu$, such that for any $C \in \mathcal{B}_0$ one can find a number ν' such that for all $m > \nu'$,

$$\mu(C) = \sum_{A^{i_k^{(m)}}_m \subset C} \mu(A^{i_k^{(m)}}_m).$$

We say that *this measure is concentrated on fundamental sequences* $(A^{i_k^{(m)}}_m)$, $k = 1, \ldots, \nu$, and the variable $\mu(A^{i_k^{(m)}}_m)$ (for sufficiently large m, it is independent of m) will be called the *value of measure on the fundamental sequence* $(A^{i_k^{(m)}}_m)$.

9°. Lemma. *Suppose that for integral measures $\mu_1(A)$ and $\mu_2(A)$, the conditions of Lemma 7° (except (2)) are satisfied. Then these measures are independent iff they are concentrated on different fundamental sequences.*

Proof. Assume that the measures μ_1 and μ_2 are independent and that $\mathcal{F}_1$ and $\mathcal{F}_2$ are σ-algebras generated by their values (they are independent, too). If $(A^{i_m}_m)$ is some random fundamental sequence on which μ_1 possesses a positive value, then the index i_m can be chosen to be $\mathcal{F}_1$-measurable (it depends only on values of the measure μ_1). Then the independence of $\mathcal{F}_1$ and $\mathcal{F}_2$ yields

$$\begin{aligned}\mathbf{P}\{\mu_2(A^{i_m}_m) > 0\} &= \sum \mathbf{P}\{i_m = k\}\mathbf{P}\{\mu_2(A^k_m) > 0\} \\ &\leq \sup_{k \leq r_m} \mathbf{P}\{\mu_2(A^k_m) > 0\} \to 0\end{aligned}$$

as $m \to \infty$, by virtue of the stochastic continuity of μ_2.

Now we are to prove the sufficiency. Let $(A^{i_k^{(m)}}_m)$, $k = 1, \ldots, \nu_1$, and $(A^{j_k^{(m)}}_m)$, $k = 1, \ldots, \nu_2$, be the fundamental sequences on which the measures μ_1 and μ_2 are concentrated, respectively. They are different by the assumption. For sufficiently large m, the sets $A^{i_1^{(m)}}_m, \ldots, A^{i_{\nu_1}^{(m)}}_m, A^{j_1^{(m)}}_m, \ldots, A^{j_{\nu_2}^{(m)}}_m$ are disjoint.

Since $\mu_1(A^j_m) = 0$ for $j\overline{\in}(i_1^{(m)},\ldots,i_{\nu_1}^{(m)})$; $\mu_2(A^j_m) = 0$ for $j\overline{\in}(j_1^{(m)},\ldots,j_{\nu_2}^{(m)})$; and $(i_1^{(m)},\ldots,i_{\nu_1}^{(m)}) \cap (j_1^{(m)},\ldots,j_{\nu_2}^{(m)}) \neq \varnothing$; we have $\mu_1(A^j_m)\mu_2(A^j_m) = 0$, $\max \mu_1(A^j_m)\mu_2(A^j_m) = 0$, and the conditions of Lemma 7° are satisfied. This lemma implies independence of μ_1 and μ_2.

□

10°. Theorem. *Suppose that there are l integral nonnegative stochastically continuous r.m.i.v. $\mu_1,\ldots,\mu_l$ such that $\mu_i(A)$ is $\mathcal{F}_A$-measurable for all $i = 1,\ldots,l$. Denote by S_i a set of random fundamental sequences on which the measure μ_i is concentrated. If the sets S_i are mutually disjoint, then the measures μ_i are independent.*

Proof. This statement can be proved by induction on l. It has been proved for $l = 2$. Assume that it is valid for $l = k$. Using the fact that the measures $\sum_{i=1}^l a_i\mu_i$ and μ_{l+1}, where a_i are integers, satisfy 9°, we find that they are independent. Hence,

$$\mathbf{M}\exp\left\{-t\sum_{i=1}^{l} a_i\mu_i(A) - s\mu_{l+1}(A)\right\}$$

$$= \mathbf{M}\exp\left\{-t\sum_{i=1}^{l} a_i\mu_i(A)\right\}\mathbf{M}\exp\{-s\mu_{l+1}(A)\}$$

$$= \prod_{i=1}^{l}\mathbf{M}\exp\{-ta_i\mu_i(A)\}\,\mathbf{M}\exp\{-s\mu_{l+1}(A)\}$$

(we have used independence of $\mu_1,\ldots,\mu_l$ (according to the induction hypothesis)). Putting $t = 1/m$, $a_i/m \to t_i$, $s = t_{l+1}$ in this equality and passing to the limit, we get

$$\mathbf{M}\exp\left\{-\sum_{i=1}^{l+1} t_i\mu_i(A)\right\} = \prod_{i+1}^{l+1}\mathbf{M}\exp\{-t_i\mu_i(A)\}\,.$$

Hence, $\mu_1(A),\ldots,\mu_{l+1}(A)$ are independent for any A; and, besides, the vectors $(\mu_1(A^i_m),\ldots,\mu_{l+1}(A^i_m))$, $i = 1,\ldots,r_m$, are independent for any m. Therefore, the measures $m_1,\ldots,\mu_{l+1}$ are independent on $\mathcal{A}_m$ for any m, which implies their independence on $\mathcal{B}_0$.

□

11°. Lemma. *Let μ be a nonnegative integral stochastically continuous r.m.i.v. concentrated on the random fundamental sequences $(A_m^{k_i^{(m)}})$, $i = 1,\ldots,\nu$, and let ξ_i be a value of the measure μ on $(A_m^{k_i^{(m)}})$. If $g(n)$ is a function from Z_+ into Z_+ for which $g(0) = 0$, then*

$$\tilde{\mu}(C) = \lim_{m\to\infty}\sum_{A_m^{k_i^{(m)}}\subset C} g(\xi_i) \tag{2.10}$$

is also a nonnegative integral stochastically continuous r.m.i.v. concentrated on that subset of fundamental sequences $(A_m^{k_i^{(m)}})$ *for which* $g(\xi_i) > 0$.

Proof. Clearly, $\tilde{\mu}(C) = 0$ whenever $\mu(C) = 0$. Therefore, $\mathbf{P}\{\tilde{\mu}(C) = 0\} \geq \mathbf{P}\{\mu(C) > 0\}$, and the stochastic continuity of μ yields the stochastic continuity of $\tilde{\mu}$. Formula (2.10) implies that the measure $\tilde{\mu}(C)$ is finitely additive on $\mathcal{B}_0$ and takes values from Z_+. Finally, the same formula yields

$$\tilde{\mu}(C) = \lim_{m\to\infty} \sum_{A_m^i \subset C} g(\mu(A_m^i))$$

(if $\mu(A_m^i) = 0$, then $g(\mu(A_m^i)) = 0$). Thus, $\tilde{\mu}(C)$ is measurable with respect to $\mathcal{F}_C$ if $\mathcal{F}_C$ is a σ-algebra generated by the variables $\{\mu(A), A \subset \mathcal{B}_0, A \subset C\}$. Since the σ-algebras $\mathcal{F}_{C_1}$ and $\mathcal{F}_{C_2}$ are independent when $C_1 \cap C_2 = \varnothing$, then $\mu(C)$ is the measure with independent values.

□

12°. Theorem. *Let* μ *be a nonnegative integral stochastically continuous r.m.i.v. and let* $\mathcal{F}_A$ *be a* σ*-algebra generated by* $\{\mu(B), B \in \mathcal{B}_0, B \subset A\}$. *There exists a sequence of mutually independent Poisson measures* $\{\mu_1, \mu_2, \ldots\}$ *such that* $\mu_k(A)$ *is* $\mathcal{F}_A$*-measurable for every* k

$$\mu(A) = \sum_{k=1}^{\infty} k\mu_k(A). \tag{2.11}$$

Proof. Let $(A_m^{k_i^{(m)}})$, $i = 1, \ldots, \nu$, be fundamental sequences on which the measure μ is concentrated, $\lim \mu(A_m^{k_i^{(m)}}) = \xi_i$. Denote by $g_k(z)$ a function from Z_+ to Z_+; $g_k(z) = 1$ for $z = k$ and $g_k(z) = 0$ for $z \neq k$. Suppose that

$$\mu_k(C) = \lim_{m\to\infty} \sum_{A_m^{k_i^{(m)}} \subset C} g_k(\xi_i).$$

By virtue of 11°, it is a nonnegative integral stochastically continuous r.m.i.v. for which $\mu_k(A)$ is $\mathcal{F}_A$-measurable; $\lim_m \max_{i\leq r_m} \mu_k(A_m^i) \leq 1$. Hence, 6° implies that μ_k is a Poisson measure. Further, the measures $\mu_1, \mu_2, \ldots, \mu_k$ are concentrated on different fundamental sequences for all n (μ_k is concentrated on those $(A_m^{k_i^{(m)}})$ for which $\xi_i = k$). Hence, by virtue of 10°, measures $\mu_1, \ldots, \mu_n$ are independent for all n; i.e., $\{\mu_k\}$ is a sequence of independent Poisson measures. Formula (2.11) follows from the equality $\xi_i = \sum k g_k(\xi_i)$;

$$\mu(A) = \lim_{m\to\infty} \sum_{A_m^{k_i^{(m)}} \subset A} g_k(\xi_i) = \lim_{m\to\infty} \sum_{A_m^{k_i^{(m)}} \subset A} \sum_k k g_k(\xi_i)$$

$$= \sum_k k \lim_{m\to\infty} \sum_{A_m^{k_i^{(m)}} \subset A} g_k(\xi_i) = \sum_k k\mu_k(A).$$

Permutation of the summation and limit transition is possible, because both the sums are finite.

□

Stochastically Continuous Nonnegative r.m.i.v.

Consider some nonnegative stochastically continuous r.m.i.v. μ on an algebra $\mathcal{B}_0 = \bigcup \mathcal{A}_m$, where $\mathcal{A}_m$ are finite algebras with atoms $A_m^1, \ldots, A_m^{r_m}$. Let us define for each $\varepsilon > o$ a set S_ε of (random) fundamental sequences $(A_m^{i_m})$ such that $\mu((A_m^{i_m})) \geq \varepsilon$. Their number does not exceed $\mu(X)/\varepsilon$.Hence, $\bigcup_{\varepsilon>0} S_\varepsilon$ is at most countable. We introduce the random variables $\xi_k = \lim_{m\to\infty} \mu(A_m^{i_m^{(k)}})$, where $(A_m^{i_m^{(k)}})$ are all fundamental sequences belonging to $\bigcup_{\varepsilon>0} S_\varepsilon$ and enumerated so that $i_m^{(k)}(w)$ are random variables. Denote by T a set of $t \geq 0$ for which $\mathbf{P}\{\xi_k = t\} = 0$ for all k. We fix a certain countable dense set $T_1 \subset T$. For every $s \in T_1$, we denote by $\mathcal{C}^s$ an algebra of subsets of $[s, \infty)$ generated by half-intervals of the form $[t_1, t_2)$, $s \leq t_1$, $t_1, t_2 \in T_1$. Let us choose finite subalgebras $\mathcal{C}_m^s$ in $\mathcal{C}^s$ such that $\mathcal{C}_m^s \subset \mathcal{C}_{m+1}^s$, $\bigcup_m \mathcal{C}_m^s = \mathcal{C}^s$, and the atoms of $\mathcal{C}_m^s$ are half-intervals; $\mathcal{C}^s \otimes \mathcal{B}_0 = \bigcup_m \mathcal{C}_m^s \otimes \mathcal{A}_m$; moreover, $\mathcal{C}_m^s \otimes \mathcal{A}_m$ are finite subalgebras of $\mathcal{C}^s \otimes \mathcal{B}_0$. We define the random measures ν_m on $\mathcal{C}^s \otimes \mathcal{A}_m$ by the equality (for $\Delta \in \mathcal{C}^s$)

$$\nu_m(\Delta \times A_m^i) = I_\Delta(\mu(A_m^i)). \tag{2.12}$$

(If $\mathcal{A}$ and $\mathcal{C}$ are finite algebras of sets with atoms $A_1, \ldots, A_r$ and $C_1, \ldots, C_l$, then $A_i \times C_k$ will be the atoms of $\mathcal{A} \otimes \mathcal{C}$; and in order to define a measure on $\mathcal{A} \otimes \mathcal{C}$, it suffices to define it on the sets $A_i \times C_k$.)

13°. Theorem. *The limit*

$$\lim_{m\to\infty} \nu_m(\Delta \times B) = \nu(\Delta \times B) \tag{2.13}$$

exists with probability 1 for all $\Delta \in \mathcal{C}^s$ and $B \subset \mathcal{B}_0$; $\nu(\Delta \times B)$ can be extended to become the function on $\mathcal{C}^s \otimes \mathcal{B}_0$ that is a Poisson r.m.i.v. stochastically continuous with respect to $\mathcal{C}_m^s \otimes \mathcal{A}_m$.

Proof. Given any $B \in \mathcal{B}_0$, we can find n such that $B \in \mathcal{A}_n$; and $\nu_m(\Delta \times B)$ is defined for $m \geq n$ by

$$\nu_m(\Delta \times B) = \sum_{A_m^i \subset B} I_\Delta(\mu(A_m^i)).$$

Let $(A_m^{i_m^{(k_l)}})$, $l = 1, 2, \ldots, r$ be all the random fundamental sequences for which $\mu(A_m^{i_m^{(k_l)}}) > s$ for all m. Since $s \in T_1$, we have $\xi_{k_l} = \lim \mu(A_m^{i_m^{(k_l)}}) > s$ (the limit equals s with probability 0). If Δ' is the boundary of Δ (this is a finite set from T_1), then $\mathbf{P}\{\xi_{k_l} \in \Delta'\} = 0$. Hence,

$$\lim I_\Delta(\mu(A_m^{i_m^{(k_l)}})) = I_\Delta(\xi_{k_l})$$

with probability 1. Since for sufficiently large m,

$$\nu_m(\Delta \times B) = \sum_{l=1}^{r} I_\Delta(\mu(A_m^{i_m^{(k_l)}})),$$

the following limit exists

$$\lim_{m\to\infty} \nu_m(\Delta \times B) = \sum_{l=1}^{r} I_\Delta(\xi_{k_l}).$$

The existence of the limit (2.13) is thus proved. Formula (2.13) allows us to define $\nu(\Delta_m \times A_m)$ on atoms of the algebra $\mathcal{C}_m^s \otimes \mathcal{A}_m$; and hence, ν can be defined on this algebra and on $\bigcup_m \mathcal{C}_m^s \otimes \mathcal{A}_m = \mathcal{C}^s \otimes \mathcal{B}_0$. Let $\Delta_m \times A_m$ be a sequence of atoms of $\mathcal{C}_m^s \otimes \mathcal{A}_m$ fundamental with respect to $\{\mathcal{C}_m^s \otimes \mathcal{A}_m\}$. Since $\nu(\Delta_m \times A_m) \le \frac{1}{s}\mu(A_m)$ and $\mu(A_m) \to 0$ in probability, ν is stochastically continuous. To verify that values of ν on mutually disjoint sets are independent, note first that for mutually disjoint $\Delta_1, \Delta_2, \ldots, \Delta_l$ and for $B \in \mathcal{B}_0$ all variables $\nu(\Delta_i \times B)$ are $\mathcal{F}_B$-measurable, they are nonnegative integral r.m.i.v. on B. The measure $\nu(\Delta_i \times B)$ is concentrated on those fundamental sequences $(A_m^{i_m^{(k_l)}})$ for which $\xi_{k_l} \subset \Delta_i$. Thus, by virtue of 10°, $\nu(\Delta_i \times B)$ are independent (for different i) r.m.i.v. on B. Hence, if Δ_i^m, $i \le q_m$, are atoms of $\mathcal{C}_m^s$, then the variables $\{\nu(\Delta_i^m \times A_m^j), i \le q_m, j \le r_m\}$ are jointly independent. This proves that ν is a r.m.i.v. on $\mathcal{C}_m^s \otimes \mathcal{B}_0$. It is clear from the construction of ν that this is a counting measure. Therefore, ν is a Poisson measure due to 5°.

□

Let us define for $s \in T_1$ a nonnegative r.m.i.v.

$$\mu^s(B) = \lim_{m\to\infty} \sum_{A_m^i \subset B} g_s(\mu(A_m^i)), \tag{2.14}$$

where

$$g_s(t) = \begin{cases} 0, & t < s, \\ t, & t \ge s. \end{cases}$$

The existence of the limit is proved just as in 13°. The finite additivity of $\mu^s(B)$ follows from (2.14). $\mu^s(B)$ is $\mathcal{F}_B$-measurable; hence, $\mu^s(B)$ is a nonnegative r.m.i.v. Clearly, $\mu^s(B) \le \mu(B)$, and $\mu(B) - \mu^s(B)$ is a nonnegative r.m.i.v., too.

14°. Lemma. *There exists a nonrandom finite additive nonnegative measure on $\mathcal{B}_0$,*

$$m_0(B) = \lim_{s \downarrow 0}(\mu(B) - \mu^s(B)).$$

Proof. (2.14) implies that $\mu^{s_1}(B) \geq \mu^{s_2}(B)$ for $0 < s_1 < s_2$, $s_1, s_2 \in T_1$, with probability 1. Hence, there exists a limit $\lim_{s\downarrow 0}(\mu(B) - \mu^s(B)) = m_0(B)$. Note that $\mu(B) - \mu^s(B)$ has the property

$$\lim_{m\to\infty} \max_{i \leq r_m}[\mu(A^i_m) - \mu^s(A^i_m)] \leq s. \tag{2.15}$$

If not, then the fundamental sequence $(A^{j_m}_m)$ can be found for which $\mu(A^{j_m}_m) - \mu^s(A^{j_m}_m) > s$. Therefore, $\mu(A^{j_m}_m) > s$ for all m, and

$$\mu^s(A^{j_k}_k) \geq \lim_{m\to\infty} g_s(\mu(A^{j_m}_m)) = \lim_{m\to\infty} g(\mu(A^{j_m}_m)),$$
$$\lim_{k\to\infty} \mu^s(A^{j_k}_k) \geq \lim_{m\to\infty} \mu(A^{j_m}_m).$$

But since $\mu^s(A^{j_k}_k) \leq \mu(A^{j_k}_k)$, we have $\lim_{m\to\infty} \mu^s(A^{j_m}_m) = \lim_{m\to\infty} \mu(A^{j_m}_m)$ and

$$\lim_{m\to\infty} [\mu(A^{j_m}_m) - \mu^s(A^{j_m}_m)] = 0,$$

which contradicts the assumption that this limit is $\geq s$. Inequality (2.15) is thus valid. Therefore,

$$\begin{aligned} 1 &= \lim_{m\to\infty} \mathbf{P}\{\max_{i\leq r_m}[\mu(A^i_m) - \mu^s(A^i_m)] \leq s\} \\ &= \lim_{m\to\infty} \prod_{i=1}^{r_m}(1 - \mathbf{P}\{[\mu(A^i_m) - \mu^s(A^i_m)] > s\}) \\ &\leq \exp\left\{- \lim_{m\to\infty} \sum \mathbf{P}\{\mu(A^i_m) - \mu^s(A^i_m) > s\}\right\}, \end{aligned} \tag{2.16}$$
$$\lim_{m\to\infty} \sum \mathbf{P}\{\mu(A^i_m) - \mu^s(A^i_m) > s\} = 0.$$

Since $m_0(B)$ is a nonnegative r.m.i.v., it is nonrandom iff $m_0(X)$ is nonrandom. For every m,

$$m_0(X) = \lim_{s\downarrow 0} \sum_{i=1}^{r_m}(\mu(A^i_m) - \mu^s(A^i_m)).$$

Let us choose a sequence $s_m \downarrow 0$ as $m \to \infty$ so that

$$\lim_{m\to\infty} \sum_{i=1}^{r_m} \mathbf{P}\{\mu(A^i_m) - \mu^{s_m}(A^i_m) > s_m\} = 0 \tag{2.17}$$

(it is possible by virtue of (2.16)). Since

$$\mathbf{P}\{\sum_{i=1}^{r_m}(\mu(A^i_m) - \mu^{s_m}(A^i_m)) \wedge s_m \neq \sum_{i=1}^{r_m}(\mu(A^i_m) - \mu^{s}(A^i_m))\}$$
$$\leq \sum_{i=1}^{r_m}\mathbf{P}\{\mu(A^i_m) - \mu^{s_m}(A^i_m) > s_m\},$$

we have

$$m_0(X) = \lim_{m\to\infty}\sum_{i=1}^{r_m}(\mu(A^i_m) - \mu^{s_m}(A^i_m)) \wedge s_m,$$

in the sense of convergence in probability. Using 13°, Chapter 1, we conclude that all the moments of the variable $\sum_i^{r_m}(\mu(A^i_m) - \mu^{s_m}(A^i_m)) \wedge s_m$ are uniformly bounded (consequently, $m_0(X)$ has all the moments); moreover,

$$\mathbf{M}(m_0(X))^k = \lim_{m\to\infty}\mathbf{M}(\sum_{i=1}^{r_m}(\mu(A^i_m) - \mu^{s_m}(A^i_m)) \wedge s_m)^k. \tag{2.18}$$

In particular,

$$\mathbf{M}m_0(X) = \lim_{m\to\infty}\sum_{i=1}^{r_m}\mathbf{M}((\mu(A^i_m) - \mu^{s_m}(A^i_m)) \wedge s_m),$$

$$\begin{aligned}\mathbf{D}m_0(X) &= \lim_{m\to\infty}\sum_{i=1}^{r_m}\mathbf{D}((\mu(A^i_m) - \mu^{s_m}(A^i_m)) \wedge s_m)\\ &\leq \lim_{m\to\infty}\sum_{i=1}^{r_m}s_m\mathbf{M}(\mu(A^i_m) - \mu^{s_m}(A^i_m)) \wedge s_m\\ &\leq \lim_{m\to\infty}s_m\mathbf{M}m_0(X) = 0.\end{aligned}$$

The equality $\mathbf{D}m_0(X) = 0$ implies that $m_0(X)$ is nonrandom. Since $m_0(A^i_m) \leq \mu(A^i_m)$, we have $m_0((A^{i_m}_m)) = 0$ for every fundamental sequence $(A^{i_m}_m)$. □

15°. Lemma. *Suppose that* $s = s^m_1 < s^m_2 < \cdots < s^m_k < \cdots$ *for all* m; $s^m_k \uparrow \infty$ *as* $k \to \infty$; $[s^m_i, s^m_{i+1}[\in \mathcal{C}^s$ *for all* i, m *and* $\lim_{m\to\infty}\max_k|s^m_{k+1} - s^m_k| = 0$. *Then*

$$\mu^s(B) = \lim_{m\to\infty}\sum_{k=1}^{\infty}s^m_k\nu([s^m_k, s^m_{k+1}[\times B). \tag{2.19}$$

Proof. Note that

$$\mu^{s^m_k}(B) - \mu^{s^m_{k+1}}(B) = \lim_{n\to\infty}\sum_{A^i_n\subset B}[g_{s^m_k}(\mu(A^i_n)) - g_{s^m_{k+1}}(\mu(A^i_n))]$$

and since for $t < u$,

$$tI_{[t,u[}(x) \leq g_t(x) - g_u(x) \leq uI_{[t,u[}(x),$$

we have

$$s_k^m \lim_{n\to\infty} \sum_{A_n^i \subset B} I_{[s_k^m, s_{k+1}^m[}(\mu(A_n^i)) \leq \mu^{s_k^m}(B) - \mu^{s_{k+1}^m}(B)$$
$$\leq s_{k+1}^m \sum_{A_n^i \subset B} I_{[s_k^m, s_{k+1}^m[}(\mu(A_n^i)),$$
$$s_k^m \nu([s_k^m, s_{k+1}^m[\times B) \leq \mu^{s_k^m}(B) - \mu^{s_{k+1}^m}(B)$$
$$\leq s_{k+1}^m \nu([s_k^m, s_{k+1}^m[\times B).$$

For sufficiently large u, $\mu^u(X) = 0$, (e.g., for $u > \mu(X)$). Thus,

$$\sum_{k=1}^{\infty} [\mu^{s_k^m}(B) - \mu^{s_{k+1}^m}(B)] = \mu^s(B).$$

Hence,

$$\sum_{k=1}^{\infty} s_k^m \nu([s_k^m, s_{k+1}^m[\times B) \leq \mu^s(B) \leq \sum_{k=1}^{\infty} s_{k+1}^m \nu([s_k^m, s_{k+1}^m[\times B),$$
$$0 \leq \mu^s(B) - \sum_{k=1}^{\infty} s_k^m \nu([s_k^m, s_{k+1}^m[\times B)$$
$$\leq \sum_{k=1}^{\infty} (s_{k+1}^m - s_k^m) \nu([s_k^m, s_{k+1}^m[\times B)$$
$$\leq \max_k |s_{k+1}^m - s_k^m| \sum_{k=1}^{\infty} \nu([s_k^m, s_{k+1}^m[\times B).$$

Since $\nu([s_{k+1}^m, s_k^m[) = 0$ for $s_k^m > \mu(X)$, there is only a finite number of nonzero terms in the sum $\sum_{k=1}^{m} \nu([s_k^m, s_{k+1}^m[\times B)$. Therefore, it is equal to $\nu([s, \infty[\times B)$. Hence,

$$0 \leq \mu^s(B) - \sum_{k=1}^{\infty} s_k^m \nu([s_k^m, s_{k+1}^m[\times B)$$
$$\leq \max_k |s_{k+1}^m - s_k^m| \nu([s, \infty[\times B),$$

and this yields (2.19).

□

The next statement, which follows from the foregoing, describes nonnegative stochastically continuous r.m.i.v.

16°. Theorem. *If $\mu(B)$ is a nonnegative stochastically continuous r.m.i.v., then one can find a nonrandom nonnegative finitely additive function $m_0(B)$, $B \subset \mathcal{B}_0$, such that $\lim_{m\to\infty} m_0(A_m^{i_m}) = 0$ for every fundamental sequence $(A_m^{i_m})$, and a stochastically continuous Poisson measure $\nu(\Delta \times B)$ defined on some sequence of algebras $\mathcal{C}^{s_m} \otimes \mathcal{B}_0$, $s_m \downarrow 0$ such that*

$$\mu(B) = m_0(B) + \lim_{s_m \downarrow 0} \lim_{\substack{s_m = s_1^n < s_2^n < \cdots, \\ [s_k^n, s_{k+1}^n[\in \mathcal{C}^{s_m}, \\ \max |s_{k+1}^n - s_k^n| \to 0}} \sum s_k^n \nu([s_k^n, s_{k+1}^n[\times B). \tag{2.20}$$

§2.2 Random Measures with Alternating Signs

We shall use the definitions of the previous section. A finitely additive function of the set $\mu(A)$, defined on an algebra $\mathcal{B}_0$, with its values being random variables given on a fixed probability space $\{\Omega, \mathcal{F}, \mathbf{P}\}$ is called a *random measure with independent values (r.m.i.v.)* if for every collection $A_1, A_2, \ldots, A_k$ of mutually disjoint sets from $\mathcal{B}_0$, the variables $\mu(A_1), \mu(A_2), \ldots, \mu(A_k)$ are independent. Denote by $\mathcal{A}_m$ a certain fixed sequence of subalgebras of $\mathcal{B}_0$ for which $\mathcal{A}_m \subset \mathcal{A}_{m+1}$ and $\mathcal{B}_0 = \bigcup \mathcal{A}_m$. The r.m.i.v. $\mu(A)$ is called *stochastically continuous* (with respect to $(\mathcal{A}_m)$) if for any fundamental sequence (A_m) (A_m is an atom of $\mathcal{A}_m$ and $A_m \supset A_{m+1}$), $\mu(A_m) \to 0$ in probability. We shall denote the atoms of $\mathcal{A}_m$ by A_m^i, $i \le r_m$. A σ-algebra of events, generated by the variables $(\mu(B), B \subset A)$, will be denoted by $\mathcal{F}_A$. The σ-algebras $\mathcal{F}_U$ and $\mathcal{F}_V$ are independent if U and $V \in \mathcal{B}_0$, $U \cap V = \varnothing$.

A r.m.i.v. is called *purely discrete* if (1) for every fundamental sequence (A_n), there exists a limit (in the sense of convergence in probability) $\lim_{n\to\infty} \mu(A_n)$ (this limit is called a value of the measure μ on the fundamental sequence (A_n) and is denoted by $\mu((A_n)))$, (2) there exists at most a countable set of fundamental sequences $(A_n^{i_n^{(k)}})$, $k = 1, 2, \ldots$, for which $\mathbf{P}(\mu((A_n^{i_n^{(k)}})) \neq 0) > 0$, and equality (2.1) holds for all $C \subset \mathcal{A}_m$.

Symmetric r.m.i.v.

Consider a symmetric r.m.i.v. $\mu(A)$, i.e., such that $\mu(A)$ has a symmetric distribution for all $A \in \mathcal{B}_0$.

17°. Theorem. *Every symmetric r.m.i.v. $\mu(A)$ can be decomposed as follows:*

$$\mu(A) = \mu_1(A) + \mu_2(A),$$

where μ_1 and μ_2 are also symmetric mutually independent r.m.i.v., μ_1 is stochastically continuous, and μ_2 is purely discrete.

Proof. Note first that for every fundamental sequence (A_n), there exists a limit $\lim_{n\to\infty}\mu(A_n)$ with probability 1. In fact,

$$\mu(A_1)=\mu(A_n)+\sum_{k=1}^{n-1}\mu(A_k\backslash A_{k+1}).$$

Since the sets $A_k\backslash A_{k+1}$ are mutually disjoint, the terms in the sum are independent. Therefore,

$$\frac{1}{2}\mathbf{P}\left\{\sum_{k=1}^{n}(\mu(A_k\backslash A_{k-1})>c)\right\}\le\mathbf{P}\left\{\sum_{k=1}^{n}\mu(A_k\backslash A_{k-1})>c,\mu(A_n)\ge 0)\right\}$$
$$\le\mathbf{P}\{\mu(A_1)>c\}.$$

(we have used the symmetry of $\mu(A_n)$). Consequently,

$$\mathbf{P}\{|\sum_{k=1}^{n-1}\mu(A_k\backslash A_{k-1})|>c\}\le 2\mathbf{P}\{|\mu(A_1)|>c\};$$

and by virtue of 19°, Chapter 1, the series $\sum_{k=1}^{\infty}\mu(A_k\backslash A_{k-1})$ converges with probability 1. Hence, the limit

$$\lim_{n\to\infty}\mu(A_n)=\mu(A_1)-\sum_{k=1}^{\infty}\mu(A_k\backslash A_{k-1})$$

exists with probability 1. Henceforth, we shall need the following:

18°. Lemma. *Let $\xi_1,\xi_2,\ldots,\xi_n$, η_n be symmetric independent random variables; $\zeta=\sum_{k=1}^{n}\xi_k+\eta_n$; $f(t)$ be a characteristic function of the variable ζ; and c be chosen so that $f(t)>1/2$ for $|t|\le c$. Then for all $\varepsilon>0$,*

$$\sum_{k=1}^{n}\mathbf{P}\{|\xi_k|>\varepsilon\}\le\lambda(\varepsilon c)\cdot\frac{1}{2c}\int_{-c}^{c}\ln\frac{1}{f(t)}dt\le\lambda(\varepsilon c)\ln 2,$$

where

$$\lambda(t)=\left(\inf_{x>t}\left(1-\frac{\sin x}{x}\right)\right)^{-1}.$$

Proof. Since

$$\prod_{k=1}^{n}\mathbf{M}e^{it\xi_k}\mathbf{M}e^{it\eta_n}=f(t)$$

and the factors on the l.h.s. are real and continuous with respect to t, they are positive for $|t| \le c$ (if at least one of them is zero, $f(t)$ is also zero). Hence,

$$f(t) \le \prod_{k=1}^{n} \mathbf{M} \cos t\xi_k = \prod_{k=1}^{n} [1 + (\mathbf{M} \cos t\xi_k - 1)]$$
$$\le \exp\left\{\sum_{k=1}^{n} (\mathbf{M} \cos t\xi_k - 1)\right\} \quad (|t| \le c)$$

and

$$-\frac{1}{2c}\int_{-c}^{c} \ln f(t)dt \ge \sum_{k=1}^{n} \frac{1}{2c}\int_{-c}^{c} \mathbf{M}(1 - \cos t\xi_k)dt$$
$$= \sum_{k=1}^{n} \mathbf{M}(1 - \frac{\sin c\xi_k}{c\xi_k}) \ge \sum_{k=1}^{n} \mathbf{M}(1 - \frac{\sin c\xi_k}{c\xi_k}) I_{\{|\xi_k| > c\}}$$
$$\ge \lambda^{-1}(\varepsilon c) \sum_{k=1}^{n} \mathbf{P}\{|\xi_k| > c\}.$$

□

Returning to the proof of the theorem, we note that Lemma 18° yields that for every $\varepsilon > 0$, there exists only a finite number of fundamental sequences (A_n) for which $\mathbf{P}\{|\lim \mu(A_n)| > \varepsilon\} > \varepsilon$. This number does not exceed $\frac{\lambda(\varepsilon c)}{\varepsilon} \ln 2$ if c is such that $\mathbf{M}e^{it\mu(X)} \ge \frac{1}{2}$ for $|t| \le c$. Consequently, there exists at most a countable set of fundamental sequences (A_n) for which $\mathbf{P}\{\mu((A_n)) > 0\} > 0$. The last part of the proof repeats the proof of Theorem 1°.

□

Decomposition of r.m.i.v.

Let $m(B)$ be some finitely additive function on $\mathcal{B}_0$. We shall call it *regular* if the following conditions are satisfied:

(1) for every $\varepsilon > 0$, there exists $n(\varepsilon)$ such that the sequence

$$k_\varepsilon(n) = \sum_{i=1}^{r_n} I_{\{\sup_{A_m^j \subset A_n^i} |m(A_m^j)| > \varepsilon\}}$$

is constant for $n \ge n(\varepsilon)$,

(2) for every fundamental sequence (A_n), there exists $\lim_{n\to\infty} m(A_n)$.

If $m(B)$ is regular, then for every $\varepsilon > 0$, there exists only a finite number of fundamental sequences (A_n) such that $|\lim_{n\to\infty} m(A_n)| > \varepsilon$ (this number does not exceed $k_\delta(n_\delta)$ if $\delta < \varepsilon$). Thus, there exists at most a countable set of sequences (A_n) for which $\lim m(A_n) \neq 0$.

Suppose that the measure $m(B)$ is regular, and $i_n^{(1)}, \ldots, i_n^{(k_\varepsilon)}$ are those numbers for which

$$\sup_{A_m^j \subset A_n^i} |m(A_m^j)| > \varepsilon, \quad i = i_n^{(1)}, \ldots, i_n^{(k_\varepsilon)}$$

when $n \geq n_\varepsilon$. Whatever fundamental sequence (A_n) we take such that A_n does not coincide with any of the sets $A_n^{i_n^{(1)}}, \ldots, A_n^{i_n^{(k_\varepsilon)}}$ (for some $n \geq n_\varepsilon$), the inequality $\overline{\lim}_{n\to\infty} |\mu(A_n)| \leq \varepsilon$ holds. Therefore, one should check condition (2) only for a countable set of fundamental sequences

$$\bigcup_{\varepsilon>0} \{(A_n^{i_n^{(1)}}), \ldots, (A_n^{i_n^{(k_\varepsilon)}})\}.$$

If μ is some random finitely additive function on $\mathcal{B}_0$, then it is regular if the set of those ω for which μ_ω is regular (ω being fixed) has probability 1. Note that the set of such ω is measurable: It coincides with the set on which limits of the sequences

$$k_\varepsilon(n, \omega) = \sum_{i=1}^{r_n} I_{\{\sup_{A_m^j \subset A_m^i} |\mu(A_m^j)| > \varepsilon\}},$$

$$\{\mu(A_n^{i_n^{(k)}(\omega,\varepsilon)})\}$$

exist, where

$$i_n^{(k)}(\omega, \varepsilon) = \sum_{l=1}^{r_n} I_{\{\sum_{i=1}^{l} I_{\{\sup_{A_m^j \subset A_n^i} |\mu(A_n^i)| > \varepsilon\}} < k\}} + 1$$

(for the sake of convenience, we consider that $(A_n^{r_n+1})$ consists of empty sets). To investigate the regularity of an additive function, it is suitable to use such a mapping of $\mathcal{B}_0$ in the algebra generated by some system of half-intervals on $[0, 1[$: We associate with each set $A_m^i \in \mathcal{B}_0$ a half-interval $[t_m^{i-1}, t_m^i[$, where $t_m^0 = 0 < t_m^1 < \cdots < t_m^{r_m} = 1$, considering the atoms A_m^i to be enumerated so that for $i < j$ and $A_n^i \subset A_n^{i_1}$, $A_n^j \subset A_n^{j_1}$, we also have $i_1 < j_1$, $[t_m^i, t_m^{i+1}[= \bigcup [t_{m+1}^j, t_{m+1}^{j+1}[$. The points t_m^i are chosen so that $\max_i |t_m^i - t_m^{i-1}| \to 0$. We write $[t_m^{i-1}, t_m^i[= \Delta_m^i$ for short. Define a function $\xi(t)$ on the set $\bigcup_m \{t_m^0, t_m^1, \ldots, t_m^{r_m}\} = S$ by assuming that

$$\xi(0) = 0, \qquad \xi(1) = \mu(X), \qquad \xi(t_m^i) - \xi(t_m^{i-1}) = \mu(A_m^i).$$

A real-valued function $a(t)$, given on some set $\Lambda \subset R$, has k ε-oscillations if there exist $k + 1$ points $t_1 < \cdots < t_{k+1} \in \Lambda$ for which $|a(t_{i+1}) - a(t_i)| \geq \varepsilon$, $i = 1, \ldots, k$, and if there are no $k + 2$ points in Λ possessing this property.

19°. Lemma. *If $\xi(t)$ for every $\varepsilon > 0$ has a finite number of ε-oscillations on S with probability 1, then the measure μ is regular.*

Proof. If some (nonrandom) function $a(t)$ on S has a finite number of ε-oscillations for every $\varepsilon > 0$, then for any increasing (decreasing) sequence $t_n \in S$, there exists $\lim a(t_n)$. Hence, if $\xi(t)$ has for all rational ε a finite number of ε-oscillations with probability 1, then for every monotonic sequence t_n, the limits $\lim \xi(t_n)$ exist on a set of full measure. Therefore, for $A_m^{i_m} \supset A_{m+1}^{i_{m+1}}$, we have

$$\lim \mu(A_m^{i_m}) = \lim[\xi(t_m^{i_m}) - \xi(t_m^{i_m-1})],$$

because the sequence $t_m^{i_m-1}$ increases and $t_m^{i_m}$ decreases ($\Delta_m^{i_m} \supset \Delta_{m+1}^{i_{m+1}}$). The number of fundamental sequences $\Delta_m^{i_m}$ for which

$$|\lim(\xi(t_m^{i_m}) - \xi(t_{m+1}^{i_{m+1}}))| > \varepsilon/2$$

is finite. If we delete from S the intervals $\Delta_m^{i_m^{(k)}}$, $k = 1, \dots, l$, for which the previous inequality holds, we get

$$\lim_{n\to\infty} \sup_{\Delta_n^j \not\subset \bigcup_{k=1}^l \Delta_m^{i_m^{(k)}}} |\xi(t_n^j) - \xi(t_n^{j-1})| < \varepsilon.$$

This means that $\xi(t)$ is regular.

□

20°. Theorem. *Symmetric r.m.i.v. are regular.*

Proof. Lemma 19° implies that it suffices to show that the function $\xi(t)$ constructed with the help of μ has a finite number of ε-oscillations. To prove this, it is sufficient to show that for every $\varepsilon > 0$, m can be found such that $\xi(t)$ has a finite number of ε-oscillations on each interval Δ_m^i. Let $\mu = \mu_1 + \mu_2$ be a decomposition of μ into stochastically continuous and discrete components, where a purely discrete component μ_2 is concentrated on the fundamental sequences $(A_n^{i_n^{(k)}})$, $k = 1, \dots,$ and let l be chosen so that $\mathbf{P}\{\lim_{k>l} \lim_{n\to\infty} \mu(A_n^{i_n^{(k)}}) > \varepsilon/4\} < 1/10$. Furthermore, we choose m so that all sets $A_m^{i_m^{(k)}}$, $k = 1, \dots, l$, are different and $\mathbf{P}\{|\mu_1(A_m^i)| > \varepsilon/4\} < \frac{1}{10}$. Now consider a number of oscillations of $\xi(t)$ on Δ_m^i. Let us denote a common point of the closed intervals $[\Delta_m^{i_m^{(k)}}]$ by t_k. There is at most one point t_k inside each interval Δ_m^i. Assume first that Δ_m^i does not contain any point t_k. Then

$$\begin{aligned}
&\mathbf{P}\{|\xi(t_m^i) - \xi(t_m^{i-1})| > \varepsilon/2\} \\
&\quad \le \mathbf{P}\{\mu_1(A_m^i) > \varepsilon/4\} + \mathbf{P}\{|\mu_2(A_m^i)| > \varepsilon/4\} \\
&\quad \le \mathbf{P}\{|\mu_1(A_m^i)| > \varepsilon/4\} + 2\mathbf{P}\{|\sum_{k>l} \lim_{n\to\infty} \mu((A_n^{i_n^{(k)}}))| > \varepsilon/4\} < 3/10
\end{aligned}$$

(we have used the fact that

$$\mu_2(A^i_m) + \sum_{\substack{k>l \\ i^{(k)}_m \neq i}} \lim_{n\to\infty} \mu(A^{i^{(k)}_n}_n) = \sum_{k>l} \lim_{n\to\infty} \mu(A^{i^{(k)}_n}_n),$$

the terms are symmetric and independent, and for these terms, $\mathbf{P}\{|\xi_1| > c\} \leq 2\mathbf{P}\{|\xi_1 + \xi_2| > c\}$). Now let $t_k \in \Delta_i$. The finiteness of a number of ε-oscillations of $\xi(t)$ on Δ_i is equivalent to the uniform boundedness of a number of ε-oscillations on the intervals $(t^{i-1}_m, t^{i^{(k)}_n - 1}_n)$ and $(t^{i^{(k)}_n}_n, t^i_m)$. Moreover,

$$\begin{aligned}
&\mathbf{P}\{|\xi(t^{i^{(k)}_n - 1}_n) - \xi(t^{j-1}_m)| > \varepsilon/2\} \\
&\leq \mathbf{P}\{|\mu_1(\bigcup_{\substack{t^{i-1}_m \leq t^{j-1}_n, \\ j \leq j^{(k)}_n - 1}} A^j_n)| > \varepsilon/4\} + \mathbf{P}\{|\mu_2(\bigcup_{\substack{t_m \leq t^{j-1}_n, \\ j \leq i^{(k)}_n - 1}} A^j_n)| > \varepsilon/4\} \\
&\leq 2\mathbf{P}\{|\mu_1(A^i_m)| > \varepsilon/4\} + 2\mathbf{P}\{|\sum_{k>l} \lim_{n\to\infty} \mu(A^{i^{(k)}_n}_n)| > \varepsilon/4\} < 4/10.
\end{aligned}$$

Similarly,

$$\mathbf{P}\{|\xi(t^i_m) - \xi(t^{i^{(k)}_n}_n)| > \varepsilon/2\} < 4/10.$$

Proof of the theorem follows from the lemma.

21°. Lemma. *Let $\xi(t)$ be a symmetric process with independent increments defined on a bounded (countable) set S with the least and the largest elements $\underline{s}$ and $\overline{s}$; moreover, $\mathbf{P}\{|\xi(\overline{s}) - \xi(\underline{s})| > \varepsilon/2\} \leq \alpha < \frac{1}{2}$. Also let ν_ε be a number of ε-oscillations of ξ on the set S. Then*

$$\mathbf{M}\nu_\varepsilon \leq \frac{1}{1-2\alpha}.$$

Proof. Suppose that S is finite: $S = \{t_1 < t_2 < \cdots < t_n\}$. Let us estimate the probability of the event $\{\nu_\varepsilon \geq j\}$. If this event happens, then for some $r \leq k$, the difference $|\xi(t_r) - \xi(t_1)|$ exceeds or equals $\varepsilon/2$ for the first time (hence, $\xi(t)$ has no ε-oscillations on the set $(t_1, \ldots, t_{r-1})$), and $\xi(t)$ has at least $j-1$ ε-oscillations on the set $(t_r, \ldots, t_n)$. Thus,

$$\begin{aligned}
\mathbf{P}\{\nu_\varepsilon \geq j\} &= \sum_r \mathbf{P}\{|\xi(t_2) - \xi(t_1)| < \varepsilon/2, \ldots, |\xi(t_{r-1}) - \xi(t_1)| < \varepsilon/2, \\
&\qquad |\xi(t_r) - \xi(t_1)| \geq \varepsilon/2, \xi(t_{r+1}) - \xi(t_r), \ldots, \xi(t_n) - \xi(t_r) \\
&\qquad \text{has at least } j-1 \ \varepsilon\text{-oscillations}\} \\
&\leq \sum_r \mathbf{P}\{|\xi(t_2) - \xi(t_1)| < \varepsilon/2, \ldots, |\xi(t_{r-1}) - \xi(t_1)| < \varepsilon/2, \\
&\qquad |\xi(t_r) - \xi(t_1)| \geq \varepsilon/2\}\mathbf{P}\{\nu_\varepsilon \geq j-1\} \\
&= \mathbf{P}\{\sup_{r\leq n} |\xi(t_r) - \xi(t_1)| > \varepsilon/2\}\mathbf{P}\{\nu_\varepsilon \geq j-1\} \\
&\leq 2\mathbf{P}\{|\xi(t_n) - \xi(t_1)| > \varepsilon/2\}\mathbf{P}\{\nu_\varepsilon \geq j-1\} \leq 2\alpha\mathbf{P}\{\nu_\varepsilon \geq j-1\}
\end{aligned}$$

(we have used 9°, Chapter 1). Hence,

$$\mathbf{P}\{\nu_\varepsilon \geq j\} \leq (2\alpha)^j, \qquad \mathbf{M}\nu_\varepsilon \leq \sum_j \mathbf{P}\{\nu_\varepsilon \geq j\} \leq \frac{1}{1-2\alpha}.$$

22°. Theorem. *Every r.m.i.v.* μ *can be represented in the form*

$$\mu = m + \mu_1 + \mu_2,$$

where m *is a nonrandom bounded additive function of sets on* $\mathcal{B}_0$, μ_1 *is a regular stochastically continuous r.m.i.v., and* μ_2 *is a purely discrete r.m.i.v.; moreover,* μ_1 *and* μ_2 *are independent.*

Proof. We use a symmetrization method. Let the probability space $\{\Omega', \mathcal{F}', \mathbf{P}'\}$ be identical to the initial one. Consider a probability space $\{\Omega\times\Omega', \mathcal{F}\otimes\mathcal{F}', \mathbf{P}\times\mathbf{P}'\}$ and define a r.m.i.v. $\tilde{\mu}(\omega,\omega',B) = \mu(\omega,B) - \mu(\omega',B)$ on it ($\mu(\omega,B)$ is a measure given on the initial probability space). Obviously, $\mu(\omega,B)$ and $\mu(\omega',B)$ are independent and equally distributed; hence, $\tilde{\mu}(\omega,\omega',B)$ is a symmetric r.m.i.v. By virtue of Theorem 20°, $\mu(\omega,B) - \mu(\omega',B)$ is regular with probability $\mathbf{P}\times\mathbf{P}' = 1$. Denote by $\mathcal{M}^r$ a set of all regular additive functions on $\mathcal{B}_0$. Then

$$\int \mathbf{P}\{\mu(\omega,B) - \mu(\omega',B) \in \mathcal{M}^r\}\mathbf{P}'(d\omega') = 1;$$

and thus, one can find $\omega_0 \in \Omega$ such that $\mathbf{P}\{\mu(\omega,B) - \mu(\omega_0,B) \in \mathcal{M}^r\} = 1$. Consider a regular r.m.i.v. $\overline{\mu}(\omega,B) = \mu(\omega,B) - \mu(\omega_0,B)$. There exists at most a countable number of fundamental sequences (A_m) for which $\overline{\mu}(\omega,A_m)$ do not approach zero in probability. Let us denote them by $(A_m^{i_m^{(k)}})$, $k = 1,2,\ldots$. Regularity implies the existence of the limits $\xi_k = \lim_{m\to\infty} \overline{\mu}(\omega, A_m^{i_m^{(k)}})$ with probability 1. These limits are independent random variables. By virtue of 20°, Chapter 1, one can find a sequence a_k such that the series $\sum(\xi_{n_i} - a_{n_i})$ converges for any subsequence n_i. Thus, the purely discrete measure is defined: $\mu_2(B) = \sum_{A_m^{i_m^{(k)}} \subset B} (\xi_k - a_k)$ for $B \subset \mathcal{A}_m$. Note now that the r.m.i.v. $\overline{\mu}(B) - \mu_2(B)$ is such that $\overline{\mu}(A_m) - \mu_2(A_m) \to 0$ on all fundamental sequences (A_m) except the fundamental sequences $(A_m^{i_m^{(k)}}), k = 1,2,\ldots$; moreover, $\lim(\overline{\mu}(A_n^{i_n^{(k)}}) - \mu_2(A_n^{i_n^{(k)}})) = -a_k$, in the sense of convergence with probability 1. On $[0,1]$, we now define a function $\varphi(t)$ without discontinuities of the second kind, so that the only discontinuities of $\varphi(t)$ are t_k, i.e., the only common points of the closed intervals $[t_m^{i_{m-1}^{(k)}}, t_m^{i_m^{(k)}}]$. If t_k does not belong to S, we put $\varphi(t_k+) - \varphi(t_k-) = a_k$; and if $t_k \in S$, it is possible that $t_k = t_l$ for some l (this means that all the intervals $\Delta_m^{i_m^{(k)}}$ and $\Delta_m^{i_m^{(l)}}$ have the common limit point and $t_k = t_n^{i_n^{(k)}} = t_n^{i_n^{(l-1)}} = t^l$), and we put $\varphi(t_k) - \varphi(t_k-) = a_k$,

$\varphi(t_l+)-\varphi(t_l)=a_l$. If we denote by $n(B)$ the additive function on $\mathcal{B}_0$ defined by the equalities $n(A^i_m)=\varphi(t^i_m)-\varphi(t^{i-1}_m)$, then $\overline{\mu}(A)+\mu_2(A)+n(A)$ is now a stochastically continuous regular r.m.i.v. Let us denote it by μ_1. Then

$$\mu_1(A)=\mu(A)-\mu(\omega_0,A)-\mu_2(A)+n(A)$$

or

$$\mu(A)=\mu(\omega_0,A)-n(A)+\mu_1(A)+\mu_2(A).$$

By writing $m(A)=\mu(\omega_0,A)-n(A)$, we obtain the required result.

□

Corollary. *Every stochastically continuous r.m.i.v. is regular.*

In fact, if μ is a stochastically continuous r.m.i.v., then according to Theorem 22°, it can be represented in the form $\mu = m+\mu_1+\mu_2$, where μ_1 is a stochastically continuous regular r.m.i.v., and μ_2 is a purely discrete one. Therefore, $m+\mu_2$ is stochastically continuous; and thus, $\mu_2(A_n^{i_n^{(k)}})+m(A_n^{i_n^{(k)}})\to 0$ for any fundamental sequence $(A_n^{i_n^{(k)}})$ for which $\lim_{n\to\infty}\mu_2(A_n^{i_n^{(k)}})\neq 0$ (the limit exists in probability). Hence, $m+\mu_2$ is a nonrandom measure. It is continuous and thus regular; μ is regular as a sum of two regular measures.

Stochastically Continuous r.m.i.v.

Let μ be a stochastically continuous r.m.i.v. It is regular, and thus we can find at most a countable set of random fundamental sequences $(A_n^{i_n^{(k)}(\omega)})$ for which the limit $\lim_{n\to\infty}\mu(A_n^{i_n^{(k)}(\omega)})=\xi_k$ differs from zero with positive probability, and for every $s>0$, there exist only a finite number of k such that $|\xi_k|\geq s$.

Consider in R_+ a set $T(\mu)$ of $t\neq 0$ for which $\sum_k \mathbf{P}\{|\xi_k|=t\}=0$. Let T_1 be some countable subset of $T(\mu)$ that is dense in R_+. For $s\in T_1$, we denote by $\widehat{\mathcal{C}}^s$ an algebra of the subsets $]-\infty,-s]\cup[s,+\infty[$ containing the intervals $[t_1,t_2[$ and $]-t_2,-t_1]$, where $t_1,t_2\in T_1$. Define on $\widehat{\mathcal{C}}^s\otimes\mathcal{B}_0$ an additive function ν by

$$\nu(U\times A^i_m)=\sum I_U(\xi_k)I_{\{i_m^{(k)}(\omega)=i\}}.$$

This is a nonnegative stochastically continuous function (with respect to the algebras $\{\widehat{\mathcal{C}}^s_m\otimes\mathcal{A}_m\}$, where $\widehat{\mathcal{C}}^s_m$ is an arbitrary sequence of finite algebras for which $\bigcup_m\widehat{\mathcal{C}}^s_m=\widehat{\mathcal{C}}^s$). In addition, ν is a r.m.i.v. (this is proved just like Theorem 13°). For $B\in\mathcal{A}_m$, we put

$$\mu^s(B)=\sum_{A_m^{i_m^{(k)}(\omega)}\subset B}\hat{g}_s(\xi_k), \tag{2.21}$$

where $\hat{g}_s(t) = t$ for $|t| \geq s$, and $\hat{g}_s(t) = 0$ for $|t| < s$. Just as in Lemma 15°, we can prove that the formula

$$\mu^s(B) = \lim_{m\to\infty}[\sum_{k=1}^{\infty}[s_k^{(m)}\nu([s_k^{(m)}, s_{k+1}^{(m)}[\times B) - \sum_{k=1}^{\infty} s_k^{(m)}\nu(] - s_{k+1}^{(m)} - s_k^{(m)}] \times B)]] \tag{2.22}$$

holds, where $s = s_1^{(m)} < s_2^{(m)} < \cdots \in T_1$, $\lim_{m\to\infty} \max_k(s_{k+1}^{(m)} - s_k^{(m)}) = 0$.

Let $\overline{\mu}_s(B) = \mu(B) - \mu_s(B)$. The measure $\overline{\mu}_s(B)$ is regular, because $\mu_s(B)$ is regular (since $\hat{g}_s(\xi_k) = 0$ for all sufficiently large k, one can find m (dependent on ω) such that the sets $A_m^{i_m^{(k)}(\omega)}$ are mutually disjoint for all k for which $\hat{g}_s(\xi_k) \neq 0$; and if $\hat{g}_s(\xi_k) = 0$ for $k > l$, then $\mu^s(A_n^i) = 0$ for $i\overline{\in}(i_n^{(1)}(\omega), \ldots, i_n^{(l)}(\omega))$). In addition, it is stochastically continuous.

23°. Theorem. *The measure $\overline{\mu}_s$ has the following properties:*

(a)
$$\lim_{n\to\infty} \max_{i\leq r_n} |\overline{\mu}_s(A_n^i)| \leq s; \tag{2.23}$$

(b) *there exists $\mathbf{M}\overline{\mu}_s(B)$, and it is a continuous additive function on $\mathcal{B}_0$;*

(c) *the following limit exists in the sense of convergence in probability,*

$$\lim_{s\downarrow 0, s\in T_1} (\overline{\mu}_s(B) - \mathbf{M}\overline{\mu}_s(B)) = \mu_0(B),$$

and μ_0 is a r.m.i.v. independent of μ_s for all $s > 0$ and such that

$$\lim_{n\to\infty} \mathbf{P}\{\max_{i\leq r_n} |\overline{\mu}_0(A_n^i)| > \varepsilon\} = 0 \tag{2.24}$$

for all $\varepsilon > 0$.

Proof. The regularity of $\overline{\mu}_s$ implies that there exists a finite number of fundamental sequences $(A_n^{i_n^{(k)}(\omega)})$, $k = 1, 2, \ldots, l(\omega)$, such that

$$\sup_{A_m^j \subset A_n^{i_n^{(k)}(\omega)}} |\overline{\mu}_s(A_m^j)| > s$$

and $|\overline{\mu}_s(A_n^j)| \leq s$ for sufficiently large n, provided that $j\overline{\in}(i_n^{(1)}(\omega), \ldots, i_n^{(l(\omega))}(\omega))$. But $\overline{\mu}_s(B)$ is constructed so that

$$\lim_{n\to\infty} |\overline{\mu}_s(A_n^{i_n^{(k)}(\omega)})| \leq s,$$

and this gives (a).

Now (a) yields

$$\lim_{n\to\infty} \mathbf{P}\{\max_{i\leq r_n} |\overline{\mu}_s(A_n^i)| \leq s_1\} = 1$$

for $s_1 > s$; and consequently,

$$\lim_{n\to\infty} \sum_{i=1}^{r_n} \mathbf{P}\{|\overline{\mu}_s(A_n^i)| > s_1\} = 0.$$

Thus, for $B \subset \mathcal{A}_m$,

$$\mathbf{P}\{\mu_s(B) \neq \sum_{A_n^i \subset B} \overline{\mu}_s(A_n^i) I_{\{|\overline{\mu}_s(A_n^i)| \le s_1\}}\}$$
$$\le \sum_{i=1}^{r} \mathbf{P}\{|\overline{\mu}_s(A_n^i)| > s_1\} \to 0 \text{ as } n \to \infty.$$

Hence,

$$\mu_s(B) = \lim_{n\to\infty} \sum_{A_n^i \subset B} \overline{\mu}_s(A_n^i) I_{\{|\overline{\mu}_s(A_n^i)| \le s_1\}},$$

in the sense of convergence in probability. Therefore, by virtue of 12°, Chapter 1, $\mu_s(B)$ has all the moments; (b) is thus proved.

In order to prove (c), we establish the fact that is of independent interest.

24°. Lemma. *If* $0 < s_1 < s_2 < \cdots < s_k \in T$, *then* $\mu^{s_k}, \mu^{s_{k-1}} - \mu^{s_k}, \ldots, \mu^{s_1} - \mu^{s_2}$, *and* $\mu - \mu^{s_1}$ *are mutually independent r.m.i.v.*

Proof. Let us first prove the independence of μ^s and $\mu - \mu^s$. The proof is analogous to the proof of Lemma 7°; just as before, it suffices to prove the independence of $\mu^s(B)$ and $\mu(B) - \mu^s(B)$ for each $B \in \mathcal{B}_0$. For the sake of convenience, we consider that $B = X$. We put

$$\xi_{mk} = \mu(A_m^k) I_{\{|\mu_s(A_m^k)| \ge s\}}, \qquad \eta_{mk} = \mu(A_m^k) - \xi_{mk},$$
$$\theta_m = \mathbf{M}\exp\left\{i\sum_{k=1}^{r_m} \xi_{mk} + iz\sum_{k=1}^{r_m} \eta_{mk}\right\}$$
$$- \mathbf{M}\exp\left\{iu\sum_{k=1}^{r_m} \xi_{mk}\right\} \mathbf{M}\exp\left\{iz\sum_{k=1}^{r_m} \eta_{mk}\right\}.$$

It suffices to show that $\lim_{m\to\infty} \theta_m = 0$. We have

$$|\theta_m| = |\prod \mathbf{M}\exp\{iu\xi_{mk} + iz\eta_{mk}\} - \prod \mathbf{M}\exp\{iu\xi_{mk}\}\mathbf{M}\exp\{iz\eta_{mk}\}|$$
$$\le \sum_{k=1}^{r_m} |\mathbf{M}\exp\{iu\xi_{mk} + iz\eta_{mk}\} - \mathbf{M}\exp\{iu\xi_{mk}\}\mathbf{M}\exp\{iz\eta_{mk}\}|$$
$$= \sum_{k=1}^{r_m} |\mathbf{M}\exp\{iu\xi_{mk}\} + \mathbf{M}\exp\{iz\eta_{mk}\} - 1$$
$$-\mathbf{M}\exp\{iu\xi_{mk}\}\mathbf{M}\exp\{iz\eta_{mk}\}|$$
$$= \sum_{k=1}^{r_m} |\mathbf{M}\exp\{iu\xi_{mk}\} - 1| \cdot |\mathbf{M}\exp\{iz\eta_{mk}\} - 1|$$

(we have used the fact that $e^{a+b} = e^a + e^b - 1$ for $ab = 0$).

Further,

$$\mathbf{M}|\exp\{iz\eta_{mk}\}-1|\le 2\mathbf{P}\{|\eta_{mk}|>\varepsilon\}+|z|\varepsilon;$$

and therefore, $\sup_k|\mathbf{M}\exp\{iz\eta_{mk}\}-1|\to 0$ as $m\to\infty$, by virtue of the stochastic continuity of μ, and

$$\sum|\mathbf{M}\exp\{iu\xi_{mk}\}-1|\le 2\sum_{k=1}^{r_m}\mathbf{P}\{|\mu(A_m^k)|>s\}.$$

Let us show that the variable on the r.h.s. is bounded. If μ' is a r.m.i.v. independent of μ with the same distributions, then $\mu-\mu'$ is symmetric, and 18° implies that $\sum_{k=1}^{r_m}\mathbf{P}\{|\mu(A_m^k)-\mu'(A_m^k)|>\varepsilon\}$ is bounded for every $\varepsilon>0$. However,

$$\mathbf{P}\{|\mu(A_m^k)|>s\}\mathbf{P}\{|\mu'(A_m^k)|\le s/2\}\le\mathbf{P}\{|\mu(A_m^k)-\mu'(A_m^k)|>s/2\},$$
$$\mathbf{P}\{|\mu'(A_m^k)|\le s/2\}\ge 1-\sup_j\mathbf{P}\{|\mu(A_m^j)|>s/2\},$$

and the r.h.s. tends to 1 as $m\to\infty$. Hence,

$$\sum\mathbf{P}\{|\mu(A_m^k)|>s\}$$
$$\le(1-\sup_j\mathbf{P}\{|\mu(A_m^j)|>s/2\})\sum\mathbf{P}\{|\mu(A_m^k)-\mu'(A_m^k)|>s/2\}.$$

Thus, $\theta_m\le\alpha_m\beta_m$, where $\alpha_m\to 0$ and β_m is bounded.

Note now that the measure μ^{s_1} defines all the measures $\mu^{s_1}-\mu^{s_2},\ldots,\mu^{s_{k-1}}-\mu^{s_k}$, e.g.,

$$\mu^{s_{k-1}}(A_m^i)-\mu^{s_k}(A_m^i)=\lim_{n\to\infty}\sum_{A_n^j\subset A_m^i}[g_{s_{k-1}}(\mu^{s_1}(A_n^j))-g_{s_k}(\mu^{s_1}(A_n^j))].$$

Therefore, $\mu-\mu^{s_1}$ is independent of the collection $\mu^{s_1}-\mu^{s_2},\ldots,\mu^{s_{k-1}}-\mu^{s_k}$. Similarly, $\mu^{s_1}-\mu^{s_2}$ is independent of the collection $\mu^{s_2}-\mu^{s_3},\ldots,\mu^{s_{k-1}}-\mu^{s_k}$, etc. Hence, $\mu-\mu^{s_1},\mu^{s_1}-\mu^{s_2},\ldots,\mu^{s_{k-1}}-\mu^{s_k}$ are independent.

□

Let us return to the proof of the theorem. It follows from the proof of (b) that $\mathbf{M}(\mu(B)-\mu^{s_k}(B))^2<\infty$. Since

$$\overline{\mu}_{s_k}(B)=\overline{\mu}_{s_1}(B)+\sum_{i=1}^{k-1}(\mu^{s_i}(B)-\mu^{s_{i-1}}(B))$$

and the terms on the r.h.s. are independent, we have

$$\mathbf{D}[\overline{\mu}_{s_k}(B)]\ge\sum_{i=1}^{k-1}\mathbf{D}[\mu^{s_i}(B)-\mu^{s_{i-1}}(B)].$$

Therefore, if $t_0 > t_1 > \cdots > t_n > \cdots$, $t_n \to 0$, we find

$$\sum_{k=0}^{\infty} \mathbf{D}[\mu^{t_k}(B) - \mu^{t_{k+1}}(B)] \leq \mathbf{D}[\overline{\mu}_{t_0}(B)],$$

and the series

$$\sum_{k=0}^{\infty}([\mu^{t_k}(B) - \mu^{t_{k+1}}(B)] - \mathbf{M}[\mu^{t_k}(B) - \mu^{t_{k+1}}(B)])$$

is thus convergent with probability 1, which follows from 15°, Chapter 1. This means that such a limit exists,

$$\begin{aligned}\lim_{n\to\infty}&[\mu(B) - \mu^{t_n}(B) - \mathbf{M}(\mu(B) - \mu^{t_n}(B))]\\ &= \mu(B) - \mu^{t_0}(B) + \sum_{k=0}^{\infty}(\mu^{t_k}(B) - \mu^{t_{k+1}}(B)) = \mu_0(B).\end{aligned}$$

Note that $\mu_0(B)$ has all the moments, $\mathbf{M}\mu_0(B) = 0$, $\mathbf{D}\mu_0(B) = \mathbf{M}\mu_0^2(B) = m_2^0(B)$ is an additive function on $\mathcal{B}_0$ (by virtue of the independence of $\mu_0(A)$ and $\mu_0(B)$ for $A \cap B = \varnothing$). Clearly,

$$\begin{aligned}\mathbf{M}[\mu_0(B) - \overline{\mu}_{t_k}(B) + \mathbf{M}\overline{\mu}_{t_k}(B)]^2 &= \sum_{i=k}^{\infty}\mathbf{D}[\mu^{t_{i+1}}(B) - \mu^{t_i}(B)]\\ &\leq \sum_{i=k}^{\infty}\mathbf{D}[\mu^{t_{i+1}}(X) - \mu^{t_i}(X)].\end{aligned}$$

Further,

$$\begin{aligned}\mathbf{P}\{\sup_{i\leq r_n}|\mu_0(A_n^i) - \overline{\mu}_{t_k}(A_n^i)| > \varepsilon\} &\leq \frac{1}{\varepsilon^2}\sum_{i=1}^{r_n}\sum_{j=k}^{\infty}\mathbf{D}[\overline{\mu}_{t_{j+1}}(A_n^i) - \overline{\mu}_{t_j}(A_n^i)]\\ &\leq \frac{1}{\varepsilon^2}\sum_{j=k}^{\infty}\mathbf{D}[\overline{\mu}_{t_{j+1}}(X) - \overline{\mu}_{t_j}(X)].\end{aligned}$$

The expression on the l.h.s. is independent of n, and we can make it arbitrarily small by choosing proper k. Further,

$$\lim_{n\to\infty}\mathbf{P}\{\sup_{i\leq r_n}|\overline{\mu}_{t_k}(A_n^i)| > \varepsilon\} = 0 \text{ for } t_k < \varepsilon.$$

Hence,

$$\overline{\lim_{n\to\infty}}\,\mathbf{P}\{\sup_{i\leq r_n}|\mu_0(A_n^i)| > \varepsilon\} \leq \frac{1}{\varepsilon^2}\sum_{j=k}^{\infty}\mathbf{D}[\overline{\mu}_{t_{j+1}}(X) - \overline{\mu}_{t_j}(X)],$$

and passing to the limit as $k \to \infty$, we obtain (2.24).

□

25°. Theorem. *Let $\mu_0(B)$ be a r.m.i.v. on $\mathcal{B}_0$ for which*

$$\lim_{n\to\infty} \mathbf{P}\{\max_{i\le r_n} |\mu_0(A_n^i)| > \varepsilon\} = 0$$

holds for every $\varepsilon > 0$. Then (a) *$\mu_0(B)$ has the Gaussian distribution, and if $a(B) = \mathbf{M}\mu_0(B)$, $b(B) = \mathbf{D}\mu_0(B)$, then $a(B)$ and $b(B)$ are additive functions on $\mathcal{B}_0$, with b being nonnegative;* (b) *$\mu_0(B)$ is continuous in the following sense:*

$$\mathbf{P}\{\lim_{n\to\infty} \max_{i\le r_n} |\mu_0(A_n^i)| = 0\} = 1. \tag{2.25}$$

Proof. Note that $\mu_0(B)$ has all the moments, as follows from statement (a) of Theorem 22°. Thus, $a(B) = \mathbf{M}\mu_0(B)$ and $b(B) = \mathbf{D}\mu_0(B)$ exist and possess the above-mentioned properties. The condition of the theorem yields

$$\lim_{n\to\infty} \max_{i\le r_n} \mathbf{P}\{|\mu_0(A_n^i)| > \varepsilon\} = 0; \tag{2.26}$$

i.e., μ_0 is stochastically continuous. Since $\mu_0(A_n^i)$ can be represented as the limit of the sum of independent variables bounded by an arbitrarily small constant, then by virtue of estimate 12°, Chapter 1 (we take $x = \varepsilon$, $c = 0$, $\alpha = 1$), we have

$$\mathbf{M}\exp\{|\mu(A_n^i)|\} - 1$$
$$\le \frac{e^{3\varepsilon}}{(1 - 4e^{2\varepsilon}\mathbf{P}\{|\mu(A_n^i)| > \varepsilon\})(1 - \mathbf{P}\{|\mu(A_n^i)| > \varepsilon\})} - 1,$$
$$\max_{i\le r_n}(\mathbf{M}\exp\{|\mu(A_n^i)|\} - 1)$$
$$\le \frac{e^{3\varepsilon}}{(1 - 4e^{2\varepsilon}\max_{i\le r_n}\mathbf{P}\{|\mu(A_n^i)| > \varepsilon\})(1 - \max_{i\le r_n}\mathbf{P}\{|\mu(A_n^i)| > \varepsilon\})} - 1.$$

This inequality, (2.26), and the arbitrariness of $\varepsilon > 0$ together imply that

$$\lim_{n\to\infty} \max_{i\le r_n}(\mathbf{M}\exp\{|\mu(A_n^i)|\} - 1) = 0.$$

Hence, $\max_{i\le r_n} |a(A_n)| \to 0$, $\max_{i\le r_n} b(A_n^i) \to 0$, and the functions $a(A)$ and $b(A)$ are continuous on $\mathcal{B}_0$. Let $\overline{\mu}_0(B) = \mu_0(B) - a(B)$. Then for all $\varepsilon > 0$,

$$\lim_{n\to\infty} \mathbf{P}\{\max_{i\le r_n} |\overline{\mu}_0(A_n^i)| > \varepsilon\} = 0;$$

and therefore,

$$\lim_{n\to\infty} \sum_{i\le r_n} \mathbf{P}\{|\overline{\mu}_0(A_n^i)| > \varepsilon\} = 0 \tag{2.27}$$

for all $\varepsilon > 0$. Thus,

$$\overline{\lim_{n\to\infty}} \mathbf{M} \sum (\overline{\mu}_0(A_n^i))^2 I_{\{|\overline{\mu}_0(A_n^i)|>\varepsilon\}}$$

$$\overline{\lim_{n\to\infty}} \mathbf{M} \sum (\frac{\lambda}{2}(\overline{\mu}_0(A_n^i))^4 + \frac{1}{2\lambda} I_{\{|\mu_0(A_n^i)|>\varepsilon\}}$$

$$\leq \overline{\lim_{n\to\infty}} \mathbf{M} \left(\left[\frac{\lambda}{2}(\overline{\mu}_0(X))^4\right] + \frac{1}{2\lambda} \sum_{i\leq r_n} \mathbf{P}\{|\overline{\mu}_0(A_n^i)| > \varepsilon\} \right)$$

$$= \frac{\lambda}{2} \mathbf{M}(\mu_0(X))^4$$

(we have used the next relation: If $\xi_1, \ldots, \xi_n$ are independent, $\mathbf{M}\xi_i = 0$, $\mathbf{M}\xi_i^4 < \infty$, then

$$\mathbf{M}(\xi_1 + \cdots + \xi_n)^4 = \sum \mathbf{M}\xi_i^4 + 3\sum_{i\neq j} \mathbf{M}\xi_i^2 \mathbf{M}\xi_j^2 \geq \sum \mathbf{M}\xi_i^4).$$

Since $\mathbf{M}(\overline{\mu}_0(X))^4 < \infty$ and λ is arbitrary, the Lindenberg condition

$$\lim_{n\to\infty} \sum_{i\leq r_n} \mathbf{M}(\mu_0(A_n^i))^2 I_{\{|\overline{\mu}_0(A_n^i)|>\varepsilon\}} = 0$$

is satisfied for the variables $\overline{\mu}_0(A_n^i)$ such that $\mathbf{M}\overline{\mu}_0(A_n^i) = 0$, $\sum_{A_n^i \subset B} \mathbf{D}\overline{\mu}_0(A_n^i) = b(B)$. Hence, $\overline{\mu}_0(B)$ and therefore $\mu_0(B)$ have the normal distribution. Let us now prove (2.25). It follows from the continuity of $a(B)$ that it suffices to consider the case $a(B) = 0$. Let $\mathbf{D}\mu_0(X) = 1$ and $t_m^i = \sum_{j<i} \mathbf{D}[\mu(A_m^j)]$, $S = \bigcup_m \{t_m^i, i \leq r_m\}$, and let $\xi(t)$ be defined on S by $\xi(t_m^i) = \sum_{j\leq i} \mu(A_m^j)$. Then the event

$$\bigcap_{i=0}^{r_m-1} \{ \sup_{\substack{s\in S\\ t_m^i \leq s \leq t_m^{i+1}}} |\xi(s) - \xi(t_m^i)| \leq \varepsilon/2\}$$

involves the following one:

$$\sup_{\substack{n\geq m\\ i\leq r_n}} |\xi(t_n^i) - \xi(t_n^{i-1})| = \sup_{\substack{n\geq m\\ i\leq r_n}} |\mu(A_n^i)| \leq \varepsilon.$$

Therefore,

$$\mathbf{P}\{\sup_{\substack{n\geq m\\ i\leq r_n}} |\mu(A_n^i)| > \varepsilon\} \leq \sum_{i=0}^{r_m-1} \mathbf{P}\{ \sup_{\substack{s\in S\\ t_m^i \leq s \leq t_m^{i+1}}} |\xi(s) - \xi(t_m^i)| > \varepsilon/2\}$$

$$\leq \sum_{i=0}^{r_m-1} 2\mathbf{P}\{|\xi(t_m^{i+1}) - \xi(t_m^i)| > \varepsilon/2\}$$

$$= 2\sum_{i=1}^{r_n} \mathbf{P}\{|\mu(A_n^i)| > \varepsilon/2\}$$

(we have used the relation

$$\mathbf{P}\{\sup_{\substack{s\in S\\ t_m^i\le s\le t_m^{i+1}}} |\xi(s)-\xi(t_m^i)|>\varepsilon/2\} = \lim_{n\to\infty}\mathbf{P}\{\sup_{t_m^i\le t_n^j\le t_m^{i+1}} |\xi(t_n^j)-\xi(t_m^i)|>\varepsilon/2\}$$

$$\le 2\mathbf{P}\{|\xi(t_m^{j+1})-\xi(t_m^i)|>\varepsilon/2\};$$

the latter arises from the symmetry of the variables $\xi(t)$ and from inequality (1.10)). It remains to note that $\sup_{\substack{i\le r_n\\ n\ge m}} |\mu(A_n^i)|$ decreases as $m\to\infty$ and to apply (2.27).

□

26°. Theorem. *Let $\mu(B)$ be a r.m.i.v. on $\mathcal{B}_0$. Then there exist* (1) *a non-random measure $m(B)$,* (2) *a purely discrete regular r.m.i.v. $\mu_1(B)$,* (3) *a stochastically continuous Poisson r.m.i.v. $\nu_1(U\times B)$ independent of $\mu_1(B)$ on the union of algebras $\bigcup_{s_k} C^{s_k}\otimes\mathcal{B}_0$, where $s_k\downarrow 0$, C^{s_k} is an algebra of the finite sums of intervals from $]-\infty,-s_k]$ and $[s_k,\infty[$, which contains the intervals $[t,s[$ and $]-s,-t]$, where $0\le t<s$, and t and s are from some countable dense set $T_1\subset R_+$, and* (4) *a continuous Gaussian r.m.i.v. $\mu_0(B)$ such that*

$$\begin{aligned}\mu(B) &= m(B)+\mu_0(B)+\mu_1(B)\\ &\quad+\lim_{\max\Delta t_i\to 0}\sum_{0=t_0<t_1<\cdots}[t_i\nu([t_i,t_{i+1}[\times B)-t_i\nu(]-t_{i+1},-t_i]\times B)]\\ &\quad+\lim_{k\to\infty}\lim_{\max\Delta t_i\to 0}\sum_{s_k\le t_0<t_1<\cdots<t_n=1} t_i[\nu([t_i,t_{i+1}[\times B)-\nu(]-t_{i+1},-t_i]\times B)\\ &\quad-\mathbf{M}\nu([t_i,t_{i+1}[\times B)+\mathbf{M}\nu(]-t_{i+1},-t_i]\times B)].\end{aligned} \tag{2.28}$$

The proof follows from 22°, 24°, and 25°.

§2.3 Stochastic Integrals and Countably Additive r.m.i.v.
Stochastic Integral with Respect to Stochastically Continuous r.m.i.v.

Let $\mathcal{B}_0$ be some countable algebra of sets, and let $\mathcal{B}_0=\bigcup_m \mathcal{A}_m$, where $\mathcal{A}_m$ is an increasing sequence of finite algebras, and $A_m^1,\dots,A_m^{r_m}$ are the atoms of $\mathcal{A}_m$. The r.m.i.v. μ is given on $\mathcal{B}_0$ and is stochastically continuous with respect to $(\mathcal{A}_m)$. Denote by S the set of fundamental sequences (A_m) such that $A_m\supset A_{m+1}$. The elements of S will be denoted by s (maybe with indices). If $s=(A_m)$, we consider that $s\in A_m$ for all m. A real-valued function $f(s)$ given on S is called *continuous* (with respect to $(\mathcal{A}_m)$) if for every $\varepsilon>0$, m can be found such that $|f(s_1)-f(s_2)|<\varepsilon$ for all $i\le r_m$ and $s_1,s_2\in A_m^i$. This definition is equivalent to the following one: For every fundamental sequence (A_m),

$$\lim_{m\to\infty}\sup_{s_1,s_2\in A_m}|f(s_1)-f(s_2)|=0.$$

We denote by $C(\mathcal{A}_m)$ a set of continuous (with respect to $(\mathcal{A}_m)$) functions. For all $f \in C(\mathcal{A}_m)$, we define a *stochastic integral* as the limit in probability, as $m \to \infty$, of the integral sums

$$\sum_{k=1}^{r_m} f(s_m^k)\mu(A_m^k), \tag{2.29}$$

where $s_m^k \in A_m^k$ are points chosen arbitrarily. Let us denote this limit by $\int f d\mu$.

The nonrandom additive function $m(B)$ on $\mathcal{B}_0$ that is continuous with respect to $(\mathcal{A}_m)$ is an example of a stochastically continuous r.m.i.v. Let us study the conditions of the existence of the integral $\int f dm$ (in the abovementioned sense) for all $f \in C(\mathcal{A}_m)$.

27°. Theorem. *The integral $\int f dm$ exists for all $f \in C(\mathcal{A}_m)$ iff the additive function m has bounded variation, i.e.,*

$$\operatorname{Var}(m) = \sup_n \sum_{i=1}^{r_n} |m(A_n^i)| < \infty. \tag{2.30}$$

Proof. Let (2.30) hold. Then assuming that $\varepsilon_m = \sup_{i;s_1,s_2 \in A_m^i} |f(s_1) - f(s_2)|$, we have for $n > m$,

$$\begin{aligned} &|\sum_{i=1}^{r_m} f(s_m^i)m(A_m^i) - \sum_{j=1}^{r_n} f(s_n^j)m(A_n^j)| \\ &\qquad \le \sum_{i=1}^{r_m} |f(s_m^i) - f(s_n^j)| I_{\{A_n^j \in A_m^i\}} |m(A_n^j)| \le \varepsilon_m \operatorname{Var}(m), \end{aligned}$$

and the r.h.s. tends to zero as $n \to \infty$, $m \to \infty$, $n > m$. Now let $\sum_{i=1}^{r_n} |m(A_n^i)| = a_n \to \infty$. We choose a subsequence n_k so that the inequality $c_k > 2\sum_{i=k+1}^{\infty} c_i$ holds for a sequence $c_k = 1/\sqrt{a_{n_k}}$.

Let $f_k(s) = \operatorname{sign} m(A_{n_k}^i)$ if $s \in A_{n_k}^i$. We put $f(s) = \sum_{k=1}^{\infty} c_k f_k(s)$. Clearly $f(s) \in C(\mathcal{A}_n)$, since the series $\sum_{k=1}^{\infty} c_k f_k(s)$ converges uniformly. If $n > n_k$, we have for $s_n^i \in A_n^i$,

$$\sum f_k(s_n^i)m(A_n^i) = \sum f_k(s_{n_k}^j)m(A_{n_k}^j) = a_{n_k}, \quad \text{where } s_{n_k}^j \in A_{n_k}^j$$

(since f_k is constant on $A_{n_k}^j$, the sum on the r.h.s. is independent of the choice of $s_{n_k}^j$). Thus,

$$\sum f(s_{n_m}^i)m(A_{n_m}^i) = \sum_{k<m} c_k a_{n_k} + \sum_i \sum_{k \ge m} c_k f_k(s_{n_m}^i)m(A_{n_m}^i).$$

If $m(A^i_{n_m}) \neq 0$, then

$$\sum_{k\geq m} c_k f_k(s^i_{n_m}) m(A^i_{n_m}) \geq c_m |m(A^i_{n_m})| - \sum_{m+1}^{\infty} c_k |m(A^i_{n_m})| \geq \frac{1}{2} c_m |m(A^i_{n_m})|.$$

Therefore, we obtain

$$\sum_i \sum_{k\geq m} c_k f_k(s^i_{n_m}) m(A^i_{n_m}) \geq \frac{1}{2}\sqrt{a_{n_m}}.$$

Thus, $\sum f(s^i_{n_m}) m(A^i_{n_m}) \geq \frac{1}{2}\sqrt{a_{n_m}} \to \infty$ as $m \to \infty$. Hence, we have constructed the function $f \in C(\mathcal{A}_m)$ for which the integral does not exist.

□

Remark 1. Let $m(B) \in \mathcal{B}_0$ be an additive continuous function with bounded variation on $\mathcal{B}_0$. We put $\operatorname{Var} m(B) = \sup_n \sum_{A^i_n \subset B} |m(A^i_n)|$. Then $\operatorname{Var} m(B)$ is a continuous additive nonnegative function.

Indeed, it is easy to see that the sums $\sum |m(A^i_n)| I_{\{A^i_n \subset B\}}$ increase with B, and their supremum thus coincides with the limit. This implies the additivity of $\operatorname{Var} m(B)$. The continuity follows from the relation

$$0 \leq \operatorname{Var} m(B) - \sum |m(A^i_n)| I_{\{A^i_n \subset B\}} \leq \operatorname{Var} m(X) - \sum |m(A^i_n)|.$$

Since the expression on the r.h.s. approaches zero,

$$\sup_i (\operatorname{Var} m(A^i_n) - |m(A^i_n)|) \to 0.$$

Remark 2. If $f(s) \in C(\mathcal{A}_m)$ and

$$\delta_m = \sup_{s,s_1 \in A^i_m} |f(s) - f(s_1)|,$$

we have

$$|f(s^i_m) m(A^i_m) - \sum f(s^j_n) m(A^j_n)| \leq \delta_m \operatorname{Var} m(A^i_m).$$

Definition. Denote by $\mathfrak{M}_0$ the set of stochastically continuous regular r.m.i.v. μ, such that for some $s \in T(\mu)$ (notation of §2.2)

$$\mathbf{M}\overline{\mu}_s(B) = \mathbf{M}[\mu(B) - \mu^s(B)] = 0 \quad \forall B \in \mathcal{B}_0.$$

28°. Theorem. *For every $f \in C(A_m)$ and the r.m.i.v. $\mu \in \mathfrak{M}_0$, there exists $\int f d\mu$.*

Proof. Consider the difference of two integral sums (2.29) for $m < n$:

$$\sum_{k=1}^{r_m} f(s_m^k)\mu(A_m^k) - \sum_{j=1}^{r_n} f(s_n^j)\mu(A_n^j)$$
$$= \sum_{k=1}^{r_n} [f(s_m^k) - f(s_n^j)] I_{\{A_m^k \supset A_n^j\}} \mu(A_n^j).$$

To prove the theorem, it suffices to show that this difference converges in probability to zero as $n \to \infty$, $m \to \infty$, $n > m$. And this is so if

$$\sum_{j=1}^{r_n} \varepsilon_n^j \mu(A_n^j) = \sum_{j=1}^{r_n} \varepsilon_n^j \overline{\mu}_s(A_n^j) + \sum_{j=1}^{r_n} \varepsilon_n^j \mu^s(A_n^j)$$

converges to zero in probability as $n \to \infty$, $\max |\varepsilon_n^j| \to 0$ (here $s \in T(\mu)$ is such that $\mathbf{M}\overline{\mu}_s(B) = 0 \quad \forall B \in \mathcal{B}_0$). We have

$$\mathbf{M}(\sum_{j=1}^{r_n} \varepsilon_n^j \overline{\mu}(A_n^j))^2 = \sum_{j=1}^{r_n} (\varepsilon_n^j)^2 \mathbf{D}(\overline{\mu}(A_n^j)) \leq \max_{j \leq r_n} (\varepsilon_n^j)^2 \mathbf{D}\mu(X) \to 0.$$

Further, (2.21) yields

$$|\sum_{j=1}^{r_n} \varepsilon_n^j \mu^s(A_n^j)| \leq \max_{j \leq r_n} |\varepsilon_n^j| \sum |\hat{g}_s(\xi_k)| = \max_{j \leq r_n} |\varepsilon_n^j| \sum_{\xi_k > s} |\xi_k|, \tag{2.31}$$

where $\xi_k = \lim_{n\to\infty} \mu(A_n^{i_n^{(k)}(\omega)})$ and $(A_n^{i_n^{(k)}(\omega)})$, $k = 1, 2, \ldots$, are all fundamental sequences (A_m) for which $\lim_{m\to\infty} \mu(A_m) \neq 0$. They form at most a countable set, and $|\xi_k| \to 0$, so the sum on the r.h.s. of (2.31) has a finite number of terms. Thus, the r.h.s. of (2.31) tends to zero as $n \to \infty$.

□

Theorems 23° and 25° imply that every stochastically continuous regular r.m.i.v. μ can be represented in the form

$$\mu = m + \tilde{\mu}, \tag{2.32}$$

where $\tilde{\mu} \in \mathfrak{M}_0$, m is an additive function on $\mathcal{B}_0$, continuous with respect to (A_m). 27° and 28° imply the following theorem.

29°. Theorem. *For the regular stochastically continuous r.m.i.v. μ, the integral $\int f d\mu$ exists for all $f \in C(\mathcal{A}_m)$ iff μ allows the representation (2.32), where m is an additive continuous function with bounded variation on $\mathcal{B}_0$, $\tilde{\mu} \in \mathfrak{M}_0$.*

We shall denote by $\mathfrak{M}_{0,v}$ a set of r.m.i.v. that can be represented in the form (2.32), where m is additive and continuous, and $\operatorname{Var} m < \infty$, $\tilde{\mu} \in \mathfrak{M}_0$. Thus, $\int f d\mu$ exists if $f \in C(\mathcal{A}_m)$, $\mu \in \mathfrak{M}_{0,v}$.

30°. Let $f \in C(\mathcal{A}_m)$, $\mu \in \mathfrak{M}_{0,v}$. For every $B \subset \mathcal{B}_0$, the integral

$$\int_B f d\mu = \int f I_B d\mu \tag{2.33}$$

is defined, and the function $\psi(B) = \int_B f d\mu$ is a stochastically continuous r.m.i.v.

Proof. To prove the existence of the integral on the r.h.s. of (2.33), it suffices to show that $I_B \in C(\mathcal{A}_m)$. Let $B \in \mathcal{A}_m$ and $B = \bigcup_{k=1}^{l} A_m^{i_k}$. Then $I_B = \sum_{k=1}^{l} I_{A_m^{i_k}}$, and hence, for $s_1, s_2 \in A_m^i$, we have $I_B(s_1) = I_B(s_2) = 1$ if $i \in (i_1, \ldots, i_l)$, and $I_B(s_1) = I_B(s_2) = 0$ if $i \overline{\in} (i_1, \ldots, i_l)$. This means exactly that $I_B \in C(\mathcal{A}_m)$. It follows from the construction of the integral that for $f_1, f_2 \in C(\mathcal{A}_m)$,

$$\int f_1 d\mu + \int f_2 d\mu = \int (f_1 + f_2) d\mu$$

with probability 1. Hence, for $B_1 \cap B_2 = \varnothing$, $B_1, B_2 \in \mathcal{B}_0$,

$$\begin{aligned}\psi(B_1 \cup B_2) &= \int_{B_1 \cup B_2} f d\mu = \int I_{\{B_1 \cup B_2\}} f d\mu = \int (I_{B_1} + I_{B_2}) f d\mu \\ &= \int_{B_1} f d\mu + \int_{B_2} f d\mu = \psi(B_1) + \psi(B_2).\end{aligned}$$

Further, if $\mathcal{F}_B$ is a σ-algebra generated by $\{\mu(A), A \subset B\}$, then $\psi(B)$ is $\mathcal{F}_B$-measurable as the limit of $\mathcal{F}_B$-measurable variables $\sum f(s_n^k)\mu(A_n^k) I_{\{A_n^k \subset B\}}$. Thus, $\psi(B)$ is a r.m.i.v.

To prove the stochastic continuity of $\psi(B)$, we represent μ in the form $\mu = m + \mu''^t + \overline{\mu}'_t$, where m is a nonrandom additive function with bounded variation, $\mu''^t \in \mathfrak{M}_0$, $t > 0$, is such that $\mathbf{M}\mu''^t = 0$ and

$$\psi(B) = \int_B f dm + \int_B f d\mu''^t + \int_B f d\overline{\mu'_t}.$$

The remark in 28° implies the continuity of $\int_B f dm$. Consider μ''^t. For sufficiently large n, there exist $l(\omega)$ (a random number) numbers $i_n^{(1)}(\omega), \ldots, i_n^{(l)}(\omega)$ such that $\mu''^t(A_n^i) = 0$ for $i \overline{\in} (i_n^{(1)}(\omega), \ldots, i_n^{(l)}(\omega))$, and $\mu''^t(A_n^{i_n^{(k)}(\omega)}) \to \xi_k$ as $n \to \infty$.

If we put $s_k(\omega) = (A_n^{i(k)(\omega)})$, then the relation

$$\int_B f d\mu'^t = \sum I_B(s_k(\omega)) f(s_k(\omega)) \xi_k \tag{2.34}$$

holds. Thus,

$$\mathbf{P}\left\{\int_{A_n^i} f d\mu'^t \neq 0\right\} \leq \mathbf{P}\{\mu'^t(A_n^i) \neq 0\} \to 0 \quad \text{as} \quad n \to \infty$$

(by virtue of the stochastic continuity of μ'^t). Formula (2.34) also implies regularity of $\int_B f d\mu'^t$. If $|f| \leq c$, we have

$$\mathbf{M}\left(\sum_{A_n^j \subset A_m^i} f(s_n^j)\overline{\mu_t'}(A_n^j)\right)^2 \leq c^2 \mathbf{D}\overline{\mu_t'}(A_m^i);$$

and thus,

$$\mathbf{M}\left(\int_{A_m^i} f d\overline{\mu_t'}\right)^2 \leq c\mathbf{D}\overline{\mu_t'}(A_m^i) \leq c \max_{i \leq r_m} \mathbf{D}\overline{\mu_t'}(A_m^i).$$

Since the r.h.s. converges to zero as $m \to \infty$, the r.m.i.v. $\int_B f d\overline{\mu_t'}$ is stochastically continuous.

□

Extension of r.m.i.v. to Countably Additive r.m.i.v. The Stochastic Integral with Respect to a Countably Additive Measure

Consider a r.m.i.v. μ on $\mathcal{B}_0$. Let $\mathcal{B}$ be the least σ-algebra containing $\mathcal{B}_0$. We say that a countably additive r.m.i.v. μ^* is given on $\mathcal{B}$ if for all $B \in \mathcal{B}$, a random variable $\mu^*(B)$ is defined; moreover, if $B_1, B_2, \ldots$ are mutually disjoint, $\mu^*(B_k)$ are independent and

$$\mu\left(\bigcup B_k\right) = \sum_k \mu(B_k)$$

(the series on the r.h.s. converges in probability and, hence, with probability 1, too; see 20°, Chapter 1). If μ^* and μ coincide on $\mathcal{B}_0$, μ^* is called a countably additive extension of μ onto $\mathcal{B}$.

31°. Theorem. *The measure μ has a countably additive extension μ^* onto $\mathcal{B}$ iff $\mu(B_n) \to 0$ in probability for every sequence $B_n \in \mathcal{B}_0$ such that $B_n \supset B_{n+1}$ and $\bigcap B_n = \varnothing$.*

Proof. If μ^* is countably additive on $\mathcal{B}$ and $B_n \supset B_{n+1}$, $\bigcap B_n = \varnothing$, then we have $B_1 = \bigcup_{k=1}^{\infty}(B_k \backslash B_{k+1})$,

$$\begin{aligned}\mu^*(B_1) &= \sum_{k=1}^{\infty} \mu^*(B_k \backslash B_{k+1}) \\ &= \sum_{k=1}^{\infty}(\mu^*(B_k) - \mu^*(B_{k+1})) = \lim_{n\to\infty}(\mu^*(B_1) - \mu^*(B_n)).\end{aligned}$$

Hence, $\lim_{n\to\infty} \mu^*(B_n) = 0$ (in the sense of the convergence in probability). The necessity of the condition of the theorem is proved.

Now let the condition of the theorem be satisfied. Consider the function

$$\chi(z, B) = \mathbf{M} \exp\{iz\mu(B)\} \tag{2.35}$$

on $\mathcal{B}_0$. For the mutually disjoint sets $B_1, B_2, \ldots, B_k$,

$$\chi(z, \bigcup_{j=1}^{k} B_j) = \prod_{j=1}^{k} \chi(z, B_j).$$

Let $c > 0$ be such that $|\chi(z, X)| \geq 1/2$ for $|z| \leq c$. Then for all B,

$$|\chi(z, B)|\,|\chi(z, X \backslash B)| = |\chi(z, X)|, \quad |\chi(z, B)| \geq |\chi(z, X)|.$$

Thus, $|\chi(z, B)| \geq 1/2$ for $|z| \leq c$. We put

$$\psi(B) = -\int_{-c}^{c} \operatorname{Re} \ln \chi(z, B) dz; \tag{2.36}$$

$\psi(B)$ is a nonnegative additive function on $\mathcal{B}_0$, for which $\psi(B_n) \downarrow 0$, provided that $B_n \supset B_{n+1}$, $B_n \in \mathcal{B}_0$, $\bigcap B_n = \varnothing$. Hence, $\psi(B)$ can be extended to become a measure on $\mathcal{B}$. By virtue of 18° (see the proof),

$$\mathbf{P}\{|\mu(B_n)| > \varepsilon\} \leq \lambda(\varepsilon c) \cdot \frac{1}{2c} \cdot \psi(B_n), \quad \text{if} \quad \psi(B_n) \to 0. \tag{2.37}$$

and thus, $\mu(B_n) \to 0$ in probability. We define $\mu(B)$ for $B \overline{\in} \mathcal{B}_0$ by monotonicity: Let $B_n \uparrow$ or $B_n \downarrow$, and $B = \lim B_n$. Assume that $\mu(B_n)$ have already been defined and, moreover, that $\mu(B_n)$ have also been defined on an algebra containing these sets. Furthermore, μ is the r.m.i.v. on this algebra, and for B' belonging to it, the equality

$$\psi(B') = -\int_{-c}^{c} \operatorname{Re} \ln \mathbf{M} e^{iz\mu(B')} dz \tag{2.38}$$

holds. Let us show that the limit $\mu(B) = \lim \mu(B_n)$ (in the sense of the convergence in probability) exists. Let $B_n \uparrow$, $B = B_1 + \bigcup_{k=2}^{\infty}(B_k \backslash B_{k-1})$. The series $\mu(B_1) + \sum_{k=2}^{\infty} \mu(B_k \backslash B_{k-1})$ consists of the independent random variables. Its partial sums are $\mu(B_n)$. For $n > m$,

$$\mathbf{P}\{|\mu(B_n) - \mu(B_m)| > \varepsilon\}\mathbf{P}\{|\mu(B_n \backslash B_m)| > \varepsilon\} \\ \leq \lambda(\varepsilon c) \cdot \frac{1}{2c} \cdot \psi(B_n \backslash B_m) \to 0$$

as $n, m \to \infty$. In addition, $\ln \chi(z, B) = \lim_{n\to\infty} \ln \chi(z, B_n)$, and thus (2.38) holds for $B' = B$. By analogy, we obtain that for $B_n \downarrow$ and $B = \bigcap B_n$, there exists $\lim \mu(B_n)$ in the sense of the convergence in probability, and if we denote it by $\mu(B')$, (2.38) holds.

Note that as a result of the limit transition, we shall again obtain the independence of values of μ on disjoint sets. Let, for example, $B^{(1)} = \bigcap B_n^{(1)}$, where $B_n^{(1)} \uparrow$ and $B^{(2)} = \bigcap B_n^{(2)}$, where $B_n^{(2)} \downarrow$ and $B^{(1)} \cap B^{(2)} = \varnothing$. Then $B_n^{(1)} \cap B^{(2)} = \varnothing$ for all n. Thus, $B^{(2)} = \bigcap(B_n^{(2)} \backslash B_m^{(1)})$, whatever m we take. Consequently, $\mu(B_m^{(1)})$ is independent of $\mu(B^{(2)}) = \lim \mu(B_n^{(2)} \backslash B_m^{(1)})$ (the variables $\mu(B_m^{(1)})$ and $\mu(B_n^{(2)} \backslash B_m^{(1)})$ are independent for all n). And since $\mu(B_m^{(1)})$ and $\mu(B^{(2)})$ are independent for all m, the variables $\mu(B^{(1)}) = \lim_{m\to\infty} \mu(B_m^{(1)})$ and $\mu(B^{(2)})$ are independent, too. Similarly, we can investigate the case when $B_n^{(1)} \downarrow$, $B_n^{(2)} \downarrow$, $B^{(1)} = \bigcap B_n^{(1)}$, $B^{(2)} = \bigcap B_n^{(2)}$, $B^{(1)} \cap B^{(2)} = \varnothing$. But if $B^{(1)} \cap B^{(2)} = \varnothing$ and $B^{(1)} = \bigcup B_n^{(1)}$, $B^{(2)} = \bigcup B_n^{(2)}$, then $B_n^{(1)} \cap B_n^{(2)} = \varnothing$, and $\mu(B_n^{(1)})$ and $\mu(B_n^{(2)})$ are independent; therefore, their limits are also independent.

Thus, by extending the r.m.i.v. μ by monotonicity, we preserve the independence of the values on disjoint sets, additivity, and (2.38). Hence, $\mu(B)$ can be extended onto the least monotonic class containing $\mathcal{B}_0$ that coincides with $\mathcal{B}$.

□

Let us consider the concrete classes of r.m.i.v. on $\mathcal{B}_0$ and establish the conditions of the countable additivity. For nonrandom additive functions of sets m, these conditions are known: m should be continuous on $\mathcal{B}_0$.

32°. Theorem. *Let μ be a purely discrete r.m.i.v., and $(A_n^{i^{(k)}_n})$, $k = 1, 2, \ldots$, be all the fundamental sequences for which $\lim_{n\to\infty} \mu(A_n^{i^{(k)}_n}) \neq 0$. Then μ can be extended to become a countably additive r.m.i.v. iff $\bigcap A_n^{i^{(k)}_n}$ is nonempty in X for all k.*

Proof. The necessity of the condition follows from the fact that for the countably additive r.m.i.v., $\bigcap A_n^{i^{(k)}_n} = \varnothing$ implies $\mu(A_n^{i^{(k)}_n}) \to 0$ in probability. Let us prove the sufficiency. Let $B_n \in \mathcal{A}_n$ and $\bigcap B_n = \varnothing$. We are to prove that $\mu(B_n) \to 0$ in probability. By the definition of a purely discrete measure,

$$\mu(B_m) = \sum_{A_m^{i_m^{(k)}} \subset B_n} \lim_{m\to\infty} \mu(A_m^{i_m^{(k)}}).$$

If there exist arbitrarily large m such that $B_m \supset A_n^{i_n^{(1)}}$, then $B_m \supset \bigcap_n A_n^{i_n^{(1)}}$ for all m; the set on the right is nonempty, and thus B_m is also nonempty. Hence, there exists m_1 such that $B_n \cap A_n^{i_n^{(1)}} = \varnothing$ for $n > m_1$. In exactly the same way, there exists m_l such that $B_n \cap A_n^{i_n^{(l)}} = \varnothing$ for $n > m_l$. Therefore, $\mu(B_m) = \sum_{k\in L_m} \lim \mu(A_n^{i_n^{(k)}})$, where $L_m \subset Z_+$ and $\bigcap L_m = \varnothing$; and thus, $\mu(B_m) \to 0$ in probability. It remains to use Theorem 31°.

□

33°. Theorem. *Let μ be a stochastically continuous r.m.i.v. for which $\lim_{n\to\infty} \sup_{i\le r_n} |\mu(A_n^i)| \le s$. Then all the moments of μ exist, and μ can be extended onto $\mathcal{B}$ as a countably additive r.m.i.v. iff the additive functions (on $\mathcal{B}_0$) $a(B) = \mathbf{M}\mu(B)$ and $b(B) = \mathbf{D}\mu(B)$ are continuous on $\mathcal{B}_0$: If $B_n \supset B_{n+1}$, $\bigcap B_n = \varnothing$, then $a(B_n) \to 0$ and $b(B_n) \to 0$.*

Proof. The sufficiency of the condition follows from the fact that $\mu(B_n) \to 0$ in probability if $a(B_n) \to 0$ and $b(B_n) \to 0$ and from Theorem 31°.

Necessity. Let μ be countably additive, μ' be independent of μ and have the same distribution as μ, $\tilde{\mu} = \mu - \mu'$. Then $\tilde{\mu}$ is also countably additive and symmetric, and $\overline{\lim}_{n\to\infty} \sup_{i\le r_n} |\tilde{\mu}(A_n^i)| \le 2s$. So $\tilde{\mu}(B)$ has all the moments; furthermore,

$$\begin{aligned}\mathbf{M}(\tilde{\mu}(B))^4 &\le \mathbf{M}(\tilde{\mu}(X))^4 \\ &= \mathbf{M}(\tilde{\mu}(B))^4 + \mathbf{M}(\tilde{\mu}(X\backslash B))^4 + 6\mathbf{M}(\tilde{\mu}(B))^2\mathbf{M}(\tilde{\mu}(X-B))^2.\end{aligned}$$

Thus, if B_k are mutually disjoint, then

$$\mathbf{D}\sum_{k=1}^{\infty} \tilde{\mu}(B_k) = \sum_{k=1}^{\infty} \mathbf{D}\tilde{\mu}(B_k).$$

Therefore, $\mathbf{D}\tilde{\mu}(B) = 2\mathbf{D}\mu(B)$ is continuous on $\mathcal{B}_0$. The sufficient condition, which has already been proved, yields that the r.m.i.v. $\mu(B) - a(B)$ is countably additive. Hence, $a(B)$ is countably additive as the difference of two countably additive r.m.i.v.

□

34°. Theorem. *Let μ be some stochastically continuous r.m.i.v., let μ^s be constructed by using μ by means of formula (2.21), and let $\nu(U \times B)$ be the Poisson measure on $C^s \otimes \mathcal{B}_0$ in terms of which μ^s is expressed in (2.22). Then μ^s can be extended up to a countably additive r.m.i.v. iff $\mathbf{M}\nu(]-\infty,-s] \cup [s,\infty[\times B)$ is continuous on $\mathcal{B}_0$.*

Proof. $\nu(]-\infty,-s]\cup[s,\infty[\times B)$ is a Poisson r.m.i.v. on $\mathcal{B}_0$. If $\mathbf{M}\nu([s,\infty[\times B)$ is continuous on $\mathcal{B}_0$, we have $\nu(]-\infty,-s]\cup[s,\infty[\times B_n) \to 0$ in probability for $B_n \in \mathcal{A}_n$, $B_n \downarrow$, $\bigcap B_n = \varnothing$; i.e., $\sum_{k=1}^{l(\omega)} I_{\{A_n^{i_n^{(k)}(\omega)} \subset B_n\}} \to 0$ in probability, where $(A_n^{i_n^{(k)}(\omega)})$, $k = 1,\dots,l(\omega)$, are all random fundamental sequences such that $\lim_{n\to\infty} \mu(A_n^{i_n^{(k)}(\omega)}) > s$. And since

$$\mu^s(B_n) = \sum_{k=1}^{l(\omega)} \xi_k I_{\{A_n^{i_n^{(k)}(\omega)} \subset B_n\}},$$

we get

$$|\mu^s(B_n)| \le \max_{k\le l(\omega)} |\xi_k| \sum_{k=1}^{l(\omega)} I_{\{A_n^{i_n^{(k)}(\omega)} \subset B_n\}} \to 0$$

in probability. The sufficiency of the conditions follows from Theorem 31°. Now let μ^s be countably additive. Introduce the r.m.i.v. for $B \subset \mathcal{A}_n$ as follows:

$$\mu^s_+(B) = \sum_{k=1}^{l(\omega)} \xi_k I_{\{\xi_k>0\}} I_{\{A_n^{i_n^{(k)}(\omega)} \subset B_n\}} = \lim_{\max(s_{k+1}-s_k)\to 0} \sum_{s=s_0<s_1<\cdots} s_k \nu([s_k, s_{k+1}[\times B),$$

$$\mu^s_-(B) = \sum_{k=1}^{l(\omega)} \xi_k I_{\{\xi_k>0\}} I_{\{A_n^{i_n^{(k)}(\omega)} \subset B_n\}} = \lim_{\max(s_{k+1}-s_k)\to 0} \sum_{s=s_0<s_1<\cdots} (-s_k)\nu(]-s_{k+1},-s_k] \times B)$$

(cf. (2.22)). The r.m.i.v. μ^s_+ and μ^s_- are independent, because they are constructed by means of independent Poisson r.m.i.v., $\mu^s = \mu^s_+ + \mu^s_-$. If $B_n \in \mathcal{B}_0$, $B_n \supset B_{n+1}$, $\bigcap B_n = \varnothing$, then $\mu^s(B_n) \to 0$ in probability. Since $\mu^s_+ \ge 0$, there exists $\lim \mu^s_+(B_n)$; and by analogy, $\lim \mu^s_-(B_n)$ exists, too; moreover, these limits are independent and their sum equals zero. Thus, $\lim \mu^s_+(B_n) \to a$, $\lim \mu^s_-(B_n) \to -a$, where $a \ge 0$. Let us show that $a = 0$. Assume that this is not true: $a > 0$. Then in the sense of the convergence in probability,

$$a = \lim_{n\to\infty} \sum [\mu^s_+(A^i_n) \wedge a] I_{\{A^i_n \subset B_n\}}.$$

Since the terms of the sums to the right are bounded by the number a and are independent, these sums has all the moments, which are uniformly bounded, and

$$\begin{aligned}
a &= \lim_{n\to\infty} \sum \mathbf{M}[\mu_+^s(A_n^i) \wedge a] I_{\{A_n^i \subset B_n\}}, \\
a^2 &= \lim_{n\to\infty} \mathbf{M}(\sum [\mu_+^s(A_n^i) \wedge a] I_{\{A_n^i \subset B_n\}})^2 \\
&= a^2 + \lim_{n\to\infty} \sum \mathbf{D}[\mu_+^s(A_n^i) \wedge a] I_{\{A_n^i \subset B_n\}} \\
&= a^2 + \lim_{n\to\infty} \sum \mathbf{M}[\mu_+^s(A_n^i) \wedge a]^2 I_{\{A_n^i \subset B_n\}} \\
&\quad - \lim_{n\to\infty} \sum (\mathbf{M}[\mu_+^s(A_n^i) \wedge a])^2 I_{\{A_n^i \subset B_n\}} \\
&\geq a^2 + s \lim_{n\to\infty} \sum \mathbf{M} \sum \mathbf{M}[\mu_+^s(A_n^i) \wedge a] I_{\{A_n^i \subset B_n\}} \\
&\quad - \lim_n \max_j \mathbf{M}[\mu_+^s(A_n^j) \wedge a] \sum \mathbf{M}[\mu_+^s(A_n^j) \wedge a] I_{\{A_n^i \subset B_n\}} \\
&\geq a^2 + sa - a \lim_{n\to\infty} \max_j \mathbf{M}[\mu_+^s(A_n^j) \wedge a] = a^2 + sa.
\end{aligned}$$

We have used the fact that $\mu_+^s(A_n^i) \geq s$ if $\mu_+^s(A_n^i) > 0$ and also that $\max_j \mathbf{M}[\mu_+^s(A_n^j) \wedge a] \leq \varepsilon + a \max_j \mathbf{P}\{\mu_+^s(A_n^j) \geq \varepsilon\}$, and by virtue of the stochastic continuity, the variable on the right can be made as small as desired for all sufficiently large n by choosing proper ε. We arrive at a contradiction. Hence, μ_+^s and μ_-^s are countably additive and

$$\nu(]-\infty, -s] \cup [s, \infty[\times B) \leq 1/s[\mu_+^s(B) - \mu_-^s(B)].$$

From the countable additivity of $\nu(]-\infty, -s] \cup [s, \infty[\times B)$, it follows, by virtue of 33°, that $\mathbf{M}\nu(]-\infty, -s] \cup [s, \infty[\times B)$ is also countably additive.

□

35°. Theorem. *Let μ be a countably additive r.m.i.v. that can be represented in the form $\mu = m + \mu_1 + \mu_2$, where μ_1 and μ_2 are stochastically continuous and independent r.m.i.v., such that for some $t > 0$,*

$$\mathbf{P}\{\overline{\lim_{n\to\infty}} \max_{i\leq r_n} |\mu_1(A_n^i)| \leq t\} = 1$$

and

$$\mathbf{M}\mu_1(B) = 0,$$

and for μ_2, there exist a finite number $l(\omega)$ of fundamental sequences $(A_n^{i_n^{(k)}(\omega)})$, $k = 1, \ldots, l(\omega)$, and the variables ξ_k such that $|\xi_k| > t$, and for $B \subset A_n$,

$$\mu_2(B) = \sum \xi_k I_{\{A_n^{i_n^{(k)}(\omega)} \subset B\}}.$$

Then m, μ_1, μ_2 are countably additive.

Proof. Let $B_n \downarrow$, $B_n \subset \mathcal{B}_0$, $\cap B_n = \varnothing$. There exist the limits $\lim \mu_1(B_n) = \mu_1(B_1) - \sum_{k=1}^{\infty} \mu_1(B_k \backslash B_{k+1})$ and $\lim \mu_2(B_n) = \sum \xi_{i_k}$, where i_k are the indices such that $A_n^{i_n^{(k)}(\omega)} \subset B_n$ for all n. Thus, the limit $\lim_{n\to\infty} m(B_n)$ exists. Furthermore,

$$0 = \lim_{n\to\infty} m(B_n) + \lim_{n\to\infty} \mu_1(B_n) + \lim_{n\to\infty} \mu_2(B_n).$$

Hence, $\lim_{n\to\infty} \mu_1(B_n)$ is constant; moreover, $\mathbf{M}\mu_1(B_n) = 0$, $\mathbf{D}\mu_1(B_n)$ is bounded, and therefore, $\lim_{n\to\infty} \mu_1(B_n) = 0$. Just as in the proof of 34°, the constancy of $\lim \mu_2(B_n)$ yields that $\lim_{n\to\infty} \mu_{2+}(B_n)$ and $\lim_{n\to\infty} \mu_{2-}(B_n)$ are also constant and thus equal to 0. Hence, $\lim_{n\to\infty} m(B_n) = 0$, too.

□

Remark. Let μ be a countably additive r.m.i.v. and $\mu = \mu_1 + \mu_2$, where μ_1 is a purely discrete r.m.i.v. concentrated on the fundamental sequences $(A_n^{i_n^{(k)}})$, $k = 1, 2, \ldots$, and μ_2 is independent of μ_1 and is stochastically continuous on these sequences. Then μ_1 and μ_2 are countably additive. This follows from the fact that $\bigcap A_n^{i_n^{(k)}}$ cannot be empty, since $\lim_{n\to\infty} \mu(A_n^{i_n^{(k)}}) = \lim_{n\to\infty} \mu_1(A_n^{i_n^{(k)}}) \neq 0$. Therefore, μ_1 is countably additive by virtue of 31°; and thus, μ_2 is also countably additive.

Stochastic Integrals with Respect to Countable Additive r.m.i.v.

We have already considered a stochastic integral with respect to additive measure. It exists for a rather restricted set of functions. Let us now consider a countably additive r.m.i.v. Our first step is to construct a stochastic integral for bounded $\mathcal{B}$-measurable functions on X. Then we shall study how to extend the domains of the definition for different classes of countably additive r.m.i.v. Let us denote by $\mathbf{B}_{(\mathcal{B})}$ a space of $\mathcal{B}$-measurable bounded real-valued functions, and by $\mathbf{B}^0_{(\mathcal{B})}$, a set of finite-valued functions from $\mathbf{B}_{(\mathcal{B})}$: For every $f(x) \in \mathbf{B}^0_{(\mathcal{B})}$, there exist a finite set of $B_1, \ldots, B_l \in \mathcal{B}$ and the numbers $c_1, \ldots, c_l$ such that $f = \sum c_k I_{B_k}$.

If μ is a r.m.i.v. given on $\mathcal{B}$, we define

$$\int f d\mu = \sum c_k \mu(B_k) \tag{2.39}$$

for $f \in \mathbf{B}^0_{(\mathcal{B})}$.

36°. Theorem. *For every $f \in \mathbf{B}_{(\mathcal{B})}$ and for a countably additive r.m.i.v. μ on $\mathcal{B}$, the integral $\int f d\mu$ is defined as the limit in probability*

$$\int f d\mu = \lim_{n\to\infty} \int f_n d\mu, \tag{2.40}$$

where $f_n \in \mathbf{B}^0_{(\mathcal{B})}$, $\|f - f_n\| \to 0$, $\|g\| = \sup_x |g(x)|$.

Proof. If μ is countably additive, and $\mu = m + \mu_1 + \mu_2$, where m is nonrandom, μ_1 is purely discrete, and μ_2 is stochastically continuous, then m, μ_1, μ_2 are countably additive. Clearly, (2.40) holds for $\mu = m$. If μ_1 is purely discrete and concentrated on $(A_n^{i_n^{(k)}})$, $k = 1, 2, \ldots$, then the sets $A^{(k)} = \bigcap A_n^{i_n^{(k)}}$ are atoms of the σ-algebra $\mathcal{B}$, each $\mathcal{B}$-measurable function f is constant on A_k, and the series $\sum_{k_i} \mu(A^{k_i})$ consisting of the independent random variables converges for an arbitrary sequence k_i (and thus, the sum is independent of the order of terms). To prove (2.40) for $\mu = \mu_1$, it is sufficient to show that for $\max_k |\varepsilon_{n_k}| \to 0$,

$$\sum_k \varepsilon_{n_k} \mu_1(A^k) \to 0$$

in probability.

Let c be chosen so that $\sum \mathbf{P}\{|\mu_1(A^k)| > c\} < \delta$. We put $\xi_k' = 0$ for $|\xi_k| \le c$, $\xi_k'' = 0$ for $|\xi_k| > c$, and $\xi_k' + \xi_k'' = \mu_1(A^k)$. Then

$$\mathbf{P}\{\sum \varepsilon_{n_k} \xi_k' \ne 0\} \le \sum \mathbf{P}\{|\mu_1(A_k)| > c\} < \delta,$$

and

$$|\mathbf{M} \sum_k \varepsilon_{n_k} \xi_k''| = |\sum_k \varepsilon_{n_k} \mathbf{M}\xi_k''| \le \sup_k |\varepsilon_{n_k}| \sum |\mathbf{M}\xi_k''|,$$

$$\mathbf{D} \sum_k \varepsilon_{n_k} \xi_k'' \le (\sup_k \varepsilon_{n_k})^2 \sum \mathbf{D}\xi_k'';$$

and the r.h.s. tend to zero as $\max_k |\varepsilon_{n_k}| \to 0$ (we have used Theorem 18°, Chapter 1).

Now let μ be a stochastically continuous r.m.i.v. We represent it in the form $\mu = \overline{\mu}_s + \mu^s$ (these r.m.i.v. were introduced above). If $\mathbf{M}\overline{\mu}_s(B) = a(B)$, $\mathbf{D}\overline{\mu}_s(B) = b(B)$, then $a(B)$ and $b(B)$ are countably additive by virtue of 33°.

Since for $f = \sum c_k I_{B_k}$, where $B_k \cap B_j = \emptyset$ if $k \ne j$, we have

$$\mathbf{M} \int f d\overline{\mu}_s = \mathbf{M} \sum c_k \overline{\mu}_s(B_k) = \sum c_k a(B_k) = \int f da,$$

$$\mathbf{D} \int f d\overline{\mu}_s = \int f^2 db,$$

then

$$\mathbf{M}\left(\int f_n d\overline{\mu}_s - \int f_m d\overline{\mu}_s\right)^2 = \left(\int (f_n - f_m)da\right)^2 + \int (f_n - f_m)^2 db$$
$$\leq \|f_n - f_m\|^2((\operatorname{Var} a)^2 + b(X)) \to 0$$

for $\|f_n - f_m\| \to 0$. Consequently, if $\mu = \overline{\mu}_s$, the limit on the r.h.s. of (2.40) exists in the sense of the mean square convergence.

Finally, consider $\int (f_n - f_m)d\mu^s$. Note that for a countably additive r.m.i.v. μ^s, the measure $\mu^s_* = \mu^s_+ - \mu^s_-$ is also countably additive; this follows from the proof of 34°. μ^s_* is a nonnegative countably additive r.m.i.v.; moreover, $|\mu^s(B)| \leq \mu^s_*(B)$. Thus, $|\sum_{k=1}^{l} c_k\mu(B_k)| \leq \sup_k |c_k|\mu_*(X)$, provided that $B_k \cap B_j = \varnothing$ for $k \neq j$. Hence,

$$\left|\int (f_n - f_m)d\mu^s\right| \leq \|f_n - f_m\|\mu^s_*(X) \to 0$$

for $\|f_n - f_m\| \to 0$. □

The extension of the integral (2.40) onto unbounded $\mathcal{B}$-measurable functions f is constructed depending on the concrete form of a measure, but the integral is always represented as the limit in probability of the integrals of bounded functions, which converge to f in some sense.

37°. Theorem. *Let the r.m.i.v. μ be such that there exist $a(B) = \mathbf{M}\mu(B)$ and $b(B) = \mathbf{D}\mu(B)$, which are countably additive functions on $\mathcal{B}$. Then $\int f d\mu$ is defined for every function f for which $\int |f|da_* + \int f^2 db < \infty$ (here $a_*(B) =$ $\operatorname{Var} a(B)$ is also a countably additive and now nonnegative measure on $\mathcal{B}$), as the mean square limit $\lim \int f_n d\mu$, where f_n is a sequence from $\mathbf{B}_{(\mathcal{B})}$ such that*

$$\lim_{n\to\infty}\left(\int |f_n - f|da_* + \int |f_n - f|^2 db\right) = 0. \tag{2.41}$$

In addition, the relations

$$\mathbf{M}\int f d\mu = \int f da, \qquad \mathbf{M}\left(\int f d\mu\right)^2 = \left(\int f da\right)^2 + \int f^2 db \tag{2.42}$$

hold.

Proof. If the f_n satisfy the condition of the theorem, then

$$\lim_{n,m\to\infty}\left(\int |f_n - f_m|da_* + \int |f_n - f_m|^2 db\right) = 0.$$

Hence,

$$\mathbf{M}\left(\int f_n d\mu - \int f_m d\mu\right)^2 = \left(\int (f_n - f_m)da\right)^2 + \int (f_n - f_m)^2 db$$
$$\leq \left(\int |f_n - f_m|da_*\right)^2 + \int |f_n - f_m|^2 db \to 0$$

as $n, m \to \infty$. Therefore, $\int f_n d\mu$ converges in mean square to some limit, which we denote by $\int f d\mu$. Formula (2.42) follows from the fact that the mean square convergence of random variables involves the convergence of their first two moments.

□

Remark. One can easily see that if f satisfies the conditions of Theorem 37°, then $\mu_f(B) = \int f I_B d\mu$ is also a countably additive r.m.i.v., for which the following functions

$$a_f(B) = \int f I_B da, \qquad b_f(B) = \int f^2 I_B db$$

are countably additive. Furthermore, if $a_{*f}(B) = \int |f| I_B da_*$ and if g is such that $\int |g| da_{*f} = \int |g|\, |f| da_* < \infty$ and $\int g^2 db_f = \int g^2 f^2 db < \infty$, then we have $\int g d\mu_f = \int g f d\mu$.

38°. Theorem. *Let μ_0 be a Gaussian countably additive r.m.i.v. on $\mathcal{B}$, and $\mathbf{M}\mu_0(B) = 0$, $\mathbf{D}\mu_0(B) = b(B)$. Then b is a countably additive function on $\mathcal{B}$. Denote by $L_2(X, b)$ a linear space of the functions $f(x)$ such that $\int f^2 db < \infty$. For all $f \in L_2(X, b)$, an integral $\int f d\mu_0$ is defined so that the following conditions are satisfied:*

(1) $\mathbf{M} \int f d\mu_0 = 0$; (2) $\mathbf{D} \int f d\mu_0 = \int f^2 db$; (3) $\int f d\mu_0$ *has the Gaussian distribution;* (4) $\int (c_1 f_1 + c_2 f_2) d\mu_0 = c_1 \int f_1 d\mu_0 + c_2 \int f_2 d\mu_0$ *(the equality for random variables holds with probability 1); and* (5) $\int I_B d\mu_0 = \mu_0(B)$. *These conditions determine the integral uniquely.*

Proof. If (4) and (5) are satisfied, then (2.39) holds for $f \in \mathbf{B}^0_{(\mathcal{B})} \subset L_2(X, b)$, and (1), (2), and (3) are satisfied for $\int f d\mu_0$. Clearly, $\mathbf{B}^0_{(\mathcal{B})}$ is dense in $L_2(X, b)$ in the sense of the norm $\|f\|_b^2 = \int f^2 db$. If we denote by H_{μ_0} the Hilbert space generated by the variables $\mu_0(B)$ with a scalar product $\langle \xi, \eta \rangle = \mathbf{M}\xi\eta$ for $\xi, \eta \in H_{\mu_0}$, then $\int f d\mu$ defines an isometric mapping of $\mathbf{B}^0_{(\mathcal{B})}$ onto some subset of H_{μ_0}, i.e., $\langle \int f d\mu, \int f d\mu \rangle = \|f\|_b^2$ for $f \in \mathbf{B}^0_{(\mathcal{B})}$. Hence, it can be extended by continuity (i.e., in the sense of the mean square convergence) onto $L_2(X, b)$, and conditions (1)–(3) remain valid (if ξ_n are Gaussian random variables and converge to ξ in probability, then ξ is also a Gaussian random variable).

□

Remark. $\int_B f d\mu_0 = \int f I_B d\mu_0$ is also a Gaussian countably additive r.m.i.v.

39°. Theorem. *Let ν be a Poisson countably additive r.m.i.v. on $\mathcal{B}$, and $\mathbf{M}\nu(B) = n(B)$ (n is a countably additive function on $\mathcal{B}$). $\int f d\nu$ is defined for any $\mathcal{B}$-measurable function $f \geq 0$ finite everywhere with respect to the* *measure $n(B)$ as a limit in the sense of the convergence in probability of the integrals $\int f \wedge c d\nu$ (as $c \to \infty$). Moreover, for every $h > 0$ and $c \geq 0$,*

$$\mathbf{P}\left\{\int f d\nu > h\right\} \leq n(\{x : f \geq c\}) + \frac{c}{h} n(X). \tag{2.43}$$

Proof. Let us first show that (2.43) holds for $f \geq 0$ from $\mathbf{B}_{(\mathcal{B})}$. We have

$$\begin{aligned}
\mathbf{P}\left\{\int f d\nu > h\right\} &= \mathbf{P}\left\{\int f \wedge c d\nu + \int (f - f \wedge c) d\nu > h\right\} \\
&\leq \mathbf{P}\left\{\int (f - f \wedge c) d\nu > 0\right\} + \mathbf{P}\left\{\int f \wedge c d\nu > h\right\} \\
&\leq \mathbf{P}\{\nu(\{x : f(x) - f \wedge c > 0\}) > 0\} + \frac{1}{h} \mathbf{M} \int f \wedge c d\nu \\
&\leq \mathbf{P}\{\nu(\{x : f > c\}) > 0\} + \frac{c}{h} \mathbf{M}\nu(X) \\
&\leq 1 - e^{-n(\{x : f > c\})} + \frac{c}{h} n(X) \leq n(\{x : f > c\}) + \frac{c}{h} n(X)
\end{aligned}$$

(we have used the Chebyshev inequality). Hence, for $c_1 < c_2$,

$$\begin{aligned}
\mathbf{P}\left\{\int f \wedge c_2 d\nu - \int f \wedge c_1 d\nu > 0\right\} &\leq n(\{x : f \wedge c_2 - f \wedge c_1 > 0\}) \\
&\leq n(\{x : f > c_1\})
\end{aligned} \tag{2.44}$$

(we use (2.43) for $c = 0$). The r.h.s. of (2.44) tends to zero as $c_1 \to \infty$. Therefore, $\int f \wedge c d\nu$ has the limit as $c \to \infty$ in the sense of the convergence in probability.

□

Remark. If f is $\mathcal{B}$-measurable and $|f|$ is finite almost everywhere with respect to the measure n, then the integral $\int f d\nu$ is defined by the equality $\int f d\nu = \int |f| d\nu - \int (|f| - f) d\nu$; moreover, $|\int f d\nu| \leq \int |f| d\nu$, and (2.43) yields

$$\mathbf{P}\left\{\left|\int f d\nu\right| > h\right\} \leq n(\{x : |f| > c\}) + \frac{c}{h} n(X). \tag{2.45}$$

Along with a Poisson r.m.i.v. ν, we shall consider a centered Poisson measure $\tilde{\nu}$

$$\tilde{\nu}(B) = \nu(B) - n(B), \qquad n(B) = \mathbf{M}\tilde{\nu}(B).$$

Then $\mathbf{D}\tilde{\nu}(B) = n(B)$. If $n(B)$ is a measure on $\mathcal{B}$, then $\int f d\tilde{\nu}$ is defined for all f such that $\int |f| dn < \infty$.

Consider now σ-finite countably additive r.m.i.v. μ. Let $X = \bigcup V_k$, where $V_k \in \mathcal{B}$, $k = 1, 2, \ldots$, and $V_j \cap V_k = \varnothing$ when $j \neq k$, and let the countably additive mutually independent r.m.i.v. μ_k be given such that $\mu_k(B) = 0$ for all $B \subset X \backslash V_k$. For every $B \in \mathcal{B}$, $B \subset V_k$, the variable

$$\mu(B) = \sum_k \mu_k(B) \tag{2.46}$$

is defined. $\mu(B)$ is called a *σ-finite countably additive r.m.i.v.*

A collection of sets $\mathcal{R}_0 \subset X$ is called a *ring* if along with each pair of sets A_1 and A_2, it contains $A_1 \cup A_2$ and $A_1 \backslash A_2$ (thus, it also contains $A_1 \cap A_2 = A_1 \cup A_2 \backslash [(A_1 \cup A_2 \backslash A_1) \cup (A_1 \cup A_2 \backslash A_2)]$). A ring $\mathcal{R}$ is called a σ-ring if along with every sequence of sets A_n, it contains $\bigcap A_n$.

We say that a r.m.i.v. is given on the ring $\mathcal{R}$ if for all $B \in \mathcal{R}$, a random variable $\mu(B)$ is given so that the following conditions are satisfied: (1) $\mu(B_1 \cup B_2) = \mu(B_1) + \mu(B_2)$ with probability 1 for $B_1, B_2 \in \mathcal{R}$, $B_1 \cap B_2 = \varnothing$, (2) if $B_1, B_2, \ldots, B_n \in \mathcal{R}$ are mutually disjoint, then $\mu(B_1), \mu(B_2), \ldots, \mu(B_n)$ are independent. If μ is given on the σ-ring $\mathcal{R}$, and if besides (1) and (2), the following condition is satisfied: (3) for a sequence (B_k) of mutually disjoint sets from $\mathcal{R}$ such that $\bigcup B_k \subset \mathcal{R}$, the equality

$$\mu(\bigcup B_k) = \sum \mu(B_k)$$

holds (the series on the r.h.s. converges in probability and, hence, with probability 1, too), then μ is a countably additive measure on the σ-ring $\mathcal{R}$.

A σ-finite countably additive r.m.i.v. is defined on a σ-ring of sets $B = \bigcup_{k=1}^n B_k$, where $B_k \in \mathcal{B}$, $B_k \subset V_k$. If $\mathcal{R}$ contains a sequence of mutually disjoint sets V_k such that $\bigcup V_k = X$, then every countably additive r.m.i.v. on $\mathcal{R}$ is a σ-finite countably additive r.m.i.v.; moreover, it can be represented in the form (2.46) if we put $\mu_k(B) = \mu(B \cap V_k)$.

40°. Theorem. *Let $\mathcal{R}_0$ be a denumerable ring such that $\bigcup_{B \subset \mathcal{R}_0} = X$, and $\mathcal{R}$ be a σ-ring generated by $\mathcal{R}_0$. A r.m.i.v. $\mu(B)$ is given on $\mathcal{R}_0$ such that* $\mathbf{M}\mu(B) = 0$, $\mathbf{D}\mu(B) = m(B)$.

I. *If $m(B)$ is continuous on $\mathcal{R}_0$ (the conditions $B_n \downarrow$, $B_n \in \mathcal{R}_0$, $\bigcap B_n = \varnothing$ imply that $m(B_n) \to 0$), then $\mu(B)$ can be extended to a σ-finite measure on the σ-algebra $\mathcal{B}$, which is generated by the ring $\mathcal{R}$.*

II. *For every function $f(x) \in L_2(X, m)$ (i.e., $\int f^2 dm < \infty$; note that m is σ-finite), the following integral is defined:*

$$\int f d\mu = \lim \int f I_{B_n} d\mu, \tag{2.47}$$

where $B_n \subset \mathcal{R}$, $B_n \uparrow$, $\bigcup B_n = X$. The limit on the right is independent of the choice of B_n; it exists in the sense of the mean square convergence; and further,

$$\mathbf{M} \int f d\mu = 0, \qquad \mathbf{D} \int f d\mu = \int f^2 dm. \tag{2.48}$$

Proof. Statement I follows from 32°. 36° implies the existence of $\int fI_{B_n}d\mu$, and the equalities

$$\mathbf{M}\int fI_{B_n}d\mu = 0, \qquad \mathbf{D}\int fI_{B_n}d\mu = \int f^2 I_{B_n}d\mu. \tag{2.49}$$

Since for $n > m$,

$$\mathbf{M}\Big(\int fI_{B_n}d\mu - \int fI_{B_m}d\mu\Big)^2 = \int f^2(I_{B_n} - I_{B_m})d\mu \to 0$$

as $n, m \to \infty$, the limit on the r.h.s. of (2.47) exists in the sense of the mean square convergence; (2.48) is obtained by the limit transition in (2.49).

□

Consider a presentation of a stochastically continuous countably additive r.m.i.v. as a stochastic integral. Let us denote by $(A_n^{i_n^{(k)}(\omega)})$, $k = 1, 2, \ldots$, all random fundamental sequences for which $\lim \mu(A_n^{i_n^{(k)}(\omega)}) = \xi_k \neq 0$ (by virtue of the regularity of μ, their number is at most countable, and for every $\varepsilon > 0$, the set of k for which $|\xi_k| > \varepsilon$ is finite).

Denote by $T(\mu) \in R_+$ a set of those t for which $\sum \mathbf{P}\{|\xi_k| = t\} = 0$, and let $T_1 \subset T(\mu)$ be some countable set dense in R_+. By $\mathcal{R}_0$ we denote a ring $\bigcup_k C^{s_k}$, where $s_k \in T_1$, $s_k \downarrow 0$, and C^{s_k} is an algebra of sets in $]-\infty, -s_k] \cup [s, \infty[$ generated by the intervals $[s, \infty[$ and $]-\infty, s]$, $s \in T_1 \cap [s_k, \infty[$.

Let us define a Poisson r.m.i.v. on the ring $\mathcal{R}_0 \otimes \mathcal{B}_0$ such that

$$\nu(U \times B) = \sum I_{\{\xi_k \in U\}} I_{\{A_n^{i_n^{(k)}(\omega)} \subset B\}} \tag{2.50}$$

for $U \in \mathcal{R}_0$, $B \subset \mathcal{A}_n$ (this is really a Poisson r.m.i.v. by virtue of 13°). If $\mu^s(B)$ is defined by (2.21), then (2.22) is valid.

41°. Theorem. *Let X be a complete separable metric space, $\mathcal{B}$ be a σ-algebra of Borel sets, and $\mathcal{B}_0$ be some countable algebra generating $\mathcal{B}$. If $\mu(B)$ is a countably additive stochastically continuous r.m.i.v., then*

(1) ν can be extended to become a countably additive r.m.i.v. on a σ-ring $\mathcal{R} \otimes \mathcal{B}$, where $\mathcal{R}$ is a σ-ring of Borel subsets of R lying at a positive distance from the origin,

(2) if $\tilde{\nu}(U \times B) = \nu(U \times B) - \mathbf{M}\nu(U \times B)$ is a centered Poisson measure, then for all $c > 0$, the integral

$$\int t \wedge c\, d\tilde{\nu} \tag{2.51}$$

exists (an integral with respect to the measure $\tilde{\nu}$ is defined as

$$\lim_{\delta \to 0} \int I_{\{|t| > \delta\}} \times f d\nu),$$

(3) *if μ^s is defined by* $\mu^s = \sum \hat{g}_s(\xi_k) I_{\{A_n^{i_n^{(k)}(\omega)} \subset B\}}$, $B \in \mathcal{A}_n$, *then*

$$\mu^s(B) = \int t I_{\{t > s\}} I_B d\nu, \tag{2.52}$$

(4) *for every* $s \in T_1$,

$$\mu(B) = \mu_0(B) + \int t I_{\{|t| < s\}} I_B d\tilde{\nu} + \int t I_{\{|t| > s\}} I_B d\nu, \tag{2.53}$$

where $\mu_0(B)$ is a Gaussian countably additive stochastically continuous function independent of ν.

Proof. By virtue of 34°, the r.m.i.v. $\nu(]-\infty, -s] \cup [s, \infty[\times B)$ is countably additive on $\mathcal{B}$; and thus, $\pi^s(B) = \mathbf{M}\nu(]-\infty, -s] \cup [s, \infty[\times B)$ is also countably additive.

To prove (1), it suffices to show that the function $\pi^s(C) = \mathbf{M}\nu(C)$ is continuous on the σ-algebra $\widehat{\mathcal{C}}^s \otimes \mathcal{B}_0$. Since $\pi^s(]-\infty, t[\times B) \leq \pi^s(t) = \pi^s(]-\infty, t[\times X)$, and $\pi^s(]-\infty, t[\times B)$ is a measure with respect to B, there exists a regular conditional probability $g(t, B)$ such that

$$\pi^s(]-\infty, t[\times B) = \int_{-\infty}^{t} g(u, B) dF^s(u)$$

and $0 \leq g(u, B) \leq 1$; $g(u, B)$ is a measure with respect to B for almost all u (with respect to the measure $dF^s(u)$). If we denote by $C\big|_t$ a cross section of the set C from $\mathcal{C}^s \otimes \mathcal{B}_0$, i.e., $C\big|_t = \{x : (t, x) \in C\}$, we obtain

$$\pi^s(C) = \int g(u, C\big|_u) dF^s(u)$$

(this is the full probability formula). If $C^n \downarrow$, $C^n \in \mathcal{C}^s \otimes \mathcal{B}_0$, $\bigcap C^n = \varnothing$, we have

$$\pi^s(C^n) = \int g(u, C^n\big|_u) dF^s(u)$$

and $g(u, C^n\big|_u) \to 0$ for almost all u. Hence, $\pi^s(C^n) \to 0$. Statement (1) is thus proved. Statement (3) follows from (2.22). Thus, for $0 < s_1 < s_2$, $s_1, s_2 \in T_1$, we have

$$\overline{\mu}_{s_2} - \overline{\mu}_{s_1}(B) = \mu^{s_1}(B) - \mu^{s_2}(B) = \int t I_{\{s_1 < |t| < s_2\}} I_B d\nu,$$

$$\overline{\mu}_{s_2}(B) - \overline{\mu}_{s_1}(B) - \mathbf{M}\overline{\mu}_{s_2}(B) + \mathbf{M}\overline{\mu}_{s_1}(B) = \int t I_{\{s_1 < t < s_2\}} I_B d\tilde{\nu},$$

$$\mathbf{D}[\overline{\mu}_{s_2}(B) - \overline{\mu}_{s_1}(B)] = \int t^2 I_{\{s_1 < t < s_2\}} I_B d\pi.$$

Since

$$\int t^2 I_{\{s_1<t<s_2\}} I_B d\pi \le \mathbf{D}\overline{\mu}_{s_2}(B),$$

by taking $B = X$, we obtain $\int t^2 I_{\{t<s_2\}} d\pi \le \mathbf{D}\overline{\mu}_{s_2}(X)$; and therefore, the integral $\int t I_{\{t<s_2\}} d\tilde{\nu}$ exists, so (2) is satisfied. We have, further, for $s_1 < s$, $s_1, s \in T(\mu)$,

$$\mu(B) = \int t I_{\{t>s\}} I_B d\nu + \mathbf{M}\overline{\mu}_s(B) + \overline{\mu}_{s_1}(B) - \mathbf{M}\overline{\mu}_{s_1}(B) + \int t I_{\{s_1<t<s_2\}} I_B d\tilde{\nu}. \tag{$*$}$$

By virtue of statement (c) in 23°, the limit $\lim_{s_1\downarrow 0}[\mathbf{M}\overline{\mu}_s(B) + \overline{\mu}_{s_1}(B) - \mathbf{M}\overline{\mu}_{s_1}(B)] = \mu_0(B)$ exists, and μ_0 is independent of μ^s (for all $s > 0$) and, hence, of ν. Passing to the limit as $s_1 \downarrow 0$ in ($*$), we get (2.53).

□

§2.4 Random Linear Functionals and Generalized Functions

Random Linear Functionals Corresponding to Countably Additive r.m.i.v.

Let X be a compact metric space, $\mathcal{B}$ be a σ-algebra of its Borel subsets, and C_X be a space of continuous functions $f(x) : X \to R$, $\|f\| = \sup_x |f(x)|$. C_X with this norm is a separable Banach space. Let μ be a countably additive r.m.i.v. on $\mathcal{B}$ with values in some complete probability space $(\Omega, \mathcal{F}, \mathbf{P})$.

In order to investigate r.m.i.v. on $\mathcal{B}$, we shall consider an algebra $\mathcal{B}_0 = \bigcup_m \mathcal{A}_m$, where $\mathcal{A}_m$ is an algebra whose atoms are constructed as follows: Let $\delta_k \to 0$; $S_1^{(k)}, \ldots, S_n^{(k)}$ be a covering of X by closed spheres of radius δ_k; $B_i^{(k)} = S_i^{(k)} \setminus \bigcup_{j<k} S_j^{(k)}$, and A_m^j have the form $B_{i_1}^{(1)} \cap B_{i_2}^{(2)} \cap \cdots \cap B_{i_m}^{(m)}$; the indices j enumerate all nonempty intersections of this kind. The diameter of the set A_m^j does not exceed δ_m, and thus for any fundamental sequence (A_m), the set $\bigcap A_m$ is either empty, and then $\mu(A_m) \to 0$ in probability for a countably additive r.m.i.v., or $\bigcap A_m$ is a single-point set, and then this fundamental sequence can be a concentration point of a purely discrete r.m.i.v. Let μ be a purely discrete countably additive r.m.i.v., and $(A_m^{i_m^{(k)}})$ be those fundamental sequences on which it is concentrated, $\{x_k\} = \bigcap_m A_m^{i_m^{(k)}}$, $\xi_k = \lim_{m\to\infty} \mu(A_m^{i_m^{(k)}})$. Then ξ_k is a sequence of independent random variables such that the series $\sum \xi_{n_k}$ converges, whatever sequence of different natural numbers is taken. The measure μ is given by

$$\mu(B) = \sum I_B(x_k)\xi_k, \tag{2.54}$$

and for a bounded measurable function f, we have

$$\int f d\mu = \sum f(x_k)\xi_k. \tag{2.55}$$

The last two statements are ascertained by the following arguments: (2.54) holds for $B = A^j_m$, and thus it is valid on $\mathcal{B}_0$; the r.h.s. of (2.54) is countably additive on $\mathcal{B}$; and therefore, two countably additive r.m.i.v. coincide on $\mathcal{B}_0$ and, hence, on $\mathcal{B}$. (2.54) implies (2.55).

For all $f \in C_X$, the variable $\mu(f) = \int f d\mu$ is defined; moreover, the following conditions are satisfied:

(1) For all $f_1, f_2 \in C_X$, $a_1, a_2 \in R$,

$$\mathbf{P}\{\mu(a_1 f_1 + a_2 f_2) = a_1\mu(f_1) + a_2\mu(f_2)\} = 1,$$

(2) $\mu(f_n) \to 0$ in probability if $\|f_n\| \to 0$,

(3) denote by $\operatorname{supp} f$ the least closed subset of X for which $f(x) = 0$ if $x \overline{\in} \operatorname{supp} f$; let G_1 and G_2 be two open sets, $G_1 \cap G_2 = \varnothing$, $f_1, f_2, \ldots, f_l, g_1, \ldots, g_m \subset C_X$, $\operatorname{supp} f_k \subset G_1$, $k = 1, \ldots, l$, $\operatorname{supp} g_j \subset G_2$, $j = 1, \ldots, m$; then the two groups of random variables $\{\mu(f_k), k = 1, \ldots, l\}$ and $\{\mu(g_j), j = 1, \ldots, m\}$ are mutually independent. Only statement (2) needs to be verified. It follows from the next statement.

41°. Lemma. *Let $\|f_n\| \to 0$; then $\int f_n d\mu \to 0$ in probability.*

Proof. The statement of the lemma is valid if μ is a nonrandom countably additive function (it has bounded variation in this case). Suppose that μ is purely discrete. Then the sequences of points x_k and of the independent random variables ξ_k (for which the series $\sum \xi_{n_k}$ converges for any sequence of different natural numbers n_k) exist, such that $\mu(B) = \sum I_B(x_k)\xi_k$ and

$$|\int f_n d\mu| = |\sum f_n(x_k)\xi_k| \le |\sum f_n(x_k)\xi_k I_{\{|\xi_k|>c\}}| + |\sum f_n(x_k) I_{\{|\xi_k|\le c\}}|.$$

The first sum on the r.h.s. consists of a finite number of terms (their number is independent of n), and each vanishes as $n \to \infty$. Therefore, this sum also vanishes as $n \to \infty$. For the second sum, we have

$$\begin{aligned}\mathbf{M}|\sum f_n(x_k)\xi_k I_{\{|\xi_k|\le c\}}|^2 &= \Big(\sum \mathbf{M}\xi_k I_{\{|\xi_k|\le c\}} f_n(x_k)\Big)^2 \\ &\quad + \sum f_n^2(x_k)\mathbf{D}\xi_k I_{\{\xi_k|\le c\}} \\ &\le \|f_n\|^2\left[\Big(\sum |\mathbf{M}\xi_k I_{\{|\xi_k|\le c\}}|\Big)^2 + \sum \mathbf{D}\xi_k I_{\{|\xi_k|\le c\}}\right].\end{aligned}$$

The coefficient of $\|f_n\|^2$ is bounded by virtue of 18°, Chapter 1, and hence this product tends to zero as $\|f_n\| \to 0$. Suppose that μ has the finite moments $\mathbf{M}\mu(B) = a(B)$, $\mathbf{D}\mu(B) = b(B)$. Then

$$\mathbf{M}(\int f_n d\mu)^2 = |\int f_n da|^2 + \int f_n^2 db \le \|f_n\|^2[(\operatorname{Var} a)^2 + b(X)] \to 0$$

for $\|f_n\| \to 0$. Finally, let μ be a stochastically continuous measure like μ^s; i.e., for this measure, a finite number of fundamental sequences $(A_n^{i_n^{(k)}(\omega)})$, $k = 1, 2, \dots, l(\omega)$ exist, such that $|\lim_{n\to\infty} \mu_n^s(A_n^{i_n^{(k)}(\omega)})| \geq s$ and $\mu^s(X - \bigcup_{k=1}^{l(\omega)} A_n^{i_n^{(k)}(\omega)}) = 0$. Then

$$|\int f_m d\mu| \leq \|f_m\| \sum |\lim \mu^s(A_m^{i_n^{(k)}(\omega)})| \to 0$$

as $\|f_m\| \to 0$. □

42°. Definition. Assume that each $f \in C_X$ is associated with a random variable $\mu(f)$ so that conditions (1)–(3) are satisfied. In this case, we say that a *random linear functional* is given on C_X. Condition (3) can be formulated more simply:

(3^1) If f_1 and f_2 are two functions for which $\operatorname{supp} f_1 \cap \operatorname{supp} f_2 = \varnothing$, then $\mu(f_1)$ and $\mu(f_2)$ are independent.

Let us show that (3) implies (3^1). Since $\operatorname{supp} f_k$ are compact sets and do not intersect, they have nonintersective neighbourhoods $G_k \supset \operatorname{supp} f_k$, $G_1 \cap G_2 = \varnothing$; and (3) yields that $\mu(f_1)$ and $\mu(f_2)$ are independent.

Suppose that (3^1) is satisfied and $\{f_k, k = 1, \dots, l\}$ and $\{g_j, j = 1, \dots, m\}$ are the same as in (3). Then the functions $\sum_{k=1}^{l} \alpha_k f_k$ and $\sum_{j=1}^{m} \beta_j g_j (\alpha_k, \beta_j \in R)$ have their supports in G_1 and G_2, respectively, and are thus independent; i.e.,

$$\mathbf{M}\left(\exp\left\{i\sum_{k=1}^{l} \alpha_k \mu(f_k)\right\}\right)\left(\exp\left\{i\sum_{j=1}^{l} \beta_j \mu(g_j)\right\}\right)$$
$$= \mathbf{M}\exp\left\{i\sum_{k=1}^{l} \alpha_k \mu(f_k)\right\} \cdot \mathbf{M}\exp\left\{i\sum_{j=1}^{l} \beta_j \mu(g_j)\right\},$$

and this means that the collections $\{\mu(f_1), \dots, \mu(f_l)\}$ and $\{\mu(g_1), \dots, \mu(g_m)\}$ are independent. Finally, (3) (or (3^1)) implies the stronger statement.

43°. *Suppose that $\mu(f)$ satisfies condition (3). For every open set G, let us denote by $\mathcal{F}_G$ a σ-algebra generated by the set of variables $\{\mu(f), \operatorname{supp} f \subset G\}$ and by all the sets of P-measure zero. Then (1) the σ-algebras $\{\mathcal{F}_{G_k}, k \geq i\}$ are independent if $G_i \cap G_k = \varnothing$ for $i \neq k$; (2) moreover, $\mathcal{F}_{\cup G_k} = \vee_k \mathcal{F}_{G_k}$.*

Proof. It suffices to consider the two sets G_1 and G_2. In fact, if $\mathcal{F}_{G_1}$ and $\mathcal{F}_{\cup_{k>1} G_k}$ are independent, then $\mathcal{F}_{G_1}$ does not depend on the collection of σ-algebras $\{\mathcal{F}_{G_k}, k > 1\}$, since $\mathcal{F}_{\cup_{k>1} G_k} \supset \bigcup_{k>1} \mathcal{F}_{G_k}$. In exactly the same way, we get that $\mathcal{F}_{G_2}$ does not depend on the collection of σ-algebras $\{\mathcal{F}_{G_k}, k > 2\}$, etc. But this yields that all the σ-algebras $\mathcal{F}_{G_k}$ are independent.

Independence of the σ-algebras $\mathcal{F}_{G_1}$ and $\mathcal{F}_{G_2}$ for $G_1 \cap G_2 = \varnothing$ follows from the fact that by virtue of (3^1), the sets of variables $\{\mu(f), \operatorname{supp} f \subset G_1\}$ and

$\{\mu(f), \operatorname{supp} f \subset G_2\}$, generating these σ-algebras, are independent. Further, if $\mathcal{F}_{G_1 \cup G_2} = \mathcal{F}_{G_1} \vee \mathcal{F}_{G_2}$, we have $\mathcal{F}_{\cup_{k=1}^n G_k} = \vee_{k=1}^n \mathcal{F}_{G_k}$, and hence,

$$\mathcal{F}_{\cup_{k=1}^\infty G_k} = \vee_{k=1}^\infty \mathcal{F}_{G_k}.$$

Since $\mathcal{F}_{G_1} \subset \mathcal{F}_{G_1 \cup G_2}$, $\mathcal{F}_{G_2} \subset \mathcal{F}_{G_1 \cup G_2}$, we have $\mathcal{F}_{G_1} \vee \mathcal{F}_{G_2} \subset \mathcal{F}_{G_1 \cup G_2}$. On the other hand, for $f \in C_X$, $\operatorname{supp} f \subset G_1 \cup G_2$, we have $f I_{G_k} \in C_X$, $\operatorname{supp} f I_{G_k} \subset G_k$, and thus,

$$\begin{aligned}
\mathcal{F}_{G_1 \cup G_2} &= \sigma\{\mu(f), f \in C_X, \operatorname{supp} f \subset G_1 \cup G_2\} \\
&\subset \sigma\{\mu(f), f \in C_X, \operatorname{supp} f \subset G_k, k = 1, 2\} \\
&\subset \sigma\{\mu(f), f \in C_X, \operatorname{supp} f \subset G_1\} \otimes \sigma\{\mu(f), f \in C_X, \operatorname{supp} f \in G_2\} \\
&= \mathcal{F}_{G_1} \otimes \mathcal{F}_{G_2} = \mathcal{F}_{G_1} \vee \mathcal{F}_{G_2}.
\end{aligned}$$

□

Remark. *If G_1 and G_2 are two open sets, then $\mathcal{F}_{G_1 \cup G_2} = \mathcal{F}_{G_1} \vee \mathcal{F}_{G_2}$.*

In fact, if $\operatorname{supp} f \subset G_1 \cup G_2$, we can always find the continuous functions φ_1 and φ_2 such that $\operatorname{supp} \varphi_1 \subset G_1$, $\operatorname{supp} \varphi_2 \subset G_2$, and $\varphi_1 + \varphi_2 = 1$ for $x \in \operatorname{supp} f$. Then $\mu(f) = \mu(\varphi_1 f_1) + \mu(\varphi_2 f_2)$, and therefore, $\mu(f)$ is measurable with respect to $\mathcal{F}_{G_1} \vee \mathcal{F}_{G_2}$, i.e., $\mathcal{F}_{G_1 \cup G_2} \subset \mathcal{F}_{G_1} \vee \mathcal{F}_{G_2}$. The inverse inclusion is obvious.

□

A Family of σ-Algebras Generated by a Random Linear Functional

44°. Theorem. *Let us define $\mathcal{F}_F = \bigcap_{G \supset F} \mathcal{F}_G$ (here F and G are, respectively, closed and open sets from X). Then* (a) *for every open set G, we have*

$$\mathcal{F}_G = \bigvee_{F \subset G} \mathcal{F}_F,$$

where F are closed sets from X; (b) *$\mathcal{F}_G$ and $\mathcal{F}_{X \setminus G}$ are independent.*

Proof. The definition of $\mathcal{F}_F$ implies that $\mathcal{F}_F \subset \mathcal{F}_G$ for $F \subset G$ and $\mathcal{F}_G \subset \mathcal{F}_F$ for $G \subset F$. We prove (a). Let $G = \bigcup S_n$, where S_n is some sequence of open spheres. Denote by S_n^ε an open sphere with the same center as S_n, but of radius shorter by ε (if the radius of S_n does not exceed ε, then S_n^ε is an empty set). $\overline{S}_n^\varepsilon$ denotes a closed sphere. Then for all ε,

$$\vee_{F \subset G} \mathcal{F}_F \supset \mathcal{F}_{\cup_{k \le n} \overline{S}_k^\varepsilon} \supset \mathcal{F}_{\cup_{k \le n} S_k^\varepsilon}.$$

Hence,

$$\vee_{F \subset G} \mathcal{F}_F \supset \vee_n \vee_{\varepsilon > 0} \mathcal{F}_{\cup_{k \le n} S_k^\varepsilon} = \vee_n \mathcal{F}_{\cup_{k \le n} S_k} = \mathcal{F}_G.$$

In order to prove (b), we note that $\mathcal{F}_{F_1}$ and $\mathcal{F}_{F_2}$ are independent for closed F_1 and F_2, for which $F_1 \cap F_2 = \varnothing$, since we can always find open G_1 and G_2 such that $G_1 \cap G_2 = \varnothing$, $F_1 \subset G_1$, $F_2 \subset G_2$, and thus $\mathcal{F}_{F_1} \subset \mathcal{F}_{G_1}$, $\mathcal{F}_{F_2} \subset \mathcal{F}_{G_2}$. Let F_k be a sequence of closed sets such that $F_k \uparrow$ and $\bigcup F_k = G$. Then $\mathcal{F}_{X\backslash G}$ and $\mathcal{F}_{F_k}$ are independent ($X\backslash G$ is closed, $(X\backslash G) \cap F_k = \varnothing$). Hence, $\mathcal{F}_{X\backslash G}$ and $\vee_k \mathcal{F}_{F_k} = \mathcal{F}_G$ are independent.

Representation of a Random Linear Functional by the Stochastic Integral

Let $\mu(f)$ be a linear random functional. There exists a function $g(t)$, $t \in R$, satisfying the following conditions: (1) $g(-t) = g(t)$, $g(0) = 0$, $g(t) \uparrow +\infty$ as $t \uparrow +\infty$, $g(t+s) \le g(t) + g(s)$ for $s, t \in R_+$; (2) $\mathbf{M}g(\mu(f)) < \infty$ for all $f \in C_X$. In fact, if

$$\varphi(c) = \sup_{\|f\| \le 1} \mathbf{P}\{|\mu(f)| > c\},$$

then $\varphi(c) \to 0$ as $c \to \infty$, because otherwise we can find such $\delta > 0$ and a sequence $f_n \in C_X$ for which $\|f_n\| \le 1$ and $\mathbf{P}\{|\mu(f_n)| > n\} \ge \delta$. But in this case, $\|\frac{1}{n} f_n\| \to 0$ and $\mu(\frac{1}{n} f_n)$ does not tend to zero in probability. Clearly, for $\|f\| \le 1$, we have

$$\mathbf{M}g(\mu(f)) = \int_0^\infty \mathbf{P}\{|\mu(f)| > c\} dg(c) \le \int_0^\infty \varphi(c) dg(c) = \int_0^\infty \varphi(c) g'(c) dc. \tag{2.56}$$

Hence, it remains to find a function $g(c)$ such that $g'(c) \ge 0$, $g'(c) \downarrow 0$, as $c \uparrow \infty$, $\int_0^\infty g'(c) dc = +\infty$; but the integral on the r.h.s. of (2.56) is finite.

45°. Lemma. *Let $g(t)$, $t \in R_+$, $g(0) = 0$, $g(t) \uparrow +\infty$, as $t \uparrow +\infty$. Then for every $\varepsilon > 0$ and $c > 0$, there exists $n(\varepsilon, c)$ such that the following inequality*

$$\sup_{\substack{|\delta_i| \le 1, \\ i=1,\dots,n}} \mathbf{M}g\left(\sum_{k=1}^n \delta_k \xi_k\right) \ge c \tag{2.57}$$

holds for $n > n(\varepsilon, c)$, whatever independent variables $\xi_1, \dots, \xi_n$ we take for which $\mathbf{M}g(|\xi_k|) > \varepsilon$.

Proof. The condition $\mathbf{M}g(|\xi|) \ge \varepsilon$ yields that for some $a > \alpha > 0$ and $\beta > 0$, $\mathbf{P}\{a \ge |\xi| \ge \alpha\} \ge \beta$; the inequality (2.57) is satisfied if

$$\mathbf{P}\{|\sum_{k=1}^n \delta_k \xi_k| \le g^{-1}(\lambda c)\} \le 1 - 1/\lambda \quad (\lambda > 1) \tag{2.58}$$

for all sufficiently large n and some δ_k.

Let $\hat{\delta}_k$ be mutually independent random variables independent of $\xi_1, \xi_2, \ldots$, for which $\mathbf{P}\{\hat{\delta}_k = 1\} = \mathbf{P}\{\hat{\delta}_k = -1\} = 1/2$. Then by setting $g^{-1}(\lambda c) = r$, we get

$$\mathbf{P}\left\{|\sum_{k=1}^{n} \hat{\delta}_k \xi_k| \le r\right\} = \sum_{\substack{\delta_k=\pm 1,\\ k=1,\ldots,n}} \mathbf{P}\left\{|\sum_{k=1}^{n} \delta_k \xi_k| \le r\right\} \cdot \frac{1}{2^n}.$$

And if the l.h.s. is less than $1 - \lambda^{-1}$, then at least for one choice of $\delta_k = \pm 1$, (2.58) is valid. For $u > 0$, we have

$$\mathbf{P}\left\{|\sum_{k=1}^{n} \hat{\delta}_k \xi_k| \le r\right\} = \mathbf{P}\{e^{-u|\sum_{k=1}^{n} \hat{\delta}_k \xi_k|} \ge e^{-ur}\}$$

$$\le e^{ur}\mathbf{M}e^{-u|\sum_{k=1}^{n} \hat{\delta}_k \xi_k|} = e^{ur}\mathbf{M}\frac{1}{\pi}\int e^{iu\sum_{k=1}^{n} \hat{\delta}_k \xi_k x} \frac{1}{1+x^2} dx$$

$$= e^{ur}\pi^{-1}\int \prod_{k=1}^{n} \mathbf{M}e^{iu\xi_k \hat{\delta}_k x} \frac{1}{1+x^2} dx = e^{ur}\pi^{-1}\int \prod_{k=1}^{n} (\mathbf{M}\cos u\xi_k x) \frac{1}{1+x^2} dx$$

$$\le e^{ur}\pi^{-1}\int \prod_{k=1}^{n} (1 + \mathbf{M}(\cos u\xi_k x - 1) I_{\{\alpha \le |\xi_k| \le a\}}) \frac{dx}{1+x^2}$$

$$\le e^{ur}\pi^{-1}\left(\int_{ua|x|<\pi} \prod_{k=1}^{n} (1 + \mathbf{M}(\cos u\xi_k x - 1) I_{\{\alpha \le |\xi_k| \le a\}}) dx + \int_{ua|x|>\pi} \frac{dx}{x^2}\right)$$

$$\le e^{ur}\pi^{-1}\left(\int_{ua|x|<\pi} [1 + (\cos ux\alpha - 1)\beta]^n dx + \frac{2ua}{\pi}\right)$$

$$\le \frac{2a}{\pi^2} ue^{ur} + \frac{e^{ur}}{\pi}\int_{ua|x|<\pi} (1 - u^2 q x^2)^n dx.$$

Here $0 < q < \frac{a^2}{\pi^2}$, Hence, for some constants $\theta_1, \theta_2, \theta_3$,

$$\mathbf{P}\left\{|\sum_{k=1}^{n} \hat{\delta}_k \xi_k| \le r\right\} \le e^{ur}\left(\theta_1 u + \theta_2 \int e^{-nu^2 q x^2} dx\right)$$

$$\le e^{ur}\left(\theta_1 u + \frac{\theta_3}{u\sqrt{n}}\right),$$

and the expression on the right can be made arbitrarily small for all $n > u^{-3}$ by choosing proper u.

□

We now define a function of the set $\lambda(B)$ by

$$\lambda(G) = \sup_{\substack{\|f\| \le 1,\\ \operatorname{supp} f \subset G}} \mathbf{M}g(\mu(f)), \qquad \lambda(B) = \inf_{G > B} \lambda(G). \tag{2.59}$$

Here G is an open set.

46°. Lemma. *The function of the set $\lambda(G)$ satisfies the following conditions:*

(1) *it is monotonic: $\lambda(B_1) \leq \lambda(B_2)$ for $B_1 \subset B_2$;*

(2) *it is semiadditive: $\lambda(\bigcup B_k) \leq \sum \lambda(B_k)$;*

(3) *if F is closed and G_n is open, $G_n \supset F$, $G_n \downarrow F$, then $\lambda(F) = \lim_{n\to\infty} \lambda(G_n)$;*

(4) *suppose that $\lambda(X) = c$. In this case, there is no sequence of mutually disjoint closed sets $F_1, F_2, \ldots, F_n$ for which $\lambda(F_k) > \varepsilon$, provided that $n > n(\varepsilon, c)$ ($n(\varepsilon, c)$ is the same as in Lemma 45°).*

Proof. Property (1) is obvious.

(2) First we assume that the G_k are open and f is such that $\operatorname{supp} f \subset \bigcup G_k$, $\|f\| \leq 1$, and

$$\lambda(\bigcup G_k) \leq \varepsilon + \mathbf{M}g(\mu(f)).$$

Since $\operatorname{supp} f$ is a compact set, $\operatorname{supp} f \subset \bigcup_{k=1}^{n} G_k$ for some n. One can easily construct the functions $f_k \subset C_X$ such that $\operatorname{supp} f_k \subset G_k$, $\sum_{k=1}^{n} f_k = f$. Then

$$\mathbf{M}g(\mu(f)) \leq \mathbf{M}g\left(\sum_{k=1}^{n} \mu(f_k)\right) \leq \sum_{k=1}^{n} \mathbf{M}g(\mu(f_k)) \leq \sum_{k=1}^{n} \lambda(G_k);$$

i.e., $\lambda(\bigcup G_k) \leq \varepsilon + \sum_{k=1}^{\infty} \lambda(G_k)$, and $\varepsilon > 0$ is arbitrary. Further,

$$\begin{aligned}\lambda(\bigcup B_k) = \inf_{G \supset \bigcup B_k} \lambda(G) &\leq \inf_{\substack{G_k \supset B_k,\\ k=1,2,\ldots}} \lambda(\bigcup G_k)\\ &\leq \inf_{\substack{G_k \supset B_k,\\ k=1,2,\ldots}} \sum \lambda(G_k) = \sum \lambda(B_k).\end{aligned}$$

Property (2) is thus proved. Let us prove (3). If $G \supset F$, then there exists n such that $G_n \subset G$ and $\lambda(G) \geq \lambda(G_n) \geq \lambda(F)$. Thus, $\lambda(G) \geq \lim_{n\to\infty} \lambda(G_n)$ for all $G \supset F$. Taking the infimum with respect to $G \supset F$, we obtain the required result.

We prove (4). Let F_i, $i = 1, \ldots, n$, be mutually disjoint, $G_i \supset F_i$ also be mutually disjoint, and $\lambda(G_i) - \lambda(F_i) \leq \delta$. The functions $f_i \subset C_X$, $\|f_i\| \leq 1$, are chosen so that $\operatorname{supp} f_i \subset G$ and $\lambda(G_i) - \mathbf{M}g(\mu(f_i)) \leq \delta$. Then $|\lambda(F_i) - \mathbf{M}g(\mu(f_i))| \leq \delta$, and we can choose δ so small that $\mathbf{M}g(\mu(f_i)) \geq \varepsilon$. The $\mu(f_i)$ are independent random variables, and

$$\sup_{|\delta_i| \leq 1} \mathbf{M}g\left(\mu(\sum_{i=1}^{n} \delta_i f_i)\right) \leq \lambda(X) = c.$$

Thus, by virtue of Lemma 45°, we have $n \leq n(\varepsilon, c)$. □

47°. Theorem. *Let G be an open set and G' be its boundary, $\lambda(G') = 0$. Then there exists a limit $\overline{\mu}(G) = \lim \mu(f_n)$, where $\operatorname{supp} f_n \subset G$, $\|f_n\| \leq 1$,*

$f_n \uparrow I_G$, and $\overline{\mu}(G)$ is $\mathcal{F}_G$-measurable, $\overline{\mu}(X) - \overline{\mu}(G)$, where $\overline{\mu}(X) = \mu(1)$, is $\mathcal{F}_{X\setminus G}$-measurable. Let $\mathcal{A}_0$ be the set of those $B \subset \mathcal{B}$ for which $\lambda(B') = 0$ (B' is the boundary of B). We put $\overline{\mu}(B) = \overline{\mu}(\operatorname{int} B)$, where $\operatorname{int} B$ is the set of internal points of B. Then $\mathcal{A}_0$ is an algebra of sets; furthermore, this algebra generates $\mathcal{B}$; $\mu(B)$ is an additive function with independent values on $\mathcal{A}_0$, and for every $f \subset C_X$, we have

$$\mu(f) = \int f d\overline{\mu}. \tag{2.60}$$

Moreover, $\overline{\mu}$ is a countably additive r.m.i.v. on $\mathcal{B}$.

Proof. If $f_n \geq 0$ and $f_n \uparrow I_G$, then $\operatorname{supp}(f_n - f_m) \subset G_1$ for all sufficiently large n and m, whatever open set $G_1 \supset G'$ we take. Thus, $\mathbf{M}g(\mu(f_n) - \mu(f_m)) \leq \lambda(G_1)$; and hence,

$$\lim_{n,m\to\infty} \mathbf{M}g(\mu(f_n) - \mu(f_m)) \leq \inf_{G_1 \supset G'} \lambda(G_1) = \lambda(G') = 0.$$

Therefore, $\mu(f_n)$ has a limit in probability that we shall denote by $\overline{\mu}(G)$. The $\mathcal{F}_G$-measurability of $\overline{\mu}(G)$ follows from the $\mathcal{F}_G$-measurability of $\mu(f_n)$ for all n. Suppose now that $\varphi_n \in C_X$, $0 \leq \varphi_n \leq 1$ and $\varphi_n \uparrow I_{X\setminus\overline{G}}$, where $\overline{G}$ is the closure of G. Then $\mu(f_n) \to \overline{\mu}(X\setminus\overline{G})$, as was already proved. Clearly, $\overline{\mu}(X\setminus\overline{G})$ is $\mathcal{F}_{X\setminus G}$-measurable. Let us show that $\overline{\mu}(X\setminus\overline{G}) = \overline{\mu}(X) - \overline{\mu}(G)$. In fact, $\mu(1) - \mu(f_n) - \mu(\varphi_n) = \mu(1 - f_n - \varphi_n)$ and $\operatorname{supp}(1 - f_n - \varphi_n) \subset G_1$ for sufficiently large n, whatever $G_1 \supset G'$ we take. Thus (as was already proved), $\mu(1 - f_n - \varphi_n) \to 0$ in probability.

The relations

$$(\bigcup_{k=1}^n B_k)' \subset \bigcup_{k=1}^n B'_k, \qquad (\bigcap_{k=1}^n B_k)' \subset \bigcup_{k=1}^n B'_k,$$

$$(X - B)' = B', \qquad \lambda(\bigcup_{k=1}^n B'_k) \leq \sum_{k=1}^n \lambda(B'_k) = 0,$$

when $\lambda(B'_k) = 0$, $k = 1, \ldots, n$, imply that $\mathcal{A}_0$ is an algebra.

Further, if G_1, G_2 are open sets, $G_1 \cap G_2 = \varnothing$, $G_1, G_2, \in \mathcal{A}_0$, and $f_n \geq 0$, $f_n \uparrow I_{G_1}$, $\varphi_n \geq 0$, $\varphi_n \uparrow G_2$, $g_n \geq 0$, $g_n \uparrow I_{G_1} + I_{G_2}$, then $\operatorname{supp}(g_n - f_n - \varphi_n) \subset G$ for sufficiently large n, whatever open set $G \supset G'_1 \cup G'_2$ we take. Hence, $\mu(g_n) - \mu(f_n) - \mu(\varphi_n) \to 0$ in probability; i.e., $\overline{\mu}(G_1 \cup G_2) = \overline{\mu}(G_1) + \overline{\mu}(G_2)$.

If B_1 and $B_2 \in \mathcal{A}_0$ and $B_1 \cap B_2 = \varnothing$, we have

$$\operatorname{int} B_1 \cap \operatorname{int} B_2 = \varnothing,$$
$$\operatorname{int}(B_1 \cup B_2) \setminus \operatorname{int} B_1 \cup \operatorname{int} B_2 \subset B'_1 \subset B'_2.$$

Since $\lambda(B'_1 \cap B'_2) = 0$, we establish again that $\overline{\mu}(\operatorname{int} B_1) + \overline{\mu}(\operatorname{int} B_2) = \overline{\mu}(\operatorname{int}(B_1 \cup B_2))$, i.e., $\overline{\mu}(B_1) + \overline{\mu}(B_2) = \overline{\mu}(B_1 \cup B_2)$.

Finally, let $f \in C_X$. Then $\Delta_t = \{x : f(x) = t\}$ is a closed set for every $t \in [\inf f, \sup f]$. For every $\varepsilon > 0$, there exist only a finite number of t such that $\lambda(\Delta_t) > \varepsilon$. Hence, $\{t : \lambda(\Delta_t) > 0\}$ is at most countable. Thus, for every $\delta > 0$, one can find the points $t_0 \le \inf f < t_1 < \cdots < t_{n-1} < \sup f < t_n$ such that $|t_k - t_{k-1}| < \delta$ and $\lambda(\Delta_{t_k}) = 0$. The function $f_\delta(x) = \sum_{k=0}^{m-1} t_k I_{\{t_k \le f(x) < t_{k-1}\}}$ is $\mathcal{A}_0$-measurable and $|f_\delta(x) - f(x)| < \delta$. Therefore, every $f \subset C_X$ is a limit of $\mathcal{A}_0$-measurable functions and is thus measurable with respect to $\sigma(\mathcal{A}_0)$. But the least σ-algebra, with respect to which all $f \in C_X$ are measurable, is $\mathcal{B}$; i.e., $\sigma(\mathcal{A}_0) = \mathcal{B}$. Further, let $\varphi_{k,n}(x)$ be functions from C_X for which $\varphi_{k,n}(x) \uparrow I_{\{t_k < f(x) < t_{k+1}\}}$. Then $|\sum t_k \varphi_{k,n}(x) - \sum f(x)\varphi_{k,n}(x)| \le \delta$. Therefore,

$$\mathbf{P}\{|\sum t_k \mu(\varphi_{k,n}) - \mu(\sum f\varphi_{k,n})| > \rho\}$$
$$= \mathbf{P}\{|\mu(\frac{1}{\delta}\sum(t_k - f(x))\varphi_{k,n}(x))| > \rho/\delta\} \le \frac{1}{g(\rho/\delta)}\lambda(X) \quad (2.61)$$

for all n. Since for sufficiently large n, the support of the function $f(x) - \sum f(x)\varphi_{k,n}(x)$ lies in an arbitrary neighbourhood of $\bigcup_{k=1}^m \Delta_{t_k}$, and $\lambda(\bigcup_{k=1}^m \Delta_{t_k}) = 0$, we get $\mu(f) - \sum_{k=1}^m \mu(f\varphi_{k,n}) \to 0$ in probability. Using $\mu(\varphi_{k,n}) \to \overline{\mu}(\{x : t_k \le f < t_{k+1}\})$, we pass to the limit as $n \to \infty$ in (2.61) and obtain

$$\mathbf{P}\{|\sum t_k \overline{\mu}(\{x : t_k \le f < t_{k+1}\}) - \mu(f)| > \rho\} \le \frac{1}{g(\rho/\delta)}\lambda(X), \quad (2.62)$$

$\sum t_k \overline{\mu}(\{x : t_k \le f < t_{k+1}\}) \to \int f d\overline{\mu}$ as $\delta \to \infty$, and the r.h.s. of (2.62) tends to zero. (2.60) is thus proved.

Let us prove at last that $\overline{\mu}$ can be extended as a countably additive r.m.i.v. onto $\mathcal{B}$. By virtue of Theorem 30°, it suffices to show that for every sequence $B_n \subset \mathcal{A}_0$, for which $B_n \supset B_{n+1}$ and $\bigcap B_n = \varnothing$, $\overline{\mu}(\operatorname{int} B_n) \to 0$ in probability. The latter is satisfied if $\lambda(\operatorname{int} B_n) \to 0$, since $\lambda(\operatorname{int} B_n) \ge \mathbf{M}g(\overline{\mu}(\operatorname{int} B_n))$. But $\operatorname{int} B_n \supset \operatorname{int} B_{n+1}$, $\cap \operatorname{int} B_n = \varnothing$. Assume that $\lambda(\operatorname{int} B_n) \ge \varepsilon > 0$. Then we can find $f_n \in C_X$, $\|f_n\| \le 1$, such that $\operatorname{supp} f_n = F_n \subset \operatorname{int} B_n$ and $\mathbf{M}g(\mu(f_n)) \ge \frac{\varepsilon}{2}$. Since $\bigcap_n \bigcup_{k=n}^\infty F_k \subset \cap \operatorname{int} B_n = \varnothing$, one can find a subsequence F_{n_k} of mutually disjoint sets. We have thus found an infinite sequence of independent variables $\mu(f_{n_k})$ for which $\mathbf{M}g(\mu(f_{n_k})) \ge \frac{\varepsilon}{2}$ and $\mathbf{M}g(\sum \mu(f_{n_k})) \le \lambda(X)$. But this contradicts Lemma 45°.

□

CHAPTER 3

Processes with Independent Increments. General Properties

§3.1 Decomposition of a Process. Properties of Sample Functions

Let us consider random processes with independent increments $\xi(t)$ defined on R_+ and taking values in R, i.e., families of random variables $\{\xi(t), t \in R_+\}$ such that the following property is satisfied: for all $0 < t_0 < \cdots < t_n$ random variables $\xi(t_0)$, $\xi(t_1) - \xi(t_0), \ldots, \xi(t_n) - \xi(t_{n-1})$ are mutually independent.

In order to define finite-dimensional distributions of the process $\xi(t)$, it suffices to know the distribution of $\xi(0)$ and the distribution of an increment $\xi(t) - \xi(s)$ for $t > s$.

In fact, to define a joint distribution of several variables, it is sufficient to know their joint characteristic function. But

$$\mathbf{M}\exp\{i\sum_{k=0}^{n} s_k\xi(t_k)\} = \mathbf{M}e^{i(s_0+\cdots+s_n)\xi(0)}\mathbf{M}e^{i(s_0+\cdots+s_n)(\xi(t_0)-\xi(0))}$$
$$\times \mathbf{M}e^{i(s_1+\cdots+s_n)(\xi(t_1)-\xi(t_0))} \times \cdots \times \mathbf{M}e^{is_n(\xi(t_n)-\xi(t_{n-1}))} \qquad (3.1)$$

(we have used the representation $\xi(t_k) = \xi(0)+(\xi(t_0)-\xi(0))+(\xi(t_1)-\xi(t_0))+ \cdots + (\xi(t_k) - \xi(t_{k-1}))$ and the independence of the increments of the process $\xi(t)$).

Are characteristic functions $f_0(s) = \mathbf{M}e^{is\xi(0)}$ and $f_{t_1,t_2}(s) = \mathbf{M}e^{is(\xi(t_2)-\xi(t_1))}$, which define finite-dimensional increments by (3.1), arbitrary or not? Clearly, $f_0(s)$ can be arbitrary, but $f_{t_1,t_2}(s)$ for $t_1 < t_2 < t_3$ satisfies the evident relation

$$f_{t_1,t_2}(s)f_{t_2,t_3}(s) = f_{t_1,t_3}(s). \qquad (3.2)$$

The description of processes with independent increments is thus reduced analytically to the description of solutions of Eq. (3.2) for characteristic functions.

Below we present the simplest examples of random processes with independent increments.

Example 1. A nonrandom function $a(t)$ satisfies the above-mentioned definition and is a process with independent increments.

Example 2. A purely discrete process with independent increments.

It is constructed as follows. Let $T \subset R_+$ be some countable set, $T = \{t_k, k = 1, 2, \dots\}$; and let a set of mutually independent random variables $\{\xi_k^+, \xi_k^-, k = 1, 2, \dots\}$ be given such that the series $\sum \xi_{n_k}^+ I_{\{t_{n_k} < T\}}$ and $\sum \xi_{n_k}^- I_{\{t_{n_k} < T\}}$ converge for every sequence of nonrepeated indices, whatever T is taken. Then the following process is defined:

$$\xi(t) = \sum \xi_k^- I_{\{t_k \le t\}} + \sum \xi_k^+ I_{\{t_k < t\}}. \tag{3.3}$$

This is a process with independent increments, which is called *purely discrete.*

Definition. A process with independent increments $\xi(t)$ is called *stochastically continuous* if $\mathbf{P}\{|\xi(t) - \xi(s)| > \varepsilon\} \to 0$ as $t \to s$ for all $s \in R_+$ and $\varepsilon > 0$.

Our first problem is to decompose the process into a sum of nonrandom, purely discrete, and stochastically continuous components. We shall use the results of Chapter 2 on the decomposition of random measures with independent values.

Each process with independent increments can be associated with some r.m.i.v. on an arbitrary segment $[a, b] \subset R_+$. Let $0 \le a < b < \infty$. For each n, we choose the points $a = t_{n0} < t_{n1} < \cdots < t_{nn} = b$ so that $\max_k(t_{n\,k+1} - t_{nk}) \to 0$, and the sets $\Lambda_n = \{t_{nk}, k = 0, \dots, n\}$ are increasing. We denote by $\mathcal{A}_n$ a σ-algebra of subsets of $[a, b]$ with atoms $[t_{nk}, t_{n\,k+1}[$, $k < n - 1$, $[t_{n\,n-1}, t_{nn}]$. We set $\mu([t_{nk}, t_{n\,k+1}[) = \xi(t_{n\,k+1}) - \xi(t_{nk})$, $\mu([t_{n\,n-1}, t_{nn}]) = \xi(t_{nn}) - \xi(t_{n\,n-1})$. Clearly, we thereby define a r.m.i.v. on the algebra $\mathcal{A} = \bigcup_n \mathcal{A}_n$.

By virtue of Theorem 22°, Chapter 2, we have the following representation:

$$\mu = m + \mu_1 + \mu_2,$$

where m is a nonrandom finitely additive function of sets on $\mathcal{A}$, μ_1 is a stochastically continuous r.m.i.v., μ_2 is a purely discrete r.m.i.v., and μ_1 and μ_2 are independent.

1°. Theorem. *Every random process with independent increments $\xi(t)$ can be represented as follows:*

$$\xi(t) = m(t) + \xi_1(t) + \xi_2(t),$$

where $m(t)$ is a nonrandom function, $\xi_1(t)$ and $\xi_2(t)$ are independent, $\xi_1(t)$ is a stochastically continuous process, and $\xi_2(t)$ is a purely discrete one.

Proof. Denote $\Lambda = \bigcup_n \Lambda_n$, where $\Lambda_n \subset [a,b]$, Λ_n are finite sets by which the σ-algebras $\mathcal{A}_n$ are constructed. For all $t \in \Lambda$, the representation

$$\xi(t) - \xi(a) = m(t) - m(a) + (\xi_1(t) - \xi(a)) + (\xi_2(t) - \xi(a))$$

holds, where $m(t) - m(a) = m([a,t[)$ is a nonrandom function; $\xi_1(t) - \xi_1(a) = \mu_1([a,t[)$, where μ_1 is a stochastically continuous r.m.i.v.; $\xi_2(t) - \xi_2(a) = \mu_2([a,t[)$, where μ_2 is a purely discrete r.m.i.v. Let $t \overline{\in} \Lambda$, $t \in [a,b]$. Then one can find a sequence k_n such that $t_{n,k_n} \le t < t_{n,k_n+1}$. Theorem 22°, Chapter 2, implies that we can choose regular μ_1 and, moreover, with a finite number of ε-oscillations on Λ (see 19°, Chapter 2). Thus, $\xi_1(t_{n,k_n})$ and $\xi_1(t_{n,k_n+1})$ have limits as $n \to \infty$. By virtue of the stochastic continuity, $\xi_1(t_{n,k_n+1}) - \xi_1(t_{n,k_n}) = \mu_1([t_{n,k_n}, t_{n,k_n+1}[) \to 0$, and therefore $\xi_1(t_{n,k_n})$ and $\xi_1(t_{n,k_n+1})$ converge to the same limit. We denote it by $\xi_1(t)$; it is thus defined for all $t \in [a,b]$. Obviously, $\xi_1(t)$ is the process with independent increments on $[a,b]$, having no jumps of the second kind, because for every $\varepsilon > 0$ it has only a finite number of ε-oscillations. $\xi_1(t)$ is a stochastically continuous process. This has already been demonstrated for $t \overline{\in} \Lambda$. If $t \in \Lambda$ and $t = t_{n,k_n}$, we have $\xi_1(t-) = \lim \xi_1(t_{n,k_n-1})$, $\xi_1(t+) = \lim \xi_1(t_{n,k_n+1})$; but by virtue of the stochastic continuity of μ_1, we have $\xi_1(t_{n,k_n+1}) - \xi_1(t_{n,k_n-1}) = \mu_1([t_{n,k_n-1}, t_{n,k_n}[) + \mu_1([t_{n,k_n}, t_{n,k_n+1}[) \to 0$ in probability. In exactly the same way, we find that $\xi_1(t)$ is continuous in probability from the left for $t = b$ and from the right for $t = a$.

Consider now the construction of the process $\xi_2(t)$. Suppose that the measure μ_2 is concentrated on the fundamental sequences $([t_{n,k_n^{(i)}}, t_{n,k_n^{(i)}+1}[)$. Denote by t_i the common point of the intervals $[t_{n,k_n^{(i)}}, t_{n,k_n^{(i)}+1}[$.

Further we shall consider the points t_i to be contained in $\bigcup_n \Lambda_n$ (more precisely, we include them in this set if they were not in it before). Evidently, this procedure does not change the stochastically continuous component. For the purely discrete component a number of fundamental sequences, on which it is concentrated, can increase. All fundamental sequences, on which μ_2 is concentrated, will be of the form $[t_{n,k_n^{(i)}}, t_i[$ or $[t_i, t_{n,k_n^{(i)}+1}[$. Let us denote

$$\xi_i^- = \lim \mu([t_{n,k_n^{(i)}}, t_i[), \qquad \xi_i^+ = \lim \mu([t_i, t_{n,k_n^{(i)}+1}[),$$

$$\xi_2(t) = \sum_{t_k \le t} \xi_k^- + \sum_{t_k < t} \xi_k^+$$

(the sums on the r.h.s. are defined since the series of $\xi_{n_k}^{\pm}$ converge for any sequence of different numbers n_k).

Clearly, for all $s \in \Lambda$,

$$\xi_2(s) = \mu_2([a,s[).$$

Finally, consider the function $m^*(t) = \xi(t) - \xi_1(t) - \xi_2(t)$. It is defined for all t, and $m^*(t) = m(t) - m(a)$ for $t \in \Lambda$; $m^*(t)$ is a process with independent increments. Since $m^*(b) = m(b) - m(a)$ is nonrandom, $\mathbf{D}(m(b) - m(a)) = 0$ and $\mathbf{D}m^*(t) \leq \mathbf{D}(m(b) - m(a))$, the function $m^*(t)$ is nonrandom. Taking an arbitrary sequence $a_n \uparrow \infty$ ($a_0 = 0$), on each interval $[a_n, a_{n+1}[$, we can write $\xi(t) - \xi(a_n) = m(t) - m(a_n) + \xi_1(t) - \xi_1(a_n) + \xi_2(t) - \xi_2(a_n)$, $t \in [a_n, a_{n+1}[$ where $m(t)$ is a nonrandom function, ξ_1 and ξ_2 are independent processes; moreover, $\xi_1(t)$ is a stochastically continuous process, and $\xi_2(t)$ is a purely discrete one. This implies the proof.

□

Properties of Sample Functions of Purely Discrete Processes with Independent Increments

We shall consider random processes like (3.3). Clearly, for each t, $\xi(t)$ is defined by (3.3) with probability 1, so any modification of $\xi(t)$ is represented by the same formula.

2°. Theorem. *The process $\xi(t)$ has a modification $\xi'(t)$ for which the limits from the left and from the right exist at each point with probability 1. Furthermore,*

(a) *$\xi'(s)$ is continuous at the point t for all $t \overline{\in} \{t_1, t_2, \ldots\}$;*

(b) *$\xi'(t_k+) - \xi'(t_k) = \xi_k^+$, $\xi'(t_k) - \xi'(t_k-) = \xi_k^-$;*

(c) *the process $\xi'(t)$ is continuous from the right iff $\mathbf{P}\{\xi_k^+ = 0\} = 1$, $k = 1, 2, \ldots$;*

(d) *denote by $\mathcal{F}_t$ a σ-algebra generated by the values of $\xi(s)$ for $s \leq t$, and also let $\mathcal{F}_{t+} = \bigcap_{u>t} \mathcal{F}_u$; the process $\xi'(t)$ is continuous from the right iff $\mathcal{F}_{t+} = \mathcal{F}_t$ for all $t \geq 0$.*

Proof. We put

$$\xi_n(t) = \sum_{k \leq n} (\xi_k^- I_{\{t_k \leq t\}} + \xi_k^+ I_{\{t_k < t\}}).$$

$\xi_n(t)$ is the process with independent increments, having only $t_1, \ldots, t_n$ as the growth points; moreover, $\xi_n(t_i+) - \xi_n(t_i) = \xi_i^+$, $\xi_n(t_i) - \xi_n(t_i-) = \xi_i^-$, $i \leq n$. Whatever T we take, the series

$$\sum_{t_k < T} (\mathbf{P}\{|\xi_k^+| > c\} + \mathbf{P}\{|\xi_k -| > c\}),$$

$$\sum_{t_k < T} (|\mathbf{M}\xi_k^+ I_{\{|\xi_k^+| \leq c\}}| + |\mathbf{M}\xi_k^- I_{\{|\xi_k^-| \leq c\}}|),$$

$$\sum_{t_k < T} (\mathbf{D}\xi_k^+ I_{\{|\xi_k^+| \leq c\}} + \mathbf{D}\xi_k^- I_{\{|\xi_k^-| \leq c\}})$$

converge for $c > 0$, by virtue of Theorem 18°, Chapter 1. Hence, for every m, one can find n_m such that

$$\sum_{\substack{t_k<m,\\ k>n_m}} (\mathbf{P}\{|\xi_k^+| > c\} + \mathbf{P}\{|\xi_k^-| > c\} + |\mathbf{M}\xi_k^+ I_{\{|\xi_k^+|\le c\}}| + |\mathbf{M}\xi_k^- I_{\{|\xi_k^-|\le c\}}| + \mathbf{D}\xi_k^+ I_{\{|\xi_k^+|\le c\}} + \mathbf{D}\xi_k^- I_{\{|\xi_k^-|\le c\}}) \le 1/m^6.$$

Then

$$\begin{aligned}
\mathbf{P}\Big\{ &\sup_{\substack{t\le m,\\ n_{m+1}\ge k>n_m}} |\sum(\xi_k^+ I_{\{t_k<t\}} + \xi_k^- I_{\{t_k\le t\}})| > m^{-2}\Big\} \\
&\le \sum_{\substack{t_k<m,\\ k>n_m}} (\mathbf{P}\{|\xi_k^+| > c\} + \mathbf{P}\{|\xi_k^-| > c\}) \\
&+ \mathbf{P}\{\sup_{t\le T} |\sum_{k\ge n_m} [(\xi_k^+ I_{\{|\xi_k^+|\le c\}} - \mathbf{M}\xi_k^+ I_{\{|\xi_k^+|\le c\}}) I_{\{t_k<t\}} \\
&+ (\xi_k^- I_{\{|\xi_k^-|\le c\}} - \mathbf{M}\xi_k^- I_{\{|\xi_k^-|\le c\}}) I_{\{t_k\le t\}}]| \\
&> m^{-2} - \sum_{\substack{k>n_m,\\ t_k\le m}} (|\mathbf{M}\xi_k^+ I_{\{|\xi_k^+|\le c\}}| + |\mathbf{M}\xi_k^- I_{\{|\xi_k^-|\le c\}}|)\} \\
&\le m^{-6} + \frac{m^{-6}}{(m^{-2} + m^{-6})^2} = O(m^{-2}).
\end{aligned}$$

Therefore, a series

$$\xi_{n_1}(t) + (\xi_{n_2}(t) - \xi_{n_1}(t)) + \cdots + (\xi_{n_{m+1}(t)} - \xi_{n_m}(t)) \tag{3.4}$$

converges with probability 1 uniformly on each finite interval. Let us denote the sum of the series (3.4) by $\xi'(t)$. It is stochastically equivalent to $\xi(t)$ and has no jumps as a sum of the uniformly convergent sequence $\xi_{n_m}(t)$. $\xi'(t)$ is continuous wherever all $\xi_{n_m}(t)$ are continuous, i.e., everywhere except points $\{t_i, i = 1, 2, \dots\}$. And since $\xi_{n_m}(t_i+) - \xi_{n_m}(t_i) = \xi_i^+$, $\xi_{n_m}(t_i) - \xi_{n_m}(t_i-) = \xi_i^-$ for $n_m \ge i$, statement (b) is valid for $\xi'(t)$; (c) clearly follows from (b). Further, $\mathcal{F}_t$ coincides with the σ-algebra generated by the variables $\{\xi_k^+, t_k < t; \xi_k^-, t_k \le t\}$, and $\mathcal{F}_{t_+}$ coincides with the σ-algebra generated by $\{\xi_k^+, t_k \le t; \xi_k^-, t_k \le t\}$, $\mathcal{F}_t = \mathcal{F}_{t_+}$ iff $\xi_k^+ = 0$ for all k.

□

Remark 1. Let us denote $f_k^+(z) = \mathbf{M}e^{iz\xi_k^+}$ and $f_k^-(s) = \mathbf{M}e^{is\xi_k^-}$. Then the characteristic function of the increment of the process is given by

$$\psi(z, s, t) = \mathbf{M}e^{iz(\xi(t)-\xi(s))} = \prod_{s<t_k\le t} f_k^-(z) \prod_{s\le t_k<t} f_k^+(z).$$

Remark 2. The process $\xi(t)$ has a modification continuous from the right iff it is stochastically continuous from the right, i.e., $\xi(t+h) \to \xi(t)$ in probability as $h \downarrow 0$ for all $t > 0$. This follows from the fact that $\psi(z, s, s+) = f_k^+(z)$ for $s = t_k$, so $\psi(z, s, s+) = 1$ iff $f_k^+(z) = 1$ for all k.

Remark 3. The process $\xi'(t_+)$ is also a purely discrete process with independent increments; moreover, $\xi'(t_+)$ coincides with $\xi'(t)$ everywhere except (possibly) a countable set of points $\{t_k\}$

$$\xi'(t) = \sum_{t_k \le t} \xi_k, \quad \text{where} \quad \xi_k = \xi_k^+ + \xi_k^-.$$

Remark 4. Construction of a purely discrete process, realized in 2°, makes possible the following decomposition of $\xi_1(t) + \xi_2(t)$ into a purely discrete component and a stochastically continuous one. We put

$$\begin{aligned} \xi_2^{(n)}(t) &= \sum_{k=1}^{n} (\xi_k^+ I_{\{t_k < t\}} + \xi_k^- I_{\{t_k \le t\}}), \\ \xi_1^{(n)}(t) &= \sum_{t_i^{(n)} \le t} [\xi(t_i^{(n)}-) - \xi(t_{i-1}^{(n)}+)] + \xi(t-) - \xi(\max\{t_i^{(n)}, t_i^{(n)} < t\}). \end{aligned} \tag{3.5}$$

Here $t_1^{(n)} < t_2^{(n)} < \cdots < t_n^{(n)}$ is a permutation of $t_1, \ldots, t_n$. Then, in the sense of convergence in probability,

$$\xi_1(t) = \lim_{n \to \infty} \xi_1^{(n)}(t), \qquad \xi_2(t) = \lim_{n \to \infty} \xi_2^{(n)}(t).$$

Properties of Sample Functions of Stochastically Continuous Processes

Let $\xi(t)$ be a stochastically continuous process with independent increments, defined in R^+. Henceforth, we shall need some general properties of processes with independent increments.

3°. Lemma. *If $\xi(t)$ is stochastically continuous on the closed interval $[a, b]$, then*

(1) *it is bounded in probability, i.e.,*

$$\lim_{c \to \infty} \sup_{t \in [a,b]} \mathbf{P}\{|\xi(t)| > c\} = 0; \tag{3.6}$$

(2) *it is uniformly stochastically continuous; for every $\varepsilon > 0$,*

$$\lim_{h \to 0} \sup_{\substack{|t-s| \le h, \\ t,s \in [a,b]}} \mathbf{P}\{|\xi(t) - \xi(s)| > \varepsilon\} = 0. \tag{3.7}$$

Proof. If (3.6) is not satisfied, then there exist $\delta > 0$ and the sequence of points $t_n \in [a, b]$ such that

$$\mathbf{P}\{|\xi(t_n)| > n\} \geq \delta. \tag{3.8}$$

We can assume that t_n converges to some point $t_0 \in [a, b]$. Then $\xi(t_n) \to \xi(t_0)$ in probability, which contradicts (3.8). If (3.7) is not valid, one can find $\varepsilon > 0$, $\delta > 0$, and two sequences t_n, s_n such that $t_n - s_n \to 0$ and

$$\mathbf{P}\{|\xi(t_n) - \xi(s_n)| > \varepsilon\} > \delta. \tag{3.9}$$

We can assume that $t_n \to t_0 \in [a, b]$. Then $\xi(t_n) \to \xi(t_0)$, $\xi(s_n) \to \xi(t_0)$, and $\xi(t_n) - \xi(s_n) \to 0$ in probability. But the latter is impossible if (3.9) is valid. □

We shall need an estimate for the number of ε-oscillations of the process $\xi(t)$ given on a finite or countable set.

4°. Lemma. *Let $\xi(t)$ be a process with independent increments, given on a finite set $\Lambda = \{t_1, \dots, t_n\}$, and $\mathbf{P}\{|\xi(t_n) - \xi(t_k)| > \varepsilon\} \leq \alpha < 1/2$ for all k. If ν is a number of 4ε-oscillations of $\xi(t)$ on Λ, then*

$$\mathbf{M}\nu \leq \frac{2\alpha}{1 - 2\alpha}. \tag{3.10}$$

Proof. The event $\{\nu \geq m\}$ involves one of the following events:

$$A_k = \{|\xi(t_2) - \xi(t_1)| \leq 2\varepsilon, \dots, |\xi(t_{k-1}) - \xi(t_1)| \leq 2\varepsilon, |\xi(t_k) - \xi(t_1)| > 2\varepsilon,$$
$$\xi(t) \text{ has at least } m-1 \text{ } 4\varepsilon\text{-oscillations on the set } \{t_k, \dots, t_n\}\}.$$

The events A_k are mutually disjoint, so

$$\mathbf{P}\{\nu \geq m\} = \sum_k \mathbf{P}\{A_k\}$$
$$= \sum_k \mathbf{P}\{|\xi(t_2) - \xi(t_1)| \leq 2\varepsilon, \dots, |\xi(t_{k-1}) - \xi(t_1)| \leq 2\varepsilon, |\xi(t_k) - \xi(t_1)| > 2\varepsilon\}$$
$$\times \mathbf{P}\{\xi(t) \text{ has at least } m-1 \text{ } 4\varepsilon\text{-oscillations on } \{t_k, \dots, t_n\}\}$$
$$\leq \mathbf{P}\{\nu \geq m-1\}\mathbf{P}\{\sup_k |\xi(t_k) - \xi(t_1)| > 2\varepsilon\}$$
$$\leq \mathbf{P}\{\nu \geq m-1\} \cdot \frac{1}{1-\alpha}\mathbf{P}\{|\xi(t_m) - \xi(t_1)| > 4\varepsilon\} \leq \mathbf{P}\{\nu \geq m-1\}\frac{\alpha}{1-\alpha}$$

(we have used 9°, Chapter 1). Hence,

$$\mathbf{P}\{\nu \geq m\} \leq \left(\frac{\alpha}{1-\alpha}\right)^m,$$

$$\mathbf{M}\nu \leq \sum_{m=1}^{\infty} \left(\frac{\alpha}{1-\alpha}\right)^m = \frac{\frac{\alpha}{1-\alpha}}{1 - \frac{\alpha}{1-\alpha}} = \frac{\alpha}{1-2\alpha}.$$

□

Corollary. *Let Λ be a countable set, $\Lambda \in [a,b]$, $a,b \in \Lambda$. Assume also that the process with independent increments $\xi(t)$ is given on Λ. If for all $t \in \Lambda$,*

$$\mathbf{P}\{|\xi(b) - \xi(t)| > \varepsilon\} \leq \alpha < 1/2,$$

then (3.10) holds for the number of 4ε-oscillations of $\xi(t)$ denoted by ν.

5°. Theorem. *For every stochastically continuous process with independent increments $\xi(t)$, there exists a modification $\xi'(t)$ that is continuous from the right and has no discontinuities of the second kind.*

Proof. Let Λ be a set of rational numbers. For every $T > 0$, the process $\xi(t)$ has a finite number of ε-oscillations for $t \in \Lambda \cap [0,T]$, whatever $\varepsilon > 0$ is taken. In fact, according to Lemma 1°, for given $\varepsilon > 0$ we can find rational h such that for $|s-t| < h$, $s,t < T$, we have $\mathbf{P}\{|\xi(t) - \xi(s)| > \varepsilon/4\} \leq \frac{1}{3}$. Then, denoting by $\nu_\varepsilon(S)$ a number of ε-oscillations of the process $\xi(t)$ on a set S, we get

$$\mathbf{M}\nu_\varepsilon(\Lambda \cap [kh,(k+1)h] \cap [0,T]) \leq \frac{2/3}{1-2/3} = 2$$

by virtue of Corollary 4°. It is obvious that

$$\nu_\varepsilon([0,T] \cap \Lambda) \leq \sum_{kh<T} \nu_\varepsilon(\Lambda \cap [kh,(k+1)h] \cap [0,T]) + T/h + 1,$$

and thus the variable on the left is finite for every $T > 0$ and $\varepsilon > 0$. Hence, there exists

$$\xi'(t) = \lim_{\substack{s \to t \\ s \in \Lambda \cap]t,\infty[}} \xi(s)$$

(with probability 1 for all t simultaneously). One can easily see that $\xi'(t)$ is continuous from the right. If $t_n \in \Lambda$, $t_n > t$, and $t_n \downarrow t$, we have $\xi'(t) = \lim \xi(t_n)$. In addition, $\xi(t_n) \to \xi(t)$ in probability, since $\xi(t)$ is stochastically continuous. Therefore, $\mathbf{P}\{\xi'(t) = \xi(t)\} = 1$; i.e., $\xi'(t)$ is a modification of $\xi(t)$. $\xi'(t)$ and $\xi(t)$ coincide on Λ with probability 1. Hence, on Λ, $\xi'(t)$ has a finite number of ε-oscillations on every finite segment. Thus, for all $t > 0$, there exists a limit $\lim_{s\in\Lambda, s\uparrow t} \xi'(s)$ which coincides with $\xi'(t-)$, since $\xi'(t)$ is continuous from the right.

□

Conditions of Continuity

6°. Theorem. *Let $\xi(t)$, $t \in R_+$, be a process with independent increments. It has a continuous modification iff for every $\varepsilon > 0$ and T the following condition*

is satisfied: whatever $0 = t_{n0} < \cdots < t_{nn} = T$ ***we take such that*** $\max_k \Delta t_{nk} \to 0$, ***where*** $\Delta t_{nk} = t_{n\,k+1} - t_{nk}$, ***we have***

$$\lim_{n\to\infty} \sum_{k=1}^{n} \mathbf{P}\{|\xi(t_{nk}) - \xi(t_{nk-1})| > \varepsilon\} = 0. \tag{3.11}$$

Proof. Let $\xi(t)$ be a continuous process. Then it is uniformly continuous on $[0,T]$; and if t_{nk} satisfy the above-mentioned conditions, then for every $\varepsilon > 0$,

$$\lim_{n\to\infty} \mathbf{P}\{\max_k |\xi(t_{nk}) - \xi(t_{nk-1})| \le \varepsilon\} = 1, \tag{3.12}$$

since

$$\max |\xi(t_{nk}) - \xi(t_{nk-1})| \to 0$$

with probability 1. (3.12) is equivalent to the next condition

$$\begin{aligned} 1 &= \lim_{n\to\infty} \prod_{k=1}^{n} \mathbf{P}\{|\xi(t_{nk} - \xi(t_{nk-1}))| \le \varepsilon\} \\ &= \lim_{n\to\infty} \prod_{k=1}^{n} (1 - \mathbf{P}\{|\xi(t_{nk} - \xi(t_{nk-1}))| > \varepsilon\}). \end{aligned} \tag{3.13}$$

The inequality $1 + x \le e^x$ yields

$$\begin{aligned} &\lim_{n\to\infty} \prod_{k=1}^{n} (1 - \mathbf{P}\{|\xi(t_{nk} - \xi(t_{nk-1}))| > \varepsilon\}) \\ &\le \lim_{n\to\infty} \exp\left\{ -\sum_{k=1}^{n} \mathbf{P}\{|\xi(t_{nk}) - \xi(t_{nk-1})| > \varepsilon\} \right\} \le 1. \end{aligned}$$

Here, taking (3.13) into account, we obtain (3.11). The necessity of the condition of the theorem is proved.

Let us prove the sufficiency. Suppose that $\xi'(t)$ is a modification of the process $\xi(t)$ continuous from the right with limits from the left (its existence follows from the fact that (3.11) clearly implies the stochastic continuity of $\xi(t)$). Denote by $\nu_\varepsilon(T)$ a number of points $t \le T$ for which $|\xi'(t) - \xi'(t-)| > \varepsilon$. Obviously,

$$\nu_\varepsilon(T) \le \varliminf_{n\to\infty} \sum_{k=1}^{n} I_{\{|\xi'(t_{nk}) - \xi'(t_{nk-1})| > \varepsilon\}};$$

and by Fatou's theorem,

$$\begin{aligned} \mathbf{M}\nu_\varepsilon(T) &\le \mathbf{M} \varliminf_{n\to\infty} \sum_{k=1}^{n} I_{\{|\xi'(t_{nk}) - \xi'(t_{nk-1})| > \varepsilon\}} \\ &\le \varliminf_{n\to\infty} \mathbf{M} \sum_{k=1}^{n} I_{\{|\xi'(t_{nk}) - \xi'(t_{nk-1})| > \varepsilon\}} \\ &= \varliminf_{n\to\infty} \sum_{k=1}^{n} \mathbf{P}\{|\xi'(t_{nk}) - \xi'(t_{nk-1})| > \varepsilon\} = 0. \end{aligned}$$

Thus, $\nu_\varepsilon(T) = 0$ with probability 1 for all $\varepsilon > 0$ and $T > 0$. And this yields the continuity of $\xi'(t)$.

□

Extension of the Results to the Vector Case

Let X be a finite-dimensional space. If $\xi(t)$ is a process with independent increments with values in X, then each coordinate $\xi^i(t)$ is a one-dimensional process with independent increments. Thus, there exists a nonrandom process $a^i(t)$ such that $\xi^i(t) - a^i(t)$ has a modification without discontinuities of the second kind and can be decomposed into the sum of a stochastically continuous process and a purely discrete one. Denote by $a(t)$ a nonrandom function with coordinates $a^i(t)$. Then $\xi(t) - a(t)$ has a modification without discontinuities of the second kind. Let $\{t_1, t_2, \dots\}$ be a set of points of stochastic discontinuities of $\xi(t)$ (it is countable, because it is the union of sets of stochastic discontinuity points, corresponding to each coordinate). Introduce random variables $\tilde{\xi}_k^+ = \xi(t_k+) - \xi(t_k)$, $\tilde{\xi}_k^- = \xi(t_k) - \xi(t_k-)$. If coordinates of these random vectors are $\tilde{\xi}_k^{+i}$ and $\tilde{\xi}_k^{-i}$, respectively, then the random processes $\xi_2^i(t) = \sum_k (\xi_k^{+i} I_{\{t_k<t\}} + \xi_k^{-i} I_{\{t_k<t\}})$ are defined. The process $\xi_2(t)$ with the coordinates $\xi_2^i(t)$ will be a purely discrete process with independent increments. $\xi^i(t) - \xi_2^i(t) = \xi_1^i(t)$ are one-dimensional stochastically continuous processes with independent increments; thus, $\xi_1(t)$ with the coordinates $\xi_1^i(t)$ is a stochastically continuous process with independent increments. To show that $\xi_1(t)$ and $\xi_2(t)$ are independent, we use Remark 4.2°. It implies that $\xi_1(t) = \lim_{n\to\infty} \xi_1^{(n)}(t)$, $\xi_2(t) = \lim_{n\to\infty} \xi_2^{(n)}(t)$, and that $\xi_1^{(n)}(t)$ and $\xi_2^{(n)}(t)$ are defined by (3.5) (except with vector variables instead of real-valued ones). In order to show the independence of $\xi_1^{(n)}(t)$ and $\xi_2^{(n)}(t)$, we represent them as $\xi_i^{(n)}(t) = \lim_{n\to\infty} \xi_i^{(n,h)}(t)$, $i = 1, 2,,$ where

$$\xi_2^{(n,h)}(t) = \sum_{k=1}^{n} ([\xi(t_k) - \xi(t_k - h)] I_{\{t_k \le t\}} + [\xi(t_k + h) - \xi(t_k)]) I_{\{t<t_k\}})$$

$$\xi_1^{(n,h)}(t) = \xi(t) - \sum_{\substack{t_k<t,\\ k\le n}} [\xi((t_k + h) \wedge t) - \xi(t_k - h) \vee 0].$$

The processes $\xi_1^{(n,h)}(t)$ and $\xi_2^{(n,h)}(t)$ are constructed in terms of the increments of $\xi(t)$ on disjoint intervals and are thus independent.

Hence, $\xi_1^{(n)}(t)$ and $\xi_2^{(n)}(t)$ are independent; therefore, $\xi_1(t)$ and $\xi_2(t)$ are also independent. Thus, the following theorem is proved:

7°. Theorem. *Let $\xi(t)$ be a process with independent increments on R_+ with values in a finite-dimensional Euclidean space X. Then it can be represented as*

$$\xi(t) = a(t) + \xi_1(t) + \xi_2(t),$$

where $a(t)$ is a nonrandom function, $\xi_1(t)$ and $\xi_2(t)$ are mutually independent process with independent increments, and moreover, $\xi_1(t)$ is a stochastically continuous process and $\xi_2(t)$ is a purely discrete one. The process $\xi_1(t)$ has a modification continuous from the right with limits from the left. The process $\xi_2(t)$ has a modification without discontinuities of the second kind.

Now consider a stochastically continuous process with values in a separable Banach space X.

8°. Theorem. *For a stochastically continuous process $\xi(t)$ with independent increments and with values in X, there exists a modification continuous from the right with limits from the left. This process has continuous modification iff for every $T > 0$ and $\varepsilon > 0$,*

$$\lim \sum_{k=0}^{n-1} \mathbf{P}\{|\xi(t_{nk}) - \xi(t_{nk-1})|_X > \varepsilon\} = 0,$$

whatever $0 = t_{n0} < t_{n1} < \cdots < t_{nn}$ we take for which $\max_k \Delta t_{nk} \to 0$ as $n \to \infty$.

The proof of Theorem 8° is completely analogous to the proof of Theorems 5° and 6°, because the proof of Lemmas 3° and 4° (on which the proof of Theorems 5° and 6° is based) remains unchanged in the case of a Banach space.

§3.2 Stochastically Continuous Processes

Let us consider stochastically continuous process with independent increments $\xi(t)$, defined on R_+ with values in a finite-dimensional Euclidean space X. Moreover, we assume that $\xi(t)$ is continuous from the right and has limits from the left. Denote by $\mathcal{B}_\varepsilon$ a ring of Borel sets in X situated outside the sphere $S_\varepsilon = \{x : |x| < \varepsilon\}$; also denote $\mathcal{B}_0 = \cup \mathcal{B}_\varepsilon$. For each $A \in \mathcal{B}_0$, the following processes are defined:

$$\xi_t(A) = \sum I_{\{\xi(s)-\xi(s-)\in A\}} I_{\{s\le t\}}(\xi(s) - \xi(s-)), \tag{3.14}$$

$$\nu_t(A) = \sum I_{\{\xi(s)-\xi(s-)\in A\}} I_{\{s\le t\}}. \tag{3.15}$$

Summation in these formulas is carried out over the discontinuity points of the process $\xi(t)$. We introduce the σ-algebras $\mathcal{F}_t$ generated by the variables $\{\xi(s), s \le t\}$, and $\mathcal{F}_t^s$, $s \le t$, generated by the variables $\{\xi(u) - \xi(s), s \le u \le t\}$.

9°. Theorem. *$\xi_t(A)$ and $\nu_t(A)$ are stochastically continuous processes with independent increments. Moreover, $\xi_t(A)$ and $\nu_t(A)$ are $\mathcal{F}_t$-measurable, whereas $\nu_t(A) - \nu_s(A)$ and $\xi_t(A) - \xi_s(A)$ for $s < t$ are $\mathcal{F}_t^s$-measurable.*

Proof. To prove $\mathcal{F}_t$-measurability of $\xi_t(A)$, it suffices to obtain a formula that represents $\xi_t(A)$ in terms of the process $\xi(t)$. Let us introduce a process $\hat{\xi}_\delta(t)$ by

$$\hat{\xi}_\delta(t) = \lim_{n\to\infty} \sum \psi_\delta(\xi(\tfrac{k}{2^n}) - \xi(\tfrac{k-1}{2^n})), \tag{3.16}$$

where $\psi_\delta(x) = I_{\{|x|>\delta\}}(x - x \cdot \frac{\delta}{|x|})$. The existence of the limit (3.16) follows from the fact that for every function $x(t)$ without discontinuities of the second kind, which is continuous from the right, we have

$$\lim_{n\to\infty} \sum \psi_\delta(x(\tfrac{k}{2^n}) - x(\tfrac{k-1}{2^n})) = \sum_{s\le t} \psi_\delta(x(s) - x(s-)) \tag{3.17}$$

(the sum on the r.h.s. is taken over jump points of the function $x(s)$, which form at most a countable set; a number of nonzero terms in this sum is finite). If $\tau_1, \tau_2, \ldots, \tau_l$ are all the points from $[0, t]$ for which $|\xi(\tau_i) - \xi(\tau_i-)| > \delta$, we have

$$\hat{\xi}_\delta(t) = \sum_{i=1}^{l} \psi_\delta(\xi(\tau_i) - \xi(\tau_{i-1})). \tag{3.18}$$

Formula (3.16) implies that $\hat{\xi}_\delta(t)$ is a process with independent increments. Moreover, $\hat{\xi}_\delta(t)$ is $\mathcal{F}_t$-measurable, whereas $\hat{\xi}_\delta(t) - \hat{\xi}_\delta(s)$ is $\mathcal{F}_t^s$-measurable. Now let us show how to construct the process $\xi_A(t)$ in terms of $\hat{\xi}_\delta(t)$ if $\delta < \varepsilon$. Since $\tau_1, \ldots, \tau_l$ are all the points in $[0, t]$ for which $|\xi(s) - \xi(s-)| > \delta$ (we use (3.18)), we get

$$\xi_t(A) = \sum_{i=1}^{l} I_{\{\xi(\tau_i)-\xi(\tau_i-)\in A\}}(\xi(\tau_i) - \xi(\tau_i-)).$$

If $\frac{k-1}{n} < t < \frac{k}{n}$ and there are no other points τ_j in the segment $[\frac{k-1}{n}, \frac{k}{n}]$, we obtain

$$\psi_\delta(\xi(\tau_i) - \xi(\tau_i-)) = \hat{\xi}_\delta(\tfrac{k}{n}t) - \hat{\xi}_\delta(\tfrac{k-1}{n}t),$$
$$\xi(\tau_i) - \xi(\tau_i-) = \varphi_\delta(\hat{\xi}_\delta(\tfrac{k}{n}t) - \hat{\xi}_\delta(\tfrac{k-1}{n}t)),$$

where $\varphi_\delta(x) = \frac{\delta+|x|}{|x|}x$ (this function possesses the following property: $\varphi_\delta(\psi_\delta(x)) = x$ for $|x| > \delta$).

Consequently,

$$\xi_t(A) = \lim_{n\to\infty} \sum_{k=1}^{n} I_{\{\varphi_\delta(\hat{\xi}_\delta(kt/n) - \hat{\xi}_\delta((k-1)t/n))\in A\}} \varphi_\delta(\hat{\xi}_\delta(\tfrac{k}{n}t) - \hat{\xi}_\delta(\tfrac{k-1}{n}t)).$$

This formula implies $\mathcal{F}_t$-measurability of $\xi_t(A)$. In exactly the same way, we can represent the increment $\xi_t(A) - \xi_s(A)$ $(t > s)$ in terms of $\hat{\xi}_\delta(t_1) - \hat{\xi}_\delta(t_2)$, where $t_1, t_2 \in [s, t]$. The above statements are now evident for the process

$\nu_t(A)$ counting the jumps of $\xi(t)$. Stochastic continuity of $\xi_t(A)$ ($\nu_t(A)$) follows from the fact that the discontinuity points of $\xi_t(A)$ are also the discontinuity points of $\xi(t)$.

□

10°. *If $A_1, A_2, \dots, A_k \in \mathcal{B}_0$ and $A_i \cap A_j = \varnothing$ for $i \neq j$, then $\nu_t(A_1), \dots, \nu_t(A_k)$ are mutually independent Poisson processes.*

Proof. It is sufficient to prove this property for the case of $t \in [0, T]$, where T is arbitrarily large. Let us choose an arbitrary countable dense set Λ on $[0, T]$ and consider an algebra $\mathcal{A}_0$ generated by the intervals $[t, T[$, $t \in \Lambda$ (we assume that $0 \in \Lambda$). We introduce the random measures with independent values ν^i defined by the equalities $\nu^i([s,t[) = \nu_t(A_i) - \nu_s(A_i)$, where $i = 1, \dots, k$, and $[s,t[\in \mathcal{A}_0$. The variables $\nu^i([s,t[)$ are $\mathcal{F}_t^s$-measurable, whereas the σ-algebras $\mathcal{F}_{t_m}^{s_m}$ are independent if the intervals $[s_m, t_m[$ are mutually disjoint. Clearly, ν^i are counting measures concentrated on different fundamental sequences, because the processes $\nu_t(A_i)$ have no common jump points. Thus, the statement of the theorem arises from 9°, Chapter 2.

□

Corollary 1. *For every t, $\nu_t(A)$ is a countably additive Poisson measure with independent values.*

In fact, countable additivity of $\nu_t(A)$ on $\mathcal{B}_0$ follows from its construction, while independence of values of $\nu_t(A)$ on disjoint sets is a consequence of 10°.

Consider a ring of sets $\mathcal{B}_0 \otimes \mathcal{B}^*_{R_+}$, where $\mathcal{B}^*_{R_+}$ is a ring of bounded Borel sets on R_+. Define on $\mathcal{B}_0 \otimes \mathcal{B}^*_{R_+}$ a random function of sets

$$\tilde{\nu}(C) = \sum I_{\{(\xi(s)-\xi(s-),s)\in C\}} \tag{3.19}$$

(summation here is taken over the discontinuity points of $\xi(s)$). Then (3.19) implies that $\tilde{\nu}(C)$ is a countably additive function of the set.

Corollary 2. *$\tilde{\nu}(C)$ is a countably additive Poisson r.m.i.v.*

In $\mathcal{B}_0$ and $\mathcal{B}^*_{R_+}$ let us choose countable rings of sets $\mathcal{C}_0$ and $\mathcal{C}^*_{R_+}$ such that their σ-closures contain $\mathcal{B}_0$ and $\mathcal{B}^*_{R_+}$, respectively. We assume that $\mathcal{C}^*_{R_+}$ is generated by a countable system of half-intervals like $[s,t[$. If $A \in \mathcal{C}_0$, $[s,t[\in \mathcal{C}^*_{R_+}$, then $\tilde{\nu}(A \times [s,t[) = \nu_t(A) - \nu_s(A)$. Theorem 10° implies that $\tilde{\nu}(A_i \times [s_i,t_i[) = \nu_{t_i}(A_i) - \nu_{s_i}(A_i)$ are mutually independent if the corresponding sets $A_1 \times [s_1,t_1[, \dots, A_k \times [s_k,t_k[$ are disjoint. Thus, $\tilde{\nu}$ is a random measure with independent values on $\mathcal{C}_0 \otimes \mathcal{C}^*_{R_+}$; and hence, this is true for the extension of $\tilde{\nu}$ onto $\mathcal{B}_0 \otimes \mathcal{B}^*_{R_+}$ (see Section 2.3).

11°. Theorem. *For every $A \subset \mathcal{B}_0$, we have*

$$\xi_t(A) = \int_A x\nu_t(dx). \tag{3.20}$$

Proof. Let A be a bounded set from $\mathcal{B}_0$. If $A = \cup_{k=1}^n A_{nk}$, where $A_{nk} \in \mathcal{B}_0$ are mutually disjoint, and $\lambda_n = \max_k \operatorname{diam} A_{nk} \to 0$ as $n \to \infty$, then choosing $x_{nk} \in A_{nk}$, we get

$$\xi_t(A) = \sum_{k=1}^n \xi_t(A_{nk}) = \sum_{k=1}^n \nu_t(A_{nk})x_{nk} + \sum_{k=1}^n (\xi_t(A_{nk}) - \nu_t(A_{nk}))x_{nk}. \tag{3.21}$$

A variable $\xi_t(A_{nk})$ is a sum of $\nu_t(A_{nk})$ variables, with the difference between each t and x_{nk} not greater than $\operatorname{diam} A_{nk}$. Thus,

$$|\sum_{k=1}^n (\xi_t(A_{nk}) - \nu_t(A_{nk})x_{nk})| \le \sum \nu_t(A_{nk}) \operatorname{diam} A_{nk} \le \lambda_n \nu_t(A).$$

The second sum on the r.h.s. of (3.21) tends to zero as $n \to \infty$, while the first sum converges to the integral on the r.h.s. of (3.20). For bounded A, (3.20) is thus proved. If A is unbounded, we have by definition (see 39°, Chapter 2)

$$\int_A x\nu_t(dx) = \lim_{n\to\infty} \int_{A_n} x\nu_t(dx),$$

where A_n is a sequence of bounded sets from $\mathcal{B}_0$ and $A_n \uparrow A$. But in this case, $\xi_t(A) = \lim_{n\to\infty} \xi_t(A_n)$ (this follows from (3.14)), and $\xi_t(A) = \xi_t(A_n)$ for sufficiently large n. Applying (3.20) to the set A_n and passing to the limit as $n \to \infty$, we complete the proof of the theorem.

□

Now consider analytic characteristics of $\nu_t(A)$ and $\xi_t(A)$. Denote by $\Pi_t(A)$ a parameter of the Poisson distribution of the variable $\nu_t(A)$

$$\mathbf{M}\nu_t(A) = \Pi_t(A), \quad A \in \mathcal{B}_0.$$

This means that

$$\mathbf{P}\{\nu_t(A) = k\} = (k!)^{-1}(\Pi_t(A))^k \exp\{-\Pi_t(A)\}.$$

Since $\nu_t(A)$ does not decrease with t, $\Pi_t(A)$ is also a nondecreasing function of t. And the stochastic continuity of $\nu_t(A)$ involves the continuity of $\Pi_t(A)$.

12°. Theorem. *The variable $\xi_t(A)$ has the characteristic function*

$$\mathbf{M}\exp\{i(z,\xi_t(A))\} = \exp\{\int_A (e^{i(z,x)} - 1)\Pi_t(A)\}. \tag{3.22}$$

If A is a bounded set from $\mathcal{B}_0$, we have

$$\begin{aligned} \mathbf{M}(z,\xi_t(A)) &= \int (z,x)\Pi_t(dx), \\ \mathbf{D}(z,\xi_t(A)) &= \int (z,x)^2\Pi_t(dx). \end{aligned} \tag{3.23}$$

Proof. Let A be a bounded set from $\mathcal{B}_0$. The proof of Theorem 11° implies that

$$(z,\xi_t(A)) = \lim_{n\to\infty} \sum_{k=1}^{n} \nu_t(A_{nk})(x_{nk},z), \tag{3.24}$$

in the sense of convergence in probability; here $\cup A_{nk} = A$, $A_{nk} \cap A_{nj} = \varnothing$ for $k \neq j$, and $\sup_k \operatorname{diam} A_{nk} \to 0$ as $n \to \infty$.

$$\begin{aligned} \mathbf{M}e^{i(z,\xi_t(A))} &= \lim_{n\to\infty} \mathbf{M}\exp\{i\sum_{k=1}^{n} \nu_t(A_{nk})(x_{nk},z)\} \\ &= \lim_{n\to\infty} \prod_{k=1}^{n} \mathbf{M}\exp\{i(x_{nk},z)\nu_t(A_{nk})\} \\ &= \lim_{n\to\infty} \prod_{k=1}^{n} \exp\{(e^{i(z,x_{nk})} - 1)\Pi_t(A_{nk})\} \\ &= \lim_{n\to\infty} \exp\{\sum_{k=1}^{n}(e^{i(z,x_{nk})} - 1)\Pi_t(A_{nk})\} \\ &= \exp\{\int_A (e^{i(z,x)} - 1)\Pi(dx)\}. \end{aligned}$$

Thus, (3.22) is ascertained for bounded sets from $\mathcal{B}_0$ and can be obtained for the others by the limit transition on A.

If $A \in \mathcal{B}_0$, then $\Pi_t(A)$ is bounded; and if A is bounded, we have $|(z,x)| \leq |z|\sup\{|x|, x \in A\} < \infty$. Thus,

$$\mathbf{M}e^{is(z,\xi_t(A))} = \exp\{\int_A (e^{is(z,x)} - 1)\Pi(dx)\}$$

is an integral analytic function of s. Using the formulas

$$\mathbf{M}\xi = \frac{1}{i}\cdot\frac{d}{ds}\ln \mathbf{M}e^{is\xi}\Big|_{s=0}, \qquad \mathbf{D}\xi = -\frac{d^2}{ds^2}(-\ln \mathbf{M}e^{is\xi})\Big|_{s=0},$$

we establish the validity of (3.23)

$$(\ln \mathbf{M}e^{is(z,\xi_t(A))} = \int_A (e^{is(z,x)} - 1)\Pi_t(dx)).$$

Decomposition of a Process into Discontinuous and Continuous Components

13°. Lemma. *Let $U_c = \{x : |x| > c\}$, $c > 0$, and $\xi_t(U_c) = \xi_c(t)$. If $\Pi_T(\{x : |x| = c\}) = 0$, $\forall T > 0$, then the processes $\xi_c(t)$ and $\xi(t)-\xi_c(t)$ are independent. The process $\xi(t)-\xi_c(t)-\xi(0)$ has all the moments, and the functions $\mathrm{M}|\xi(t)-\xi_c(t)-\xi(0)|^r$ are locally bounded for all $r > 0$.*

Proof. Let U_c, $c > 0$, satisfy the condition of the theorem. Denote by $\tau_1, \ldots, \tau_l$ all points $s \in [0,t]$ for which $|\xi(s) - \xi(s-)| > c$. Then $\sup_{\substack{s<t,\\ s\neq\tau_1,\ldots,s\neq\tau_l}} |\xi(s) - \xi(s-)| < c$; and therefore, one can find $\delta > 0$ such that $|\xi(t_2) - \xi(t_1)| < c$ for $0 < t_2 - t_1 < \delta$, $t_2 \leq t$, if $]t_1, t_2] \cap \{\tau_1, \ldots, \tau_l\} = \varnothing$, and such that $|\xi(t_2) - \xi(t_1)| > c$ otherwise. Hence,

$$\xi_c(t) = \lim_{n\to\infty} \sum_{k=1}^{n-1} I_{\{|\xi(kt/n)-\xi((k-1)t/n)|>c\}}[\xi(\tfrac{k}{n}t) - \xi(\tfrac{k-1}{n}t)],$$

$$\xi(t) - \xi_c(t) = \lim_{n\to\infty} \sum_{k=0}^{n-1} I_{\{|\xi(kt/n)-\xi((k-1)t/n)|<c\}}[\xi(\tfrac{k}{n}t) - \xi(\tfrac{k-1}{n}t)] + \xi(0) \tag{3.25}$$

with probability 1. We can establish the independence of $\xi_c(t)$ and $\xi(t)-\xi_c(t)$ by proving that the variables $\xi_c(t) - \xi_c(s)$ and $\xi(t) - \xi_c(t) - \xi(s) + \xi_c(s)$ are independent for all $s < t$ (see the proof of Lemma 7°, Chapter 2; we should take into account that these variables are $\mathcal{F}_t^s$-measurable). Independence of the above-mentioned increments is established in exactly the same way as the independence of $\xi_c(t)$ and $\xi(t) - \xi_c(t) - \xi(0)$ (i.e., for $s = 0$). We put for $x, y \in X$,

$$\xi_{nk} = I_{\{|\xi(kt/n)-\xi((k-1)t/n)|>c\}}(\xi(\tfrac{k}{n}t) - \xi(\tfrac{k-1}{n}t), x),$$
$$\eta_{nk} = I_{\{|\xi(kt/n)-\xi((k-1)t/n)|<c\}}(\xi(\tfrac{k}{n}t) - \xi(\tfrac{k-1}{n}t), y),$$
$$\theta_n = \mathrm{M}\exp\{i\sum_{k=1}^{n}\xi_{nk} + i\sum_{k=1}^{n}\eta_{nk}\} - \mathrm{M}\exp\{i\sum_{k=1}^{n}\xi_{nk}\}\mathrm{M}\exp\{i\sum_{k=1}^{n}\eta_{nk}\}.$$

It suffices to show that $\lim_{n\to\infty}\theta_n = 0$, and this is proved just as in Lemma 24°, Chapter 2.

Formula (3.25), the boundedness of the terms on the r.h.s., and 13°, Chapter 1, together imply the second statement of the theorem.

Henceforth, we shall need a continuous analogue of Kolmogorov's inequality (5°, Chapter 1).

14°. Lemma. *Let $\xi(t)$ be a process continuous from the right with independent increments for which $\mathrm{M}\xi(t) = 0$ and $\mathrm{M}|\xi(t)|^2 < \infty$. Then*

$$\mathrm{P}\{\sup_{t\in[a,b]} |\xi(t)| > c\} \leq \frac{1}{c^2}\mathrm{M}|\xi(b)|^2. \tag{3.26}$$

Proof. Let $t_{nk} = a + (k/n)(b-a)$. Then

$$\mathbf{P}\{\sup_{t\in[a,b]} |\xi(t)| > c\} = \lim_{n\to\infty} \mathbf{P}\{\sup_{k\le n} |\xi_n(t_{nk})| > c\}.$$

By virtue of 5°, Chapter 1, we get

$$\mathbf{P}\{\sup_{k\le n} |\xi_n(t_{nk})| > c\} \le \frac{1}{c^2}\mathbf{M}|\xi(b)|^2.$$

□

15°. Theorem. *Let $c_k \downarrow 0$ be a sequence such that $\Pi_T(\{x : |x| = c_k\}) = 0$ for all $T > 0$ and k. Then the representation*

$$\xi(t) = \xi_{c_1}(t) + \sum_{k=1}^{\infty}([\xi_{c_{k+1}}(t) - \xi_{c_k}(t)] - \mathbf{M}[\xi_{c_{k+1}}(t) - \xi_{c_k}(t)]) + \xi^0(t) \quad (3.27)$$

holds, where $\xi^0(t)$ is a process that has a continuous modification. Furthermore, the process $\xi^0(t)$ is independent of $\nu_s(A)$, $s \le t$, $A \in \mathcal{B}_0$.

Proof. Note that $\xi(t) - \xi_{c_k}(t)$ and $\xi_{c_1}(t) - \xi_{c_k}(t)$ are independent processes; this follows from Lemma 13° (one should apply it to the process $\xi(t) - \xi_{c_1}(t) = \eta(t)$ and take c_k to be c. Then $\eta_{c_k}(t) = \xi_{c_1}(t) - \xi_{c_k}(t)$ and $\eta(t) - \eta_{c_k}(t) = \xi(t) - \xi_{c_k}(t)$). Hence, for $x \in X$,

$$\mathbf{D}(\xi(t) - \xi_{c_1}(t), x) = \mathbf{D}(\xi_{c_1}(t) - \xi_{c_k}(t), x) + \mathbf{D}(\xi(t) - \xi_{c_1}(t), x),$$

$$\sum_{i=1}^{k-1} \mathbf{D}(\xi_{c_i}(t) - \xi_{c_{i+1}}(t), x) = \mathbf{D}(\xi_{c_1}(t) - \xi_{c_k}(t), x) \le \mathbf{D}(\xi(t) - \xi_{c_1}(t), x).$$

Thus, $\sum_{i=1}^{\infty} \mathbf{D}(\xi_{c_i}(t) - \xi_{c_{i+1}}(t), x) < \infty$ and, by virtue of 15°, Chapter 1, the series $\sum_{i=1}^{\infty}(\xi_{c_i}(t) - \xi_{c_{i+1}}(t) - \mathbf{M}[\xi_{c_i}(t) - \xi_{c_{i+1}}(t)])$ converges with probability 1. Note that

$$|\xi_{c_i}(t) - \xi_{c_{i+1}}(t)| \le |c_i|\,|\xi_t(U_{c_{i+1}} - U_{c_i})| \le c_i \nu_T(U_{c_{i+1}} - U_{c_i}), \quad t \le T$$

In addition, $\xi_{c_i}(t) - \xi_{c_{i+1}}(t)$ is a stochastically continuous process. Thus, $\mathbf{M}[\xi_{c_i}(t) - \xi_{c_{i+1}}(t)]$ is a continuous function. The process $\xi^0(t)$ is a limit as $n \to \infty$ of the processes $\xi(t) - \xi_{c_n}(t) - \mathbf{M}[\xi_{c_1}(t) - \xi_{c_n}(t)] = \xi^{c_n}(t)$, which have no jumps exceeding c_n in modulus. If c_n is chosen so that the series on the r.h.s. of (3.26) converges uniformly, then $\xi^{c_n}(t)$ also converges uniformly and a limit function (it is a modification of $\xi^0(t)$) is continuous. Lemma 14° yields

$$\begin{aligned}\mathbf{P}\{\sup_{t\le T} |\xi_{c_k}(t) - \xi_{c_{k-1}}(t) - \mathbf{M}(\xi_{c_k}(t) - \xi_{c_{k-1}}(t))| \ge 2^{-k}\}\\ \le 2^{2k}\mathbf{M}|\xi_{c_k}(T) - \xi_{c_{k+1}}(T) - \mathbf{M}(\xi_{c_k}(T) - \xi_{c_{k-1}}(T))|^2.\end{aligned} \quad (3.28)$$

For every T, one can choose a sequence c_k so that the series of the variables from the r.h.s. of (3.28) converges. In fact, since for $c_k \downarrow 0$,

$$\sum_{k=1}^{\infty} \mathbf{M}|\xi_{c_{k+1}}(t) - \xi_{c_k}(t) - \mathbf{M}(\xi_{c_{k+1}}(t) - \xi_{c_k}(t))|^2 < \infty,$$

we can choose the increasing subsequence n_k such that

$$\sum_{i=n_k}^{\infty} \mathbf{M}|\xi_{c_{i+1}}(t) - \xi_{c_i}(t) - \mathbf{M}(\xi_{c_{i+1}}(t) - \xi_{c_i}(t))|^2 \le \varepsilon_k,$$

where ε_k is a sequence of numbers given beforehand. Then

$$\begin{aligned} \mathbf{M}|\sum_{i=n_k}^{n_{k+1}} &(\xi_{c_{i+1}}(t) - \xi_{c_i}(t) - \mathbf{M}(\xi_{c_{i+1}}(t) - \xi_{c_i}(t)))|^2 \\ &= \mathbf{M}|\xi_{c_{n_{k+1}}}(t) - \xi_{c_{n_k}}(t) - \mathbf{M}(\xi_{c_{n_{k+1}}}(t) - \xi_{c_{n_k}}(t))|^2 \\ &= \sum_{i=n_k}^{n_{k+1}-1} \mathbf{M}|\xi_{c_{i+1}}(t) - \xi_{c_i}(t) - \mathbf{M}(\xi_{c_{i+1}}(t) - \xi_{c_i}(t))|^2 \le \varepsilon_k \end{aligned}$$

(we have used the fact that $\mathbf{M}(\xi, \eta) = (\mathbf{M}\xi, \mathbf{M}\eta)$ for independent ξ and $\eta \in X$ such that $\mathbf{M}\xi$ and $\mathbf{M}\eta$ exist). Clearly, ε_k can be chosen so that $\sum \varepsilon_k 2^{2k} < \infty$. □

16°. Theorem. *Let $\xi^0(t)$ be a continuous process with independent increments with probability 1. Then its increments have normal distributions. There exist the continuous function $a(t)$ with values in X and the continuous monotonic function $B(t)$ with values in a space $L_+(X)$ (of nonnegative symmetric linear operators from X into X), such that for $t_1 < t_2$, the characteristic function of the increment $\xi^0(t_2) - \xi^0(t_1)$ is defined by*

$$\mathbf{M}\exp\{i(z, \xi_0(t_2) - \xi_0(t_1))\} = \exp\{i(z, a(t_2) - a(t_1)) - \tfrac{1}{2}((B(t_2) - B(t_1))z, z)\}. \tag{3.29}$$

Proof. By virtue of Theorem 8°, for every ε, we have

$$\lim_{n\to\infty} \sum_{k=1}^{n} \mathbf{P}\{|\xi^0(\tfrac{k}{n}t) - \xi^0(\tfrac{k-1}{n}t)| > \varepsilon\} = 0;$$

and thus, we can find a sequence $\varepsilon_n \downarrow 0$ such that

$$\lim_{n\to\infty} \sum_{k=1}^{n} \mathbf{P}\{|\xi^0(\tfrac{k}{n}t) - \xi^0(\tfrac{k-1}{n}t)| > \varepsilon_n\} = 0.$$

Then

$$\xi^0(t) - \xi^0(0) = \lim_{n\to\infty} \sum_{k=1}^{n} I_{\{|\xi^0(kt/n) - \xi^0((k-1)t/n)| \le \varepsilon_n\}} (\xi^0(\tfrac{k}{n}t) - \xi^0(\tfrac{k-1}{n}t)), \tag{3.30}$$

since

$$\mathbf{P}\{\xi^0(t) - \xi^0(0) \ne \sum_{k=1}^{n} I_{\{|\xi^0(kt/n) - \xi^0((k-1)t/n)| \le \varepsilon_n\}} (\xi^0(\tfrac{k}{n}t) - \xi^0(\tfrac{k-1}{n}t))\} \le \sum_{k=1}^{n} \mathbf{P}\{|\xi^0(\tfrac{k}{n}t) - \xi^0(\tfrac{k-1}{n}t)| > \varepsilon_n\}.$$

On the r.h.s. of (3.30) we have the sums of independent random variables bounded by the number ε_n, $\varepsilon_n \downarrow 0$. These sums have limit distributions that coincide with the distributions of $\xi^0(t) - \xi^0(0)$. But the central limit theorem can be applied to the variables on the r.h.s. of (3.30); and thus, the distribution of $\xi^0(t) - \xi^0(0)$ is normal. Let

$$a(t) = \mathbf{M}(\xi^0(t) - \xi^0(0)),$$
$$(B(t)z, z) = \mathbf{M}(z, \xi^0(t) - \xi^0(0) - a(t))^2.$$

Then $B(t) \in L_+(X)$ and

$$\mathbf{M} \exp\{i(z, \xi(t) - \xi(0))\} = \exp\{i(z, a(t)) - \tfrac{1}{2}(B(t)z, z)\}. \tag{3.31}$$

Monotonicity of $B(t)$ follows from the equalities (for $t_1 < t_2$)

$$\mathbf{M}(z, \xi^0(t_2) - \xi^0(0) - a(t_2))^2 = \mathbf{M}(z, \xi^0(t_1) - \xi^0(0) - a(t_1))^2 + \mathbf{M}(z, \xi^0(t_2) - \xi^0(t_1) - a(t_2) + a(t_1))^2;$$
$$((B(t_2) - B(t_1))z, z) = \mathbf{M}(z, \xi^0(t_2) - \xi^0(t_1) - a(t_2) + a(t_1))^2 \ge 0.$$

We have used the independence of $\xi^0(t_1) - \xi^0(0)$ and $\xi^0(t_2) - \xi^0(t_1)$. Using this independence again, we can write

$$\mathbf{M} \exp\{i(z, \xi^0(t_2) - \xi^0(t_1))\} = \mathbf{M} \exp\{i(z, \xi^0(t_2) - \xi^0(0))\} \times (\mathbf{M} \exp\{i(z, \xi^0(t_1) - \xi^0(0))\})^{-1}. \tag{3.32}$$

Therefore, (3.31) implies (3.29).

Further, since $\xi(t)$ is a stochastically continuous process, the r.h.s. of (3.31) is a continuous function of t; hence, $(z, a(t))$ and $(B(t)z, z)$ are continuous for all $z \in X$. This yields the continuity of $a(t)$ and $B(t)$.

□

We can now find a general form of the characteristic function of the stochastically continuous process $\xi(t)$ with independent increments taking values in the finite-dimensional Euclidean space X.

17°. Theorem. *For every stochastically continuous process with independent increments $\xi(t)$ defined on R_+ and taking values in X, the following exist:*

(1) a function $\Pi_t(A)$ defined for $A \in \mathcal{B}_0$, $t \in R_+$, which is a nonnegative measure and a continuous nondecreasing function of t for $A \in \mathcal{B}_0$, such that

$$\lim_{\varepsilon \to 0} \int_{|x|>\varepsilon} \frac{|x|^2}{1+|x|^2} \Pi_t(dx) = \int \frac{|x|^2}{1+|x|^2} \Pi_t(dx)$$

exists and is a continuous function of t;

(2) a continuous function $a(t)$ with values in X; and

(3) a continuous nondecreasing function $B(t)$ with values in $L_+(X)$ such that for $t_1 < t_2$, the characteristic function of the increment $\xi(t_2) - \xi(t_1)$ is defined by

$$\mathbf{M} \exp\{i(z, \xi(t_2) - \xi(t_1))\} = \exp\{i(z, a(t_2) - a(t_1)) - \tfrac{1}{2}((B(t_2) - B(t_1))z, z) + \int (e^{i(z,x)} - 1 - i(z,x) I_{\{|x| \le c\}}) \Pi_t(dx)\} \tag{3.33}$$

($c > 0$ is some number for which $\Pi_t(\{x : |x| = c\}) = 0$ for all $t > 0$; the value of $a(t)$ depends on the choice of c).

Proof. Let us use (3.27). Since the terms on the r.h.s. are independent, we have

$$\begin{aligned} \mathbf{M} \exp\{i(z, \xi(t) - \xi(0))\} = {} & \mathbf{M} \exp\{i(z, \xi_{c_1}(t))\} \\ & \times \prod_{k=1}^{\infty} \mathbf{M} \exp\{i(z, \xi_{c_{k+1}}(t) - \xi_{c_k}(t) \\ & - \mathbf{M}(\xi_{c_{k+1}}(t) - \xi_{c_k}(t)))\} \\ & \times \mathbf{M} \exp\{i(z, \xi^0(t) - \xi^0(0))\}. \end{aligned} \tag{3.34}$$

The first factor on the r.h.s. of (3.34) has the form

$$\exp\left\{ \int_{|x|>c_1} (e^{i(z,x)} - 1) \Pi_t(dx) \right\}. \tag{3.35}$$

Further,

$$\begin{aligned} & \mathbf{M} \exp\{i(z, \xi_{c_{k+1}}(t) - \xi_{c_k}(t) - \mathbf{M}\xi_{c_{k+1}}(t) + \mathbf{M}\xi_{c_k}(t))\} \\ & = \mathbf{M} \exp\{i(z, \xi_{c_{k+1}}(t) - \xi_{c_k}(t))\} \exp\left\{ i \left(z, \int_{c_{k+1} < |x| \le c_k} x \Pi_t(dx) \right) \right\} \\ & = \exp\left\{ \int_{c_{k+1} < |x| \le c_k} (e^{i(z,x)} - 1 - i(z,x)) \Pi_t(dx) \right\}. \end{aligned}$$

Hence,

$$\mathbf{M}\exp\{\sum_{k=1}^{\infty} i(z, \xi_{c_{k+1}}(t) - \xi_{c_k}(t) - \mathbf{M}\xi_{c_{k+1}}(t) + \mathbf{M}\xi_{c_k}(t))\}$$

$$= \exp\left\{\lim_{k\to\infty}\int_{c_k<|x|\le c_1}(e^{i(z,x)} - 1 - i(z,x))\Pi_t(dx)\right\}.$$

Note that the above-mentioned limit exists; and thus, for all $z \in X$,

$$\lim_{k\to\infty}\int_{c_k<|x|\le c_1}(z,x)^2\Pi_t(dx) < \infty.$$

Therefore,

$$\int_{|x|\le c_1}|x|^2\Pi_t(dx) = \lim_{k\to\infty}\int_{c_k<|x|\le c_1}|x|^2\Pi_t(dx) < \infty. \tag{3.36}$$

Hence, (3.33) is obtained if we put $c = c_1$ and take the functions, which exist according to Theorem 16°, to be $a(t)$ and $B(t)$. It remains to note that

$$\int\frac{|x|^2}{1+|x|^2}\Pi_t(dx) \le \int_{|x|\le c_1}|x|^2\Pi_t(dx) + \frac{c^2}{1+c^2}\Pi_t(U_c) < \infty.$$

□

Remark 1. The distribution in X, which has the characteristic function of the form (3.33), i.e.,

$$\varphi(z) = \exp\{i(z,a) - \tfrac{1}{2}(Bz,z) + \int(e^{i(z,x)} - 1 - i(z,x)I_{\{|x|\le c\}})\Pi(dx)\}, \tag{3.37}$$

is called *infinitely divisible*. The vector $a \in X$, the operator $B \in L_+(X)$, and the measure Π on $\mathcal{B}_0$, for which $\int\frac{|x|^2}{1+|x|^2}\Pi(dx) < \infty$, are determined by φ uniquely.

(1) $(Bz,z) = -\lim \frac{2}{s^2}\ln\varphi(sz)$;

(2) if $\psi(z) = \ln\varphi(z) + 1/2(Bz,z)$, then by choosing some basis $\{e_k\}$, $k = 1,\dots,d$, in X, we obtain

$$\frac{1}{2h}\int_{-h}^{h}\sum_{k=1}^{d}[-\psi(z+the_k) - \psi(z-the_k) + 2\psi(z)]dt$$

$$= \int\sum_{k=1}^{d}\left(1 - \frac{\sin(h(x,e_k))}{h(x,e_k)}\right)\Pi(dx)e^{i(z,x)}.$$

We can thus define the measure $\sum_{k=1}^{d}\left(1-\frac{\sin(h(x,e_k))}{h(x,e_k)}\right)\Pi(dx)$ and, therefore, the measure Π;

(3) after this, a is determined trivially.

Remark 2. If we have some representation of a characteristic function of a stochastically continuous process with independent increments, then by using it, we can determine an operator $B(t)$ (it is called a *diffusion operator*) and a measure Π_t (it is called a *Lévy measure* (or a *jump measure*)) uniquely.

Remark 3. A process with independent increments $\xi(t)$ is called *homogeneous* if $\xi(0) = 0$ and the distribution of $\xi(t+h) - \xi(t)$ for $h > 0$ coincides with the distribution of $\xi(h)$. For the stochastically continuous homogeneous process $\xi(t)$, the following relations:

$$a(t+h) = a(t) + a(h), \qquad B(t+h) = B(t) + B(h),$$
$$\Pi_{t+h}(A) = \Pi_t(A) + \Pi_h(A)$$

hold. They imply that $a(t) = ta$, $B(t) = tB$, and $\Pi_t(A) = t\Pi(A)$, where a, B, and Π are the same as in Remark 1. Consequently, the characteristic function of the process with independent increments is defined by

$$\begin{aligned}&\mathbf{M}\exp\{i(z,\xi(t))\}\\&= \exp\left\{t\left[i(a,z) - \tfrac{1}{2}(Bz,z) + \int (e^{i(z,x)} - 1 - i(z,x)I_{\{|x|\le c\}})\Pi(dx)\right]\right\},\end{aligned} \tag{3.38}$$

where $a \in X$, $B \in L_+(X)$, and Π is the measure on $\mathcal{B}_0$ for which $\int \frac{|x|^2}{1+|x|^2}\Pi(dx) < \infty$.

§3.3 Properties of Sample Functions

Step Processes (Composed Poisson Processes)

A function $x(t)$ given on R_+ with values in a finite-dimensional Euclidean space X is called a *step function* if there exists a sequence $t_k \uparrow \infty$ such that $x(t) = x(t_k)$ for $t_k \le t < t_{k+1}$. A process with independent increments $\xi(t)$ is called a *step process* if its sample functions are step functions. Let us consider stochastically continuous processes with independent increments.

18°. Theorem. *A process $\xi(t)$ continuous from the right is a step process with independent increments iff its characteristic function has the form*

$$\mathbf{M}\exp\{i(z,\xi(t)-\xi(0))\} = \exp\left\{\int (e^{i(z,x)-1})\Pi_t(dx)\right\}, \tag{3.39}$$

where Π_t is a finite measure on a σ-algebra $\mathcal{B}$ of Borel sets in X.

Proof. If $\xi(t)$ is a step process, then by using the notations from the previous section, we can write $\xi(t) - \xi(0) = \lim_{c\to 0} \xi_t(U_c)$ and, therefore,

$$\mathbf{M}\exp\{i(z,\xi(t)-\xi(0))\} = \lim_{c\to 0} \exp\left\{\int_{U_c} (e^{i(z,x)}-1)\Pi_t(dx)\right\}$$
$$= \exp\left\{\int (e^{i(z,x)}-1)\Pi_t(dx)\right\}.$$

Denote a number of jumps of the process $\xi(t)$ by ν_t. Then $\nu_t = \lim_{c\to 0}\nu_t(U_c)$. Thus, ν_t is also a Poisson random variable, and $\mathbf{M}\nu_t(U_c) \to \mathbf{M}\nu_t$ as $c \to 0$. Hence, $\lim_{c\to 0}\Pi_t(U_c) = \mathbf{M}\nu_t$. Extending the definition of the measureΠ_t onto $\{0\}$ by the equality $\Pi_t(\{0\}) = 0$, we get a finite measure on $\mathcal{B}$ for which (3.39) is valid.

Assume now that (3.39) is satisfied and that the measure $\Pi_t(dx)$ is finite. Without loss of generality, we can consider that $\Pi_t(\{0\}) = 0$. As follows from Remark 3 (17°), $\Pi_t(A)$ is a jump measure of the process $\xi(t)$. Hence,

$$\mathbf{M}e^{i(z,\xi_t(U_c))} = \exp\left\{\int_{U_c} (e^{i(z,x)}-1)\Pi_t(dx)\right\},$$
$$\mathbf{M}\nu_t(U_c) = \Pi_t(U_c) \le \Pi_t(X).$$

The number of jumps of the process $\xi_t(U_c)$ is thus bounded uniformly in t by the variable $\nu_t = \sup_c \nu_t(U_c)$, and $\nu_t = \lim_{c\to 0}\nu_t(U_c)$ is such that $\mathbf{M}\nu_t = \Pi_t(X)$. The limit $\lim_{c\to 0}\xi_t(U_c)$ exists, because

$$\mathbf{P}\{|\xi_t(U_{c_2}) - \xi_t(U_{c_1})| > 0\} \le \mathbf{P}\{\nu_t(U_{c_2}\backslash U_{c_1}) > 0\} \le 1 - e^{-\Pi_t(U_{c_2}\backslash U_{c_1})} \to 0$$

for $c_2 < c_1$, $c_2, c_1 \to 0$. Since ν_t is a number of jumps of the process $\lim_{c\to 0}\xi_t(U_c)$, it is a step process. But

$$\mathbf{M}\exp\{i(z, \lim_{c\to 0}\xi_t(U_c))\} = \lim_{c\to 0}\mathbf{M}\exp\{i(z,\xi_t(U_c))\}$$
$$= \exp\left\{\int (e^{i(z,x)}-1)\Pi_t(dx)\right\}$$
$$= \mathbf{M}\exp\{i(z,\xi(t)-\xi(0))\}.$$

Thus, the distributions of $\xi(t) - \xi(0)$ are identical with the distributions of the step process $\lim_{c\to 0}\xi_t(U_c)$.

□

A step process with independent increments is defined completely if the sequence of pairs of random variables is given $(\tau_k, \xi(\tau_k))$, $k = 1, \ldots,$ where $\tau_0 = 0$ and τ_k is a moment of the kth jump of the process $\xi(t)$ $(0 = \tau_0 < \tau_1 < \tau_2 < \cdots)$.

19°. Theorem. *The joint distribution of the variables $(\tau_k, \xi(\tau_k))$ is given by*

$$\mathbf{P}\{\tau_k - \tau_{k-1} > s \mid \tau_1, \dots, \tau_{k-1}, \xi(\tau_1), \dots, \xi(\tau_{k-1})\} \\ = \exp\{-\Pi_{\tau_{k-1}+s}(X) + \Pi_{\tau_{k-1}}(X)\}; \tag{3.40}$$

$$\mathbf{P}\{\xi(\tau_k) - \xi(\tau_{k-1}) \in A \mid \tau_1, \dots, \tau_{k-1}, \tau_k, \xi(\tau_1), \dots, \xi(\tau_{k-1})\} \\ = (\Pi_{\tau_k}(A) - \Pi_{\tau_{k-1}}(A))(\Pi_{\tau_k}(X) - \Pi_{\tau_{k-1}}(X))^{-1}. \tag{3.41}$$

Proof. First let us find the joint distribution of the variables τ_1 and $\xi(\tau_1) - \xi(0)$. Clearly, for $s > 0$, $A \in \mathcal{B}_0$, we have

$$\mathbf{P}\{\tau_1 > s, \xi(\tau_1) - \xi(0) \in A\} \\ = \lim_{n\to\infty} \sum_{m>ns} \mathbf{P}\{\nu(\tfrac{k}{n}) - \nu(\tfrac{k-1}{n}) = 0, k = 1, 2, \dots, m, \xi(\tfrac{m+1}{n}) - \xi(\tfrac{m}{n}) \in A\} \\ = \lim_{n\to\infty} \sum_{m>ns} \prod_{k=1}^{m} \mathbf{P}\{\nu(\tfrac{k}{n}) - \nu(\tfrac{k-1}{n}) = 0\}\mathbf{P}\{\xi(\tfrac{m+1}{n}) - \xi(\tfrac{m}{n}) \in A\},$$

where $\nu(t)$ is a number of jumps of $\xi(s)$ up to the time t. Since $\nu(t)$ is a Poisson process with a parameter $\Pi_t(X)$, we have

$$\mathbf{P}\{\nu(\tfrac{k}{n}) - \nu(\tfrac{k-1}{n}) = 0\} = \exp\{-\Pi_{k/n}(X) + \Pi_{(k-1)/n}(X)\}.$$

Further, (3.39) yields

$$\mathbf{M}\exp\{i(z, \xi(\tfrac{m+1}{n}) - \xi(\tfrac{m}{n}))\} \\ = \exp\{-\Pi_{(m+1)/n}(X) + \Pi_{m/n}(X)\} \\ \times \sum_{r=0}^{\infty} \frac{1}{r!} \left(\int e^{i(z,x)}(\Pi_{(m+1)/n}(dx) - \Pi_{m/n}(dx)) \right).$$

Hence,

$$\mathbf{P}\{\xi(\tfrac{m+1}{n}) - \xi(\tfrac{m}{n}) \in A\} \\ = \sum_{r=1}^{\infty} \frac{1}{r!} \exp\{-(\Pi_{(m+1)/n}(X) - \Pi_{m/n}(X))\} \\ \times (\Pi_{(m+1)/n}(X) - \Pi_{m/n}(X))^r \mu_r(A),$$

where

$$\mu_1(A) = [\Pi_{(m+1)/n}(A) - \Pi_{m/n}(A)][\Pi_{(m+1)/n}(X) - \Pi_{m/n}(X)]^{-1}$$

and μ_k is a k-times iterated convolution of the measure μ_1. Therefore,

$$\mathbf{P}\{\xi(\tfrac{m+1}{n}) - \xi(\tfrac{m}{n}) \in A\} \\ = \Pi_{(m+1)/n}(A) - \Pi_{m/n}(A) + O((\Pi_{(m+1)/n}(X) - \Pi_{m/n}(X))^2).$$

Thus,

$$\begin{aligned}
&\mathbf{P}\{\tau_1 > s, \xi(\tau_1) - \xi(0) \in A\} \\
&\quad = \lim_{n\to\infty} \sum_{m>ns} \exp\{-\Pi_{m/n}(X)\}(\Pi_{(m+1)/n}(A) - \Pi_{m/n}(A)) \\
&\quad = \int_s^\infty \exp\{-\Pi_u(X)\} d\Pi_u(A).
\end{aligned}$$

Taking $\mathbf{P}\{\tau_1 > s\} = \exp\{-\Pi_s(X)\}$ into account, we get

$$\mathbf{P}\{\xi(\tau_1) - \xi(0) \in A/\tau_1\} = \Pi_{\tau_1}(A)(\Pi_{\tau_1}(X))^{-1}.$$

Let us now find a conditional distribution of $\tau_k - \tau_{k-1}$ if $\tau_1, \ldots, \tau_{k-1}$ and $\xi(0), \ldots, \xi(\tau_{k-1})$ are given.

Denote by C some event determined by the variables $\tau_1, \ldots, \tau_{k-1}$ and $\xi(0), \ldots, \xi(\tau_{k-1})$. Then

$$\begin{aligned}
&\mathbf{P}\{C \cap \{\tau_k - \tau_{k-1} > s\}\} \\
&\quad = \lim_{n\to\infty} \sum_m \mathbf{P}\{C \cap \{\tau_k - \tau_{k-1} > s\} \cap \{\tau_{k-1} \in [\tfrac{m-1}{n}, \tfrac{m}{n}]\}\} \\
&\quad = \lim_{n\to\infty} \sum_m \mathbf{P}\{C \cap \{\tau_{k-1} \in [\tfrac{m-1}{n}, \tfrac{m}{n}]\}\} \\
&\qquad \cap \{\nu(\tfrac{m+1}{n}) - \nu(\tfrac{m}{n}) = 0, \ldots, \nu(\tfrac{m+n_s}{n}) - \nu(\tfrac{m+n_s-1}{n}) = 0\},
\end{aligned}$$

where n_s is an integral part of ns. The events $C \cap \{\tau_{k-1} \in [\frac{m-1}{n}, \frac{m}{n}]\}$ are determined by the behaviour of the process before the time m/n, and $\nu(\frac{i+1}{n}) - \nu(\frac{i}{n})$, $i \geq m$, are determined by the increments of the process after the time m/n. Hence,

$$\begin{aligned}
&\mathbf{P}\{C \cap \{\tau_{k-1} \in [\tfrac{m-1}{n}, \tfrac{m}{n}]\} \cap \{\nu(\tfrac{i+1}{n}) - \nu(\tfrac{i}{n}) = 0, i = m, \ldots, m + n_s\}\} \\
&\quad = \mathbf{P}\{C \cap \{\tau_{k-1} \in [\tfrac{m-1}{n}, \tfrac{m}{n}]\}\} \exp\{-\Pi_{(m+n_s)/n}(X) + \Pi_{m/n}(X)\}.
\end{aligned}$$

Therefore,

$$\begin{aligned}
&\mathbf{P}\{C \cap \{\tau_k - \tau_{k-1} > s\}\} \\
&\quad = \lim_{n\to\infty} \sum_m \mathbf{P}\{C \cap \{\tau_{k-1} \in [\tfrac{m-1}{n}, \tfrac{m}{n}]\}\} \\
&\qquad \times \exp\{-\Pi_{(m+n_s)/n}(X) + \Pi_{m/n}(X)\} \\
&\quad = \int_0^\infty \mathbf{P}\{C \cap \{\tau_{k-1} \in du\}\} \exp\{-\Pi_{s+u}(X) + \Pi_u(X)\}.
\end{aligned} \tag{3.42}$$

This implies (3.40). Let us show the validity of (3.41). Let C be the same as before; let us find a conditional joint distribution of $\xi(\tau_k) - \xi(\tau_{k-1})$ and

$\tau_k - \tau_{k-1}$. We have

$$\begin{aligned}
&\mathbf{P}\{C \cap \{\tau_k - \tau_{k-1} > s\} \cap \{\xi(\tau_k) - \xi(\tau_{k-1}) \in A\}\} \\
&\quad = \lim_{n\to\infty} \sum_m \sum_{l>ns} \mathbf{P}\{C \cap \{\tau_{k-1} \in [\tfrac{m-1}{n}, \tfrac{m}{n}]\} \\
&\qquad\qquad \cap \{\nu(\tfrac{i+1}{n}) - \nu(\tfrac{i}{n}) = 0, i = m, \dots, m+l-1\} \\
&\qquad\qquad \cap \{\xi(\tfrac{m+l+1}{n}) - \xi(\tfrac{m+l}{n}) \in A\}\} \\
&\quad = \lim_{n\to\infty} \sum_m \sum_{l>ns} \mathbf{P}\{C \cap \{\tau_{k-1} \in [\tfrac{m-1}{n}, \tfrac{m}{n}]\} \\
&\qquad\qquad \times \exp\{-\Pi_{(m+l)/n}(X) + \Pi_{m/n}(X)\} \\
&\qquad\qquad \times [\Pi_{(m+l+1)/n}(A) - \Pi_{(m+l)/n}(A)] \\
&\quad = \int_0^\infty \mathbf{P}\{C \cap \{\tau_{k-1} \in du\}\} \int_s^\infty \exp\{-\Pi_{u+v}(X) + \Pi_v(X)\} d_v \Pi_{u+v}(A).
\end{aligned} \tag{3.43}$$

This formula yields (3.41).

□

Monotonic Processes in R and X

A process $\xi(t)$ is *monotonically nondecreasing* if $\xi(t_2) \geq \xi(t_1)$ with probability 1 for $t_1 < t_2$. If $\xi(t)$ is continuous from the right, it is sufficient that this inequality hold with probability 1 for each pair t_1, t_2.

20°. Theorem. *The stochastically continuous real-valued process with independent increments $\xi(t)$, $t \in R_+$, that is continuous from the right, is monotonically nondecreasing iff its characteristic function allows the representation*

$$\mathbf{M} e^{iz(\xi(t)-\xi(0))} = \exp\left\{ iz\gamma(t) + \int (e^{izx} - 1)\Pi_t(dx) \right\}, \tag{3.44}$$

where $\gamma(t)$ is a continuous nondecreasing function, and $\Pi_t(A)$ is a σ-finite measure on $]0, \infty]$ such that $\Pi_t(A) < \infty$ for $A \in]\varepsilon, \infty]$, $\Pi_t(A)$ is a continuous nondecreasing function of t, and $\int_0^1 x\Pi_t(dx) < \infty$ for all $t > 0$.

Proof. Assume that $\xi(t)$ is a stochastically continuous nondecreasing process. Then this is also true for $\xi_c(t)$ (because all jumps of $\xi(t)$ are nonnegative) and for $\xi(t) - \xi_c(t)$, which is obtained by exclusion of all jumps exceeding c from $\xi(t)$. Since $\xi_c(t)$ increases with c, $\xi(t) - \xi_c(t)$ decreases and $\xi(t) -$

$\xi_c(t) \geq 0$. Hence, there exists $\lim_{c\to 0}[\xi(t) - \xi_c(t)] = \xi^0(t)$, and $\xi^0(t)$ is a nondecreasing process with independent increments. By Theorem 16°, the variable $\xi^0(t_2) - \xi^0(t_1)$ has the normal distribution; it is also nonnegative with probability 1, and thus $\mathbf{D}(\xi^0(t_2) - \xi^0(t_1)) = 0$. Therefore, $\xi^0(t) - \xi^0(0)$ is a nonrandom nondecreasing continuous function. We denote it by $\gamma(t)$. Then $\xi(t) - \xi(0) = \gamma(t) + \lim_{c\to 0}\xi_c(t)$. Since $\xi(t) - \xi(t-) \geq 0$ for all t, we have $\Pi_t(]-\infty, 0[) = 0$ and, thus,

$$\mathbf{M}e^{iz(\xi(t)-\xi(0))} = e^{iz\gamma(t)} \lim_{c\to 0} \mathbf{M}e^{iz\xi_c(t)} = e^{iz\gamma(t)} \exp\left\{\int_0^\infty (e^{izx} - 1)\Pi_t(dx)\right\}.$$

The existence of $\int_0^1 x\Pi_t(dx)$ is necessary to make the integral on the r.h.s. of (3.44) meaningful.

Assume that the characteristic function of $\xi(t)$ is given by (3.44). Then the characteristic function of the process with independent increments $\xi(t) - \xi(0) - \gamma(t) = \eta(t)$ has the form

$$\mathbf{M}e^{iz\eta_t} = \exp\left\{\int_0^\infty (e^{izx} - 1)\Pi_t(dx)\right\}. \tag{3.45}$$

By virtue of Remark 2 (17°), $\Pi_t(dx)$ is a jump measure for the process $\eta(t)$, and $\Pi_t(dx) = 0$ on $]-\infty, 0[$. Hence,

$$\mathbf{M}\exp\{iz\eta_c(t)\} = \exp\left\{\int_c^\infty (e^{izx} - 1)\Pi_t(dx)\right\}. \tag{3.46}$$

The process $\eta_c(t)$ is a sum of jumps of the process $\eta(t)$, which exceed c in modulus. They are nonnegative with probability 1, and $\eta_c(t)$ is thus a nondecreasing process. Together, (3.45) and (3.46) imply that $\eta(t)$ coincides with $\lim_{c\to 0}\eta_c(t)$. Hence, $\eta(t)$ is also a nondecreasing process. It follows from the equality $\xi(t) = \xi(0) + \gamma(t) + \eta(t)$ that $\xi(t)$ is nondecreasing, too ($\gamma(t)$ and $\eta(t)$ are nondecreasing).

□

Let $K \subset X$ be some cone, i.e., a convex closed set in X containing the point λx, $\lambda \geq 0$, along with each point $x \in K$. A function $x(t)$ with values in X is called *K-monotonic* if $x(t_2) - x(t_1) \in K$ for $t_1 < t_2$. The definition of a *K-monotonic process* is obvious. We say that the cone K is *nondegenerate* if it does not contain any straight line.

21°. Theorem. *If a cone K is nondegenerate, then a stochastically continuous process $\xi(t)$ with independent increments in a finite-dimensional Euclidean* *space X is K-monotonic iff the characteristic function of this process allows the representation*

$$\mathbf{M}\exp\{i(z,\xi(t)-\xi(0))\} = \exp\left\{i(z,a(t)) + \int_K (e^{i(z,x)}-1)\Pi_t(dx)\right\}, \quad (3.47)$$

where $a(t)$ is a K-monotonic continuous function and the measure Π is such that $\int_{|x|\le 1}|x|\Pi_t(dx) < \infty$, $\Pi_t(X\backslash K) = 0$, $t \ge 0$.

Proof. Let $\xi(t)$ be a K-monotonic process. Then $\xi(s)-\xi(s-) \in K$ for all $s > 0$. Thus, the measure $\Pi_t(dx)$ is concentrated on the cone K. Processes $\xi_c(t)$ and $\xi(t)-\xi_c(t)$ are also K-monotonic for all $c > 0$. Besides, $\xi_{c_1}(t)-\xi_{c_2}(t) \in K$ for $c_1 < c_2$. Note that every sequence $x_n \in K$ with $x_{n-1} - x_n \in K$ has a limit $\lim_n x_n = x_0$, $x_0 \in K$ (the K-decreasing K-bounded sequence has a limit). Thus, there exists $\lim_{c\to 0}(\xi(t) - \xi_c(t)) = \xi^0(t)$, which is a continuous K-monotonic process. Since $\xi^0(t) - \xi^0(0)$ has the normal distribution, and $\xi^0(t)-\xi^0(0) \in K$ with probability 1, where K is nondegenerate, the diffusion of $\xi^0(t)$ equals 0. Thus, $\xi^0(t)-\xi^0(0)$ is a nonrandom K-monotonic continuous function. Denote it by $a(t)$. We have $\xi(t) - \xi(0) = a(t) + \lim_{c\to 0}\xi_c(t)$ and $\mathbf{M}e^{i(z,a(t)+\xi_c(t))} = \exp\left\{i(z,a(t)) + \int_{K\cap\{|x|>c\}}(e^{i(z,x)}-1)\Pi_t(dx)\right\}$. By passing here to the limit as $c \to 0$, we obtain (3.47). For a nondegenerate cone K, there exist z_0 and $\alpha > 0$ such that $(z_0,x) \ge \alpha|x|$ for $x \in K$. Hence, the existence of the integral

$$\int_{K\cap\{|x|<c\}}(e^{i(z_0,x)}-1)\Pi_t(dx)$$

implies the existence of

$$\int_{K\cap\{|x|\le c\}}(z_0,x)\Pi_t(dx)$$

and

$$\int_{K\cap\{|x|\le c\}}|x|\Pi_t(dx).$$

Suppose now that the characteristic function of the process is given by (3.47), where $a(t)$ is a K-monotonic continuous function. Clearly, it suffices to prove that the process $\xi(t) - \xi(0) - a(t) = \eta(t)$ is K-monotonic. Since the jump measure for $\eta(t)$ is concentrated on K, $\eta_c(t)$ is a K-monotonic process for all $c > 0$. $\eta_{c_2}(t) - \eta_{c_1}(t) \in K$ for $c_2 < c_1$. Hence, the limit $\lim_{c\to 0}\eta_c(t)$ exists. Since the characteristic function of the process $\eta(t) - \eta_c(t)$ is given by

$$\mathbf{M}\exp\{i(z,\eta(t)-\eta_c(t))\} = \exp\left\{\int_{K\cap\{|x|\le c\}}(e^{i(z,x)}-1)\Pi_t(dx)\right\}$$

and since the function to the right converges to 1 as $c \to 0$, we have $\eta(t) = \lim_{c\to 0}\eta_c(t)$. Thus, $\eta(t)$ is also K-monotonic as a limit of K-monotonic processes.

□

Processes with Bounded Variation

A process $\xi(t)$, $t \in R_+$, with values in a finite-dimensional Euclidean space X *has bounded variation* if for all $t \in R_+$, the variable

$$\sup_{0=t_0<t_1<\cdots<t_n\le t} \sum_{k=1}^{n} |\xi(t_k) - \xi(t_{k-1})| = \operatorname*{Var}_{[0,t]} \xi \tag{3.48}$$

is bounded with probability 1. Let us examine stochastically continuous processes continuous from the right with independent increments. Then in order to calculate a variation of a process, one can use (3.48), taking the decompositions such that the points t_k belong to a fixed countable dense (in R_+) set.

22°. Lemma. *If $\xi(t)$ is a stochastically continuous (continuous from the right) process with independent increments and bounded variation, then* $\operatorname{Var}_{[0,t]} \xi = \eta(t)$ *is a real-valued nondecreasing stochastically continuous process with independent increments.*

Proof. Clearly, for $0 < s < t$,

$$\operatorname*{Var}_{[0,t]} \xi = \operatorname*{Var}_{[0,s]} \xi + \operatorname*{Var}_{[s,t]} \xi.$$

Since $\operatorname{Var}_{[s,t]} \xi$ can be expressed in terms of the increments of ξ on $[s,t]$, it is $\mathcal{F}_t^s$-measurable. Here $\mathcal{F}_t^s$ is a σ-algebra generated by the variables $\{\xi(u) - \xi(s), u \in [s,t]\}$. In particular, $\eta(t)$ is $\mathcal{F}_t^0$-measurable. Since the σ-algebras $\mathcal{F}_t^s$ and $\mathcal{F}_t^0$ are independent, $\eta(t) - \eta(s)$ does not depend on $\{\eta(u)$ for $u \le s\}$. □

23°. Lemma. *If $x(t)$ is a function continuous from the right with bounded variation, then for all $t > 0$,*

$$\sum_{s\le t} |x(s) - x(s-)| < \infty, \tag{3.49}$$

and the function

$$x^0(t) = x(t) - \sum_{s\le t} (x(s) - x(s-)) \tag{3.50}$$

is continuous and has bounded variation; moreover,

$$\operatorname*{Var}_{[0,t]} x = \operatorname*{Var}_{[0,t]} x^0 + \sum_{s\le t} |x(s) - x(s-)|. \tag{3.51}$$

Proof. Since the function $x(t)$ has no discontinuities of the second kind, a set of discontinuity points of $x(t)$ is at most countable. If $\{s_k\}$ are all discontinuity points of $x(t)$, then

$$\operatorname*{Var}_{[0,t]} x \geq \lim_{h\to 0} \sum_{k\leq n} |x(s_k) - x(s_k - h)| I_{\{s_k \leq t\}} = \sum_{k=1}^{n} |x(s_k) - x(s_k-)| I_{\{s_k\leq t\}}$$

for all n. Passing to the limit as $n \to \infty$, we obtain (3.49). The function

$$\hat{x}(t) = \sum_{s\leq t} (x(s) - x(s-))$$

is defined (the series converges uniformly); it has the same points and values of discontinuities as $x(t)$, and thus $x(t) - \hat{x}(t)$ is a continuous function. (3.51) follows easily from the definition of variation.

□

Remark. The first term on the r.h.s. of (3.51) is continuous with respect to t (as a variation of a continuous function), and the second one is purely discontinuous, so this formula gives a decomposition of the increasing function of t (of the variation) into the sum of continuous and purely discontinuous functions.

24°. Theorem. *A stochastically continuous process $\xi(t)$ continuous from the right with independent increments has bounded variation iff its characteristic function has the form*

$$\mathbf{M} \exp\{i(z, \xi(t) - \xi(0))\} = \exp\left\{ i(z, a(t)) + \int (e^{i(z,x)} - 1)\Pi_t(dx) \right\}, \quad (3.52)$$

where $a(t)$ is a continuous function with bounded variation, and $\Pi_t(dx)$ is such that $\int_{|x|\leq c} |x|\Pi_t(dx) < \infty$.

Proof. Assume that $\xi(t)$ has bounded variation. Denoting it by $\eta(t)$, we obtain

$$\mathbf{M} \exp\{i\lambda\eta(t)\} = \exp\left\{ i\lambda\gamma(t) + \int_0^\infty (e^{i\lambda x} - 1)G_t(dx) \right\},$$

since $\eta(t)$ is a stochastically continuous nondecreasing process with independent increments. On the other hand, by virtue of (3.51), $\gamma(t)$ is a variation of the continuous component of $\xi(t)$, and

$$\eta(t) - \gamma(t) = \int |x|\nu_t(dx),$$

where $\nu_t(A)$ is a number of jumps of $\xi(s)$ on $[0,t]$ that hit the set $A \in \mathcal{B}_0$. Hence, the process

$$\lim_{c\to 0} \xi_t(U_c) = \lim_{c\to 0} \int_{|x|>c} x\nu_t(dx) = \int x\nu_t(dx) = \xi_0(t)$$

is defined, and $\xi^0(t) = \xi(t) - \xi_0(t)$ is a continuous process with bounded variation. Since $|\xi^0(t) - \xi^0(0)| \leq \gamma(t)$, where $\gamma(t) < \infty$, and $\xi^0(t)$ has the normal distribution, we have $B(t) = 0$ and thus $\xi^0(t) = a(t)$, where $a(t)$ is a continuous function with bounded variation. The process $\int_{|x|\leq c} |x|\nu_t(dx)$ has jumps bounded by c; therefore, it has all the moments, and thus

$$\mathbf{M} \int_{|x|\leq c} |x|\nu_t(dx) = \int_{|x|\leq c} |x|\Pi_t(dx) < \infty.$$

Then (3.52) follows from the representation $\xi(t) = a(t) + \int x\nu_t(dx)$ and from (3.20).

Assume that $\xi(t)$ has the characteristic function (3.52). It suffices to examine the case $a(t) \equiv 0$. Then $\xi(t) - \xi_c(t)$ and $\xi_c(t)$ are independent, and

$$\mathbf{M} \exp\{i(z, \xi_c(t))\} = \exp\left\{\int_{|x|>c} (e^{i(z,x)} - 1)\Pi_t(dx)\right\},$$

so (3.52) yields

$$\mathbf{M} \exp\{i(z, \xi(t) - \xi_c(t))\} = \exp\left\{\int_{|x|\leq c} (e^{i(z,x)} - 1)\Pi_t(dx)\right\}.$$

The r.h.s. tends to 1 as $c \to 0$; thus,

$$\xi(t) = \lim_{c\to 0} \xi_c(t),$$

$$\xi(t) = \int x\nu_t(dx),$$

$$\operatorname*{Var}_{[0,t]} \xi = \int |x|\nu_t(dx) = \int_{|x|\leq c} |x|\nu_t(dx) + \int_{|x|>c} |x|\nu_t(dx) < \infty$$

(the first integral has finite mathematical expectation, whereas the second one is always finite).

□

§3.4 Locally Homogeneous Processes with Independent Increments

A stochastically continuous process with independent increments with values in X (X is a finite-dimensional Euclidean space) is called *locally homogeneous* if its characteristics $a(t)$, $B(t)$, and $\Pi_t(A)$ are absolutely continuous, i.e., there exist $a^*(t)$, $B^*(t)$, and $\Pi_t^*(A)$ such that

$$a(t) = \int_0^t a^*(s)ds, \qquad B(t) = \int_0^t B^*(s)ds, \qquad \Pi_t(A) = \int_0^t \Pi_s^*(A)ds.$$

We naturally assume the functions $a^*(t)$, $B^*(t)$, and $\Pi_t^*(A)$ to be measurable with respect to t and locally integrable. Clearly, $a^*(s) \in X$, $B^*(s) \in L_+(X)$. Suppose that $\Pi_s^*(A)$ is a finite measure on $\mathcal{B}_0$ (with respect to A) and that $\int \frac{|x|^2}{1+|x|^2}\Pi_s^*(dx) < \infty$.

In this case, the characteristic function of the process $\xi(t)$ allows the representation

$$\begin{aligned} &\mathbf{M}\exp\{i(z, \xi(t) - \xi(0))\} \\ &\quad = \exp\left\{\int_0^t \left[(z, a^*(s)) - \frac{1}{2}(B^*(s)z, z)\right.\right. \\ &\qquad \left.\left. + \int (e^{i(z,x)} - 1 - (z,x)I_{\{|x|\le c\}})\Pi_s^*(dx)\right] ds\right\}. \end{aligned} \tag{3.53}$$

Integro-Differential Equations for a Process

25°. Theorem. *Let $a^*(s)$, $B^*(s)$, and $\Pi_s^*(A)$ be functions continuous in s and let*

$$\lim_{\varepsilon\to 0} \sup_{0\le s\le t} \int_{|x|\le\varepsilon} (|x|^2)\Pi_s^*(dx) = 0$$

for all $t > 0$. If $f(x)$ is a twice continuously differentiable function from X to R bounded with its derivatives, then the function

$$u_t(s,x) = \mathbf{M}f(x + \xi(t) - \xi(s)), \quad 0 < s < t, \tag{3.54}$$

satisfies the integro-differential equation with the boundary condition

$$\frac{\partial}{\partial s}u_t(s,x) + L_s[u_t(s,x)] = 0, \quad \lim_{s\uparrow t} u_t(s,x) = f(x), \tag{3.55}$$

where

$$\begin{aligned} L_s\varphi &= (a^*(s), \varphi'(x)) + \frac{1}{2}\operatorname{Sp} B^*(s)\varphi''(x) \\ &\quad + \int [\varphi(x+y) - \varphi(x) - (\varphi'(x), y)I_{\{|y|\le c\}}]\Pi_s^*(dy) \end{aligned} \tag{3.56}$$

(φ' is a vector from X for which $(\varphi'(x), z) = \frac{d}{dh}\varphi(x+hz)\big|_{h=0}$; $\varphi''(x)$ is an operator from $L(X)$ for which $(\varphi''(x)u, v) = \frac{\partial^2}{\partial h \partial \theta}\varphi(x+hu+\theta v)\big|_{h=0,\theta=0}$; the trace $\operatorname{Sp} AB$ is defined for $A, B \in L(X)$ by $\Sigma(ABe_k, e_k)$, where $\{e_k\}$ is an orthonormal basis in X). If the basis is chosen and if x^i are the coordinates of x, $a_i^*(s)$ are the coordinates of $a^*(s)$, and $b_{ij}^*(s)$ are the matrix elements of the operator $B^*(s)$ in this basis, then

$$\begin{aligned} L_s\varphi(x) &= \sum a_i^*(s)\frac{\partial \varphi}{\partial x^i} + \frac{1}{2}\sum_{i,j} b_{ij}^*(s)\frac{\partial \varphi}{\partial x^i \partial x^j}(x) \\ &\quad + \int [\varphi(x+y) - \varphi(x) - I_{\{|y|\le c\}}\sum \frac{\partial \varphi}{\partial x^i}(x) y^i]\Pi_s(dy). \end{aligned} \tag{3.57}$$

Proof. Note that the functions $u_t(s,x)$ and $f(x)$ have the same number of continuous derivatives with respect to x. The second relation of (3.55) is obviously valid. The first relation of (3.55) is thus equivalent to the equality

$$f(x) - u_t(s,x) + \int_s^t L_\tau[u_t(\tau,x)]d\tau = 0. \tag{3.58}$$

Let $f_n(x)$ be some sequence of functions satisfying the following conditions: The functions f_n' and f_n'' exist and are continuous; the functions f_n, f_n', f_n'' are jointly bounded and converge uniformly on each compact set to f, f', and f'', respectively. Denote

$$u_t^{(n)}(s,x) = \mathbf{M} f_n(x + \xi(t) - \xi(s)).$$

Then $u_t^{(n)}(s,x)$, $u_t^{(n)\prime}(s,x)$, and $u_t^{(n)\prime\prime}(s,x)$ are uniformly bounded and converge uniformly on each compact set to $u_t(s,x)$, $u_t'(s,x)$, and $u_t''(s,x)$, respectively. Hence, $L_\tau[u_t^{(n)}(\tau,x)]$ are also uniformly bounded and converge uniformly on each compact set to $L_\tau[u_t(\tau,x)]$. In addition, if $u_t^{(n)}(s,x)$ satisfies the relation

$$f_n(x) - u_t^{(n)}(s,x) + \int_s^t L_\tau[u_t^{(n)}(\tau,x)]d\tau = 0, \tag{3.59}$$

then by passing to the limit in it as $n \to \infty$, we get (3.58). Assume that the function $f(x)$ allows the representation

$$f(x) = \int e^{i(z,x)}\theta(z)dz, \tag{3.60}$$

where $\theta(z)$ is such that $\int(1+|z|^2)|\theta(z)|dz < \infty$. Then $f(x)$, $f'(x)$, and $f''(x)$ are continuous and bounded. Let us find

$$u_t(s,x) = \mathbf{M}\int \exp\{i(z, x+\xi(t)-\xi(s))\}\theta(z)dz = \int e^{i(z,x)}\chi(s,t,z)\theta(z)dz,$$

and

$$\chi(s,t,z) = \exp\left\{\int_s^t Q_u^*(z)du\right\},$$

where

$$Q_s^*(z) = [i(a^*(s),z) - \tfrac{1}{2}(B^*(s)z,z)] + \int (e^{i(z,y)} - 1 - i(z,y)I_{\{|y|\le c\}})\Pi_s^*(dy).$$

Then

$$\frac{\partial}{\partial s}u_t(s,x) = -\int e^{i(z,x)}\chi(s,t,z)Q_s^*(z)\theta(z)dz.$$

Let us examine $L_s^*[u_t(s,x)]$. We have

$$(u_t'(s,x),a^*(s)) = \int i(z,a^*(s))e^{i(z,x)}\chi(s,t,z)\theta(z)dz,$$

$$\operatorname{Sp} B^*(s)u_t''(s,x) = -\int (B^*(s)z,z)e^{i(z,x)}\chi(s,t,z)\theta(z)dz,$$

$$u_t(s,x+y) - u_t(s,x) = (u_t'(s,x),y)I_{\{|y|\le c\}}$$
$$= \int [e^{i(z,y)} - 1 - i(z,y)I_{\{|y|\le c\}}]e^{i(z,x)}\chi(s,t,x)\theta(z)dz$$

Thus,

$$L_\tau^*[u_t(\tau,x)] = \int e^{i(z,x)}\chi(s,t,z)Q_s^*(z)\theta(z)dz.$$

Hence, for all $f(x)$ of the form (3.59), Eq. (3.58) is satisfied.

Suppose now that $f(x)$ satisfies the condition of the theorem. We put

$$f_n(x) = \exp\left\{-\frac{1}{n}|x|^2\right\}\left(2\pi\cdot\frac{1}{n}\right)^{-d/2}\int f(y)\exp\left\{-\frac{n|x-y|^2}{2}\right\}dy.$$

The function $f_n(x)$ is infinitely differentiable and decreases as $|x|\to\infty$ faster than any power. Thus, it can be represented in the form (3.60), where $|\theta(z)||z|^m$ is integrable for all m. Clearly, $f_n(z)$ has two derivatives uniformly bounded in n, which converge uniformly on each compact set to the corresponding derivatives of f. Since (3.59) holds for f_n, (3.58) is valid for f.

□

Consider the following special cases of Eq. (3.56):

I. $\xi(t)$ is a composed Poisson process with the characteristic function

$$\mathbf{M}\exp\{i(z,\xi(t)-\xi(0))\} = \exp\left\{\int_0^t\int (e^{i(z,y)} - 1)\Pi_s^*(dy)ds\right\}.$$

Then Eq. (3.56) has the form

$$\frac{\partial}{\partial s}u_t(s,x) + \int [u_t(s,x+y) - u_t(s,x)]\Pi_s^*(dy) = 0.$$

II. $\xi(t)$ is a composed Poisson process with drift, having the characteristic function

$$\mathbf{M}\exp\{i(z,\xi(t)-\xi(0))\} = \exp\left\{\int_0^t \left[i(a^*(s),z) + \int (e^{i(z,y)} - 1)\Pi_s^*(dy)\right] ds\right\}.$$

Eq. (3.56) has the form

$$\frac{\partial}{\partial s}u_t(s,x) + (a^*(s), u_t'(s,x)) + \int [u_t(s,x+y) - u_t(s,x)]\Pi_s^*(dy) = 0.$$

III. $\xi(t)$ is a Gaussian process with independent increments, having the characteristic function

$$\mathbf{M}\exp\{i(z,\xi(t)-\xi(0))\} = \exp\left\{\int_0^t \left[i(a^*(s),z) - \frac{1}{2}(B^*(s)z,z)\right] ds\right\}.$$

In this case, Eq. (3.56) has the form

$$\frac{\partial}{\partial s}u_t(s,x) + (a^*(s), u_t'(s,x)) + \frac{1}{2}\operatorname{Sp} B^*(s)u_t''(s,x) = 0.$$

Equation for the Distribution of Functionals in the Integral Form

The integro-differential operator (3.56) is closely connected with the process, having the characteristic function (3.53); it appears when distributions of some functionals of a random process are considered. Let us examine the functionals

$$\int_a^b \psi(s,\xi(s))ds, \tag{3.61}$$

where ψ is a continuous function. In order to determine the distribution of the variable (3.61), it is convenient to use the function

$$v_\lambda(t,x) = \mathbf{M}\exp\left\{\lambda\int_t^b \psi(s,\xi(s)-\xi(t)+x)ds\right\}. \tag{3.62}$$

So $\mathbf{M}v_{iz}(t,\xi(t))$ is a characteristic function of the variable considered. If ψ is a bounded function, then $v_\lambda(t,x)$ is an integral analytic function of λ. If $\psi \geq 0$, it is analytic in the half-plane $\operatorname{Re}\lambda < 0$ and continuous in the closed half-plane $\operatorname{Re}\lambda \leq 0$. And if ψ is real, then $v_\lambda(t,x)$ is defined and continuous for imaginary λ. Clearly, for bounded ψ, the functions $v_\lambda(t,x)$ and ψ have the same number of derivatives with respect to x.

26°. Theorem. *If ψ is bounded and has continuous bounded derivatives $\psi'(x)$ and $\psi''(x)$, then for $t \le b$, the function $v_\lambda(t,x)$ satisfies the equation*

$$\frac{\partial}{\partial t} v_\lambda(t,x) + L_t v_\lambda(t,x) + \lambda\psi(t,x) v_\lambda(t,x) = 0 \tag{3.63}$$

with the boundary condition $v_\lambda(b,x) = 1$.

Proof. We have

$$\begin{aligned}
v_\lambda(t,x) - v(t-\Delta,x) &= \mathbf{M}\exp\left\{\lambda \int_t^b \psi(\xi(s)-\xi(t)+x)ds\right\} \\
&\quad - \mathbf{M}\exp\left\{\lambda \int_{t-\Delta}^b \psi(s,\xi(s)-\xi(t-\Delta)+x)ds\right\} \\
&= \mathbf{M}\exp\left\{\lambda \int_t^b \psi(s,\xi(s)-\xi(t)+x)ds\right\} \\
&\quad - \mathbf{M}\exp\left\{\lambda \int_t^b \psi(s,\xi(s)-\xi(t-\Delta)+x)ds\right\} \\
&\quad + \mathbf{M}\exp\left\{\lambda \int_t^b \psi(s,\xi(s)-\xi(t-\Delta)+x)ds\right\} \\
&\quad \times \left(1 - \exp\left\{\lambda \int_{t-\Delta}^t \psi(s,\xi(s)-\xi(t-\Delta)+x)ds\right\}\right).
\end{aligned}$$

Since the variable $\frac{1}{\Delta}\left(1-\exp\left\{\lambda\int_{t-\Delta}^t \psi(s,\xi(s)-\xi(t-\Delta)+x)ds\right\}\right)$ is uniformly bounded and tends to $-\lambda\psi(x)$ in probability uniformly in t on each finite interval, we have

$$\begin{aligned}
&\lim_{\Delta\to 0} \mathbf{M}\exp\left\{\lambda \int_t^b \psi(s,\xi(s)-\xi(t-\Delta)+x)ds\right\} \\
&\qquad \times \frac{1}{\Delta}\left(1-\exp\left\{\lambda\int_{t-\Delta}^t \psi(s,\xi(s)-\xi(t-\Delta)+x)ds\right\}\right) \\
&\quad = -\lambda\psi(s,x)\mathbf{M}\exp\left\{\lambda\int_t^b \psi(s,\xi(s)-\xi(t)+x)ds\right\} = -\lambda\psi(s,x)v_\lambda(t,x).
\end{aligned}$$

Further,

$$\mathbf{M}\exp\left\{\lambda\int_t^b \psi(s,\xi(s)-\xi(t)+x)ds\right\}$$

$$-\mathbf{M}\exp\left\{\lambda\int_t^b \psi(s,\xi(s)-\xi(t)+\xi(t)-\xi(t-\Delta)+x)ds\right\}$$

$$= v_\lambda(t,x)-\mathbf{MM}(\exp\left\{\lambda\int_t^b \psi(s,\xi(s)-\xi(t)+\xi(t)-\xi(t-\Delta)+x)ds\right\}\mid \mathcal{F}_t)$$

$$= v_\lambda(t,x)-\mathbf{M}v_\lambda(t,\xi(t)-\xi(t-\Delta)+x).$$

The function $v_\lambda(t,x)$ is twice continuously differentiable. Putting for $s<t$,

$$\mathbf{M}v_\lambda(t,\xi(t)-\xi(s)+x)=\tilde{v}(s,x)$$

and using (3.58), we can write

$$v_\lambda(t,x)-\mathbf{M}v_\lambda(t,\xi(t)-\xi(t-\Delta)+x)=\tilde{v}(t,x)-\tilde{v}(t-\Delta,x)$$
$$=-\int_{t-\Delta}^t L_\tau[\tilde{v}(\tau,x)]d\tau.$$

Since $L_\tau[\tilde{v}(\tau,x)]$ is continuous in τ,

$$\lim\frac{1}{\Delta}\int_{t-\Delta}^t L_\tau[\tilde{v}(\tau,x)]d\tau=-L_t[\tilde{v}(t,x)]=-L_t v_\lambda(t,x)$$

uniformly in t on each bounded interval. Hence,

$$\lim_{\Delta\to 0}\frac{v_\lambda(t-\Delta,x)-v_\lambda(t,x)}{\Delta}=-L_t[v_\lambda(t,x)]-\lambda\psi(t,x)v_2(t,x)$$

uniformly in t on each bounded interval. This relation implies (3.63).

□

Remark. Let us put

$$g_m(t,x)=\mathbf{M}\left(\int_t^b \psi(s,x+\xi(s)-\xi(t))ds\right)^m. \tag{3.64}$$

Then

$$v_\lambda(t,x)=\sum_{m=0}^{\infty}\frac{\lambda^m}{m!}g_m(t,x), \tag{3.65}$$

and the functions $g_m(t,x)$ satisfy a system of integro-differential equations

$$\frac{\partial}{\partial t}g_m(t,x)+L_t g_m(t,x)+\psi(t,x)g_{m-1}(t,x)=0. \tag{3.66}$$

This system can be obtained in exactly the same way as in Theorem 26°. On the other hand, one should only show that the series (3.65) can be differentiated term-by-term and compare coefficients of the same powers of λ.

The following theorem can be established by analogy with 26°.

27°. Theorem. *Let the functions $\psi(x)$ and $f(x)$ be twice continuously differentiable and bounded with both their derivatives. Suppose that*

$$z(t,x) = \mathbf{M}\exp\left\{\int_t^b \psi(s,\xi(s)-\xi(t)+x)ds\right\} f(\xi(b)-\xi(t)+x). \quad (3.67)$$

Then for $t < B$, the function $z(t,x)$ satisfies the equation

$$\frac{\partial}{\partial t}z(t,x) + L_t z(t,x) + \psi(t,x)z(t,x) = 0$$

with the boundary condition $z(b,x) = f(x)$.

Probability of the Process Staying in a Certain Region

Let $G(t,x)$ be some real-valued function on $R_+ \times X$, continuous in t and smooth in x. Assume that for all $t \in [a,b]$, the equation $G(t,x) = 0$ defines a connected closed smooth surface in X. We shall denote it by Γ_t, and the closed region bounded by this surface will be denoted by D_t. We are interested in the probability

$$\mathbf{P}\{\xi(t) \in D_t, a \le t \le b\}.$$

To find it we use the function

$$Q(t,x) = \mathbf{P}\{\xi(s) - \xi(t) + x \in D_s, t \le s \le b\} \quad (3.68)$$

and construct the twice continuously differentiable nonnegative function $\psi(t,x)$ satisfying the condition $\psi(t,x) = 0$ for $x \in D_t$. Then for $x \overline{\in} D_t$, the functions

$$\mathbf{M}\exp\{-n\int_t^b \psi(s,\xi(s)-\xi(t)+x)ds\} = Q_n(t,x)$$

satisfy the condition $\lim_{n\to\infty} Q_n(t,x) = 0$.

Let $\psi(t)$ be twice continuously differentiable. Then by virtue of Theorem 26°,

$$\frac{\partial}{\partial t}Q_n(t,x) + L_t Q_n(t,x) - \psi(t,x)Q_n(t,x) = 0, \quad (3.69)$$
$$Q_n(b,x) = 1.$$

28°. Theorem. *Let D_t^0 be the interior of D_t. Then*

$$Q(t,x) = \lim_{n\to\infty} Q_n(t,x), \quad x \in D_t^0.$$

Proof. Note that

$$\exp\left\{-n\int_t^b \psi(s,\xi(s)-\xi(t)+x)ds\right\}=1$$

if the event $\bigcap_{t\le s\le b}\{\xi(s)-\xi(t)+x\in D_s\}$ takes place, and that

$$\lim_{n\to\infty}\exp\left\{-n\int_t^b \psi(s,\xi(s)-\xi(t)+x)ds\right\}=0$$

if $\psi(\tau,\xi(\tau)-\xi(t)+x)>0$ for some $\tau\in[t,b]$ (since $\int_t^b\psi(s,\xi(s)-\xi(t)+x)ds>0$). Hence,

$$\lim_{n\to\infty}\mathbf{M}\exp\left\{-n\int_t^b \psi(s,\xi(s)-\xi(t)+x)ds\right\}$$
$$=\mathbf{M}I_{\{\bigcap_{t\le s\le b}\{\xi(s)-\xi(t)\in D_s\}\}}=Q(t,x).$$

□

Remark 1. Suppose that we can pass to the limit as $n\to\infty$ in Eq. (3.69). Then $Q(t,x)$ satisfies the equation

$$\frac{\partial}{\partial t}Q(t,x)+(a^*(t),Q'_x(t,x))+\frac{1}{2}\operatorname{Sp}B^*Q''_{xx}(t,x)$$
$$+\int_{x+y\in D_t}[Q(t,x+y)-Q(t,x)-(Q'_x(t,x),y)I_{\{|y|\le c\}}]\Pi^*_s(dy)=0, \tag{3.70}$$

for $t<b$, $x\in D_t^0$, $Q(t,x)=0$, $x\in\Gamma_t$, $Q(b,x)=1$, $x\in D_b^0$.

Remark 2. Eq. (3.70) is valid for a composed Poisson process. In this case, it takes the form

$$\frac{\partial}{\partial t}Q(t,x)+\int_{x+y\in D_t}[Q(t,x+y)-Q(t,x)]\Pi_t(dy)=0, \tag{3.71}$$
$$x\in D_t^0,\qquad Q(b,x)=1 \text{ for } x\in D_t.$$

To show this, let us rewrite (3.69) as follows:

$$\int_t^b\int[Q_n(t,x+y)-Q_n(s,x)]\Pi_s(dy)+1-Q_n(t,x)=0.$$

In this relation, the limit transition is possible; one should take into account that $Q_n(s,x)\to 0$ for $x\overline{\in}D_t$.

CHAPTER 4

HOMOGENEOUS PROCESSES

§4.1 General Properties

Let us consider homogeneous processes with independent increments defined on R_+ with values in a finite-dimensional Euclidean space X, i.e., processes with independent increments $\xi(t)$ such that $\xi(0) = 0$ and for $h > 0$, $t \geq 0$, $\xi(t+h) - \xi(t)$ has the same distribution as $\xi(h)$.

A characteristic function of a homogeneous process is given by

$$\varphi_t(z) = \mathbf{M} \exp\{i(z, \xi(t))\} = \exp\{tK(z)\}, \tag{4.1}$$

where

$$K(z) = i(a, z) - \frac{1}{2}(Bz, z) + \int (e^{i(z,x)} - 1 - i(z, x)\mathbf{I}_{\{|x| \leq c\}})\Pi(dx), \tag{4.2}$$

$c > 0$, $a \in X$, $B \in L_+(X)$, Π is a σ-finite measure on $\mathcal{B}_X$ for which

$$\Pi(\{0\}) = 0, \qquad \int \frac{|x|^2}{1 + |x|^2} \Pi(dx) < \infty.$$

The function $K(z)$ is called a *cumulant* of the process.

1°. *Let $f(z)$ be a twice continuously differentiable function bounded and continuous with its derivatives. Then*

$$\lim_{t \to 0} \frac{1}{t}[\mathbf{M} f(x + \xi(t)) - f(x)] = Af(x), \tag{4.3}$$

where

$$Af(x) = (a, f'(x)) + \frac{1}{2} \operatorname{Sp} B f''(x) + \int (f(x+y) - f(x) - (f'(x), y)\mathbf{I}_{\{|y| \leq c\}})\Pi(dy). \tag{4.4}$$

The operator A is called a generator (or a generating operator) of the process.

Proof. The conditions of Theorem 25°, Chapter 3, are clearly satisfied in this case. Thus, if

$$u_t(s,x) = \mathbf{M}f(\xi(t) - \xi(s) + x), \quad s < t,$$

then

$$\frac{\partial}{\partial s} u_t(s,x) + L_s u_t(s,x) = 0, \qquad u_t(t,x) = f(x). \tag{4.5}$$

But

$$L_s f = Af, \qquad \begin{aligned} u_t(s,x) &= \mathbf{M}f(\xi(t) - \xi(s) + x) \\ &= \mathbf{M}f(\xi(t-s) + x) = v(t-s,x), \end{aligned}$$

where v is defined by the last equality in the chain. The function $v(t,x)$ satisfies the equation

$$\frac{\partial}{\partial t} v(t,x) = Av(t,x), \qquad v(0,x) = f(x). \tag{4.6}$$

Obviously, $v'(t,x) = \mathbf{M}f'(\xi(t)+x)$, $v''(t,x) = \mathbf{M}f''(\xi(t)+x)$ and therefore $v'(t,x) \to f'(x)$, $v''(t,x) \to f''(x)$ as $t \to 0$ for all x, hence $Av(t,x) \to Af(x)$ as $t \to 0$ for all x. Equation (4.6) yields

$$\frac{1}{t}[v(t,x) - f(x)] = \frac{1}{t}\int_0^t Av(s,x)\,ds.$$

Passing to the limit as $t \to 0$, we get (4.3).

□

2°. *Let $f(x)$ be a twice continuously differentiable function integrable on X with its derivatives, and let $\tilde{f}(z) = \int e^{i(z,x)} f(x)\,dx$. Then*

$$\int e^{i(z,x)} Af(x)\,dx = K(z)\tilde{f}(z). \tag{4.7}$$

Proof. The following formulas are valid:

$$\int e^{i(z,x)}(a, f'(x))\,dx = -i(a,z)\tilde{f}(z), \tag{4.8}$$

$$\frac{1}{2}\int e^{i(z,x)} \mathrm{sp}\,Bf''(x)\,dx = -\frac{1}{2}(Bz,z)\tilde{f}(z), \tag{4.9}$$

$$\int e^{i(z,x)} \int [f(x+y) - f(x) - (y, f'(x))\mathrm{I}_{\{y\}}]\Pi(dy)dx$$
$$= \int [e^{i(z,y)} - 1 - i(y,z)\mathrm{I}_{\{|y|\le c\}}]\Pi(dy)\cdot \tilde{f}(z). \tag{4.10}$$

In fact,

$$\int e^{i(z,x)} \frac{1}{h}[f(x+ha) - f(x)dx = \int e^{i(z,x)} f(x)\frac{1}{h}(e^{i(a,z)h} - 1)dx;$$

The r.h.s. tends to the r.h.s. of (4.8) as $h \to 0$, and the l.h.s. can be rewritten as follows:

$$\int e^{i(z,x)} \frac{1}{h}\int_0^h (a, f'(x+ta))dt\, dx = \int (a, f'(y))\frac{1}{h}\int_0^h e^{i(z,y-ta)}dt$$
$$= \int e^{i(z,y)}(a, f'(y))\left[\frac{1}{h}\int_0^h e^{it(a,z)}dt\right] dy.$$

Since $\left|\frac{1}{h}\int_0^h e^{it(a,z)}dt\right| \le 1$, $\frac{1}{h}\int_0^h e^{it(a,z)}dt \to 1$ and $|(a, f'(y))|$ is integrable, we find

$$\lim_{h\to 0} \int e^{i(z,y)}(a, f'(y))\left[\frac{1}{h}\int_0^h e^{it(a,z)}dt\right] dy = \int e^{i(z,y)}(a, f'(y))dy.$$

Thus, (4.8) is proved, and it implies

$$\int e^{i(z,x)}(f''(x)b, a)dx = \int e^{i(z,x)}((f'(x), b)', a)dx$$
$$= i(a,z)\int e^{i(a,z)}(f'(x), b)dx$$
$$= -(a,z)(b,z)\int e^{i(z,x)} f(x)dx.$$

If $\{e_k\}$ is an orthonormal basis in X, we have

$$\int e^{i(z,x)}\mathrm{Sp} f''(x)B\, dx = \int e^{i(z,x)}\sum_{k,j}(Be_k, e_j)(f''(x)e_k, e_j)dx$$
$$= -\sum (Be_k, e_j)(z, e_k)(z, e_j)\tilde{f}(z) = -(Bz, z)\tilde{f}(z).$$

Thus, (4.9) is proved; (4.10) also follows from (4.8). Summing up the relations (4.8), (4.9), and (4.10), we get (4.7).

□

Consider the family of transformations $T_t f$ defined on C_X with values in C_X,

$$T_t f(x) = \mathbf{M} f(x + \xi(t)), \quad t \geq 0. \tag{4.11}$$

3°. *$T_t f$ is a semi-group of transformations, i.e., $T_{t+s} f = T_t T_s f$ for $t > 0$, $s > 0$. T_t transforms the subspaces C_X^0 (of uniformly continuous bounded functions on X) and $C_X^{(k)}$ (of k-times continuously differentiable functions) into themselves; furthermore, $T_t f$ has the same modulus of continuity as f, and $\sup_x |((T_t f)', a)| \leq \sup_x |(f', a)|$; similar inequalities are valid for all derivatives. $T_t f$ is continuous in t for every $x \in X$, $t \geq 0$. If $f \in C_X^{(2)}$, then $\frac{d}{dt} T_t f = A T_t f$.*

Proof. (4.11) clearly allows term-by-term differentiation. This implies estimates of the derivatives of $T_t f$ in terms of the derivatives of f. (4.11) also implies the conservation of the continuity modulus of $T_t f$ (strictly speaking, the continuity modulus of $T_t f$ is not greater than the continuity modulus of f). The continuity of $T_t f$ with respect to t follows from the stochastic continuity of $\xi(t)$.

Let $t > 0$, $s > 0$. Then

$$\begin{aligned} T_{t+s} f(x) &= \mathbf{M} f(\xi(t+s) + x) = \mathbf{M} f(\xi(t+s) - \xi(t) + \xi(t) + x) \\ &= \mathbf{M}\mathbf{M}(f(\xi(t+s) - \xi(t) + \xi(t) + x)/\xi(t)) \\ &= \mathbf{M}[(\mathbf{M} f(\xi(t+s) - \xi(t) + y))_{y=\xi(t)+x}] \\ &= \mathbf{M}[(\mathbf{M} f(\xi(s) + y))_{y=\xi(t)+x}] = \mathbf{M}[T_s f(y)_{y=\xi(t)+x}] \\ &= \mathbf{M} T_s f(\xi(t) + x) = T_t T_s f(x). \end{aligned}$$

□

Let us define the operator R_λ acting from C_X into C_X according to the formula

$$R_\lambda f(x) = \int_0^\infty e^{-\lambda t} T_t f(x) dt.$$

These operators are defined for $\operatorname{Re} \lambda > 0$. If we know $R_\lambda f$ for all λ such that $\operatorname{Re} \lambda > 0$ (for all $\lambda > 0$ or on the straight line $\operatorname{Re} \lambda = \alpha > 0$), we can reconstruct $T_t f(x)$ for all $t > 0$.

4°. Theorem. *Let $r_\lambda(dx)$ be a measure such that*

$$R_\lambda f(x) = \int f(x + y) r_\lambda(dx). \tag{4.12}$$

Then

(a) $\int e^{i(z,y)} r_\lambda(dy) = \frac{1}{\lambda - K(z)}$;

(b) *$\lambda r_\lambda(dy)$ is a probability distribution; it is infinitely divisible and its characteristic function can be represented as*

$$\frac{\lambda}{\lambda - K(z)} = \exp\left\{\int (e^{i(z,x)} - 1)\Pi_\lambda(dx)\right\}, \tag{4.13}$$

where

$$\begin{gathered}\Pi_\lambda(A) = \int_0^\infty \frac{1}{t} e^{-\lambda t}\mathbf{P}\{\xi(t) \in A\}dt, \qquad 0\overline{\in}A,\\ \Pi_\lambda(\{0\}) = 0\end{gathered} \tag{4.14}$$

Proof. Denote $F_t(dy) = P\{\xi(t) \in dy\}$. Then

$$T_t f(z) = \mathbf{M} f(x + \xi(t)) = \int f(x+y)F_t(dy),$$

$$R_\lambda f(x) = \int f(x+y)\int_0^\infty e^{-\lambda t}F_t(dy)dt;$$

hence, (4.12) holds for

$$r_\lambda(dx) = \int_0^\infty e^{-\lambda t}F_t(dx)dt,$$

$r_\lambda(A)$ is a bounded measure, $r_\lambda(X) = \frac{1}{\lambda}$, and thus, $\lambda r_\lambda(A)$ is a probability measure. We now find the Fourier transform of r_λ:

$$\begin{aligned}\int e^{i(z,x)} r_\lambda(dx) &= \lambda \int_0^\infty \left[\int e^{i(z,x)}F_t(dx)\right]e^{-\lambda t}dt\\ &= \int_0^\infty e^{tK(z)-\lambda t}dt = \frac{1}{\lambda - K(z)}.\end{aligned}$$

Let us show that the measure $\Pi_\lambda(A)$ is finite for all $A \in B_0$ (B_0 is a ring of Borel sets lying at a positive distance from the point 0). Let $U_\varepsilon = \{|x| : |x| > \varepsilon\}$. Then

$$\begin{aligned}\mathbf{P}\{\xi(t) \in U_\varepsilon\} &= \mathbf{P}\{|\xi(t)| > \varepsilon\}\\ &= \mathbf{P}\{|\xi'(t) + \xi''(t)| > \varepsilon\},\end{aligned}$$

where $\xi'(t) = \int_{|x|>1} x\nu_t(dx)$, $\xi''(t) = \xi(t) - \xi'(t)$, and $\nu_t(A)$ is the number of jumps of $\xi(s)$ on $[0,t]$ which hit the set A. Thus,

$$\begin{aligned}\mathbf{P}\{\xi(t) \in U_\varepsilon\} &\le \mathbf{P}\{\xi'(t) \neq 0\} + \mathbf{P}\{|\xi''(t)| > \varepsilon\}\\ &\le 1 - e^{-t\Pi(U_\varepsilon)} + \frac{M|\xi''(t)|^2}{\varepsilon^2}.\end{aligned}$$

Note that $\xi''(t)$ is also a homogenous process having all the moments. Hence,

$$\begin{aligned}\mathbf{M}|\xi''(t)|^2 &= |\mathbf{M}\xi''(t)|^2 + \mathbf{M}|\xi''(t) - \mathbf{M}\xi''(t)|^2 \\ &= t^2|\mathbf{M}\xi''(1)|^2 + t\mathbf{M}|\xi''(1) - \mathbf{M}\xi''(1)|^2.\end{aligned}$$

Therefore, for $t \leq 1$,

$$\mathbf{P}\{\xi(t) \in U_\varepsilon\} \leq qt, \tag{4.15}$$

where q is some constant; and thus,

$$\Pi_\lambda(U_\varepsilon) \leq q + \int_1^\infty \frac{1}{t}e^{-\lambda t}\mathbf{P}\{\xi(t) \in U_\varepsilon\}dt \leq q + \frac{e^{-\lambda}}{\lambda}.$$

Let us now show that

$$\int_{|x|\leq 1} |x|\Pi_\lambda(dx) = \int_0^\infty \frac{1}{t}e^{-\lambda t}\int_{|x|\leq 1} |x|F_t(dx)dt < \infty.$$

Let us estimate the expression

$$\int_{|x|\leq 1} |x|F_t(dx) = \mathbf{M}|\xi(t)|\mathbf{I}_{\{|\xi(t)|\leq 1\}} = \mathbf{M}|\xi'(t) + \xi''(t)|\mathbf{I}_{\{|\xi(t)|\leq 1\}}.$$

Let A_t be the event $\nu(t, U_1) > 0$. Then

$$\begin{aligned}|\xi'(t) + \xi''(t)|I_{\{|\xi(t)|\leq 1\}} &\leq |\xi''(t)| + I_{A_t}, \\ \mathbf{M}|\xi(t)|\mathbf{I}_{\{|\xi(t)|\leq 1\}} &\leq \mathbf{P}\{A_t\} + \mathbf{M}|\xi''(t)| \\ &\leq 1 - \exp\{-t\Pi(U_i)\} + \sqrt{\mathbf{M}|\xi''(t)|^2}.\end{aligned}$$

Therefore, there exists q_1 such that

$$\int_{|x|\leq 1} |x|F_t(dx) \leq q_1\sqrt{t}; \tag{4.16}$$

then

$$\int_{|x|\leq 1} |x|\Pi_\lambda(dx) \leq \int_0^\infty \frac{e^{-\lambda t}}{t}q_1\sqrt{t}\,dt = \int_0^\infty \frac{q_1}{\sqrt{t}}e^{-\lambda t}dt < \infty.$$

Now, by using the formula

$$\int_0^\infty \frac{e^{-\alpha t} - e^{-\beta t}}{t}dt = \ln\frac{\beta}{\alpha},$$

we obtain

$$\begin{aligned}\frac{\lambda}{\lambda - K(z)} &= \exp\left\{\ln \frac{\lambda}{\lambda - K(z)}\right\} = \exp\left\{\int_0^\infty \frac{e^{-\lambda t + tK(z)} - e^{-\lambda t}}{t} dt\right\} \\ &= \exp\left\{\int_0^\infty \frac{e^{-\lambda t}}{t}(e^{tK(z)} - 1)dt\right\} \\ &= \exp\left\{\int_0^\infty \frac{e^{-\lambda t}}{t}[\int e^{i(z,x)}F_t(dx) - 1]dt\right\} \\ &= \exp\left\{\lim_{\varepsilon \to 0} \int_\varepsilon^\infty \frac{e^{-\lambda t}}{t} \int (e^{i(z,x)} - 1)F_t(dx)dt\right\} \\ &= \exp\left\{\lim_{\varepsilon \to 0} \int (e^{i(z,x)} - 1) \int_\varepsilon^\infty \frac{e^{-\lambda t}}{t} F_t(dx)\right\}.\end{aligned}$$

To complete the proof of the theorem, it remains to show that

$$\lim_{\varepsilon \to 0} \int (e^{i(z,x)} - 1) \int_\varepsilon^\infty \frac{e^{-\lambda t}}{t} F_t(dx)dt = \int (e^{i(z,x)} - 1)\Pi_\lambda(dx)$$

or that

$$\lim_{\varepsilon \to 0} \int (e^{i(z,x)} - 1) \int_0^\varepsilon \frac{e^{-\lambda t}}{t} F_t(dx)dt = 0. \tag{4.17}$$

However,

$$\lim_{\varepsilon \to 0} | \int_{|x|>1} (e^{i(z,x)} - 1) \int_0^\varepsilon \frac{e^{-\lambda t}}{t} F_t(dx)dt \le 2 \lim_{\varepsilon \to 0} \int_0^\varepsilon \frac{e^{-\lambda t}}{t} F_t(U_1)dt = 0$$

by virtue of (4.15), and

$$\lim_{\varepsilon \to 0} | \int_{|x|\le 1} (e^{i(z,x)} - 1) \int_0^\varepsilon \frac{e^{-\lambda t}}{t} F_t(dx)| \le |z| \lim_{\varepsilon \to 0} \int_0^\varepsilon \frac{e^{-\lambda t}}{t} \int_{|x|\le 1} |x| F_t(dx) = 0$$

by virtue of (4.16). Hence, (4.17) is proved.

□

Now consider the moments of the integrals

$$\eta_x(f) = \int_0^\infty e^{-\lambda t} f(x + \xi(t))dt.$$

Clearly, $M\eta_x(f) = R_\lambda f(x)$. But all the other moments can also be expressed in terms of the resolvent.

5°. Lemma.

$$\mathbf{M}[\eta_x(f)]^n = n!R_{n\lambda}[fR_{(n-1)\lambda}[fR_{(n-2)\lambda}[f\dots fR_\lambda f]\dots](x). \tag{4.18}$$

Proof. We have

$$\left(\int_0^\infty e^{-\lambda t}\varphi(t)dt\right)^n$$
$$= n!\int_{0<t_1<\cdots<t_n<\infty} e^{-\lambda t_1-\lambda t_2\cdots-\lambda t_n}\varphi(t_1)\varphi(t_2)\dots\varphi(t_n)dt_1\,dt_2\dots dt_n.$$

Consequently,

$$\mathbf{M}[\eta_x(f)]^n = n!\int_{0<t_1<\cdots<t_n<\infty} e^{-\lambda t_1-\cdots-\lambda t_n}\mathbf{M}f(x+\xi(t_1))\dots$$
$$f(x+\xi(t_{n-1}))\mathbf{M}(f(x+\xi(t_n))/\xi(t_{n-1}))dt_1\dots dt_n$$
$$= n!\int_{0<t_1<\cdots<t_{n-1}} e^{-\lambda t_1-\cdots-\lambda t_{n-1}}\mathbf{M}f(x+\xi(t_1))\dots$$
$$f(x+\xi(t_{n-1}))\int_{t_{n-1}}^\infty e^{-\lambda t_n T_{t_n-t_{n-1}}}f(x+\xi(t_{n-1}))\,dt_n$$
$$= n!\int_{0<t_1<\cdots<t_{n-1}} e^{-\lambda t_1-\cdots-\lambda t_{n-1}}\mathbf{M}f(x+\xi(t_1))\dots$$
$$f(x+\xi(t_{n-1}))e^{-\lambda t_{n-1}}R_\lambda f(x+\xi(t_{n-1}))dt_1\dots dt_{n-1}$$
$$= n!\int_{0<t_1<\cdots<t_{n-2}} e^{-\lambda t_1-\cdots-\lambda t_{n-2}}\mathbf{M}f(x+\xi(t_1))\dots f(x+\xi(t_{n-2}))$$
$$\times\int_{t_{n-2}}^\infty e^{-2\lambda t_{n-1}}\mathbf{M}(f(x+\xi(t_{n-1}))R_\lambda f(x+\xi(t_{n-1}))/$$
$$\xi(t_{n-2}))dt_{n-1}dt_1\dots dt_{n-2}$$
$$= n!\int_{0<t_1<\cdots<t_{n-2}} e^{-\lambda t_1-\cdots-\lambda t_{n-2}}e^{-2\lambda t_{n-2}}\mathbf{M}f(x+\xi(t_1)\dots$$
$$f(x+\xi(t_{n-2}))R_{2\lambda}[fR_\lambda f](x+\xi(t_{n-2}))dt_1\dots dt_{n-2} = \dots$$
$$= n!R_{n\lambda}[fR_{(n-1)\lambda}[fR_{(n-2)\lambda}[\dots fR_\lambda f]\dots](x).$$

6°. Theorem. *Let $f(x)$ be a continuous bounded function. We put*

$$v_t(x) = \mathbf{M}\exp\left\{\int_0^t f(\xi(s)+x)ds\right\}.$$

Then $v_t(x)$ satisfies the integral equation

$$v_t(x) = 1+\int_0^t T_s[fv_{t-s}](x)ds$$
$$= \iint_0^t f(x+y)v_{t-s}(x+y)F_s(dy)ds. \tag{4.19}$$

Proof. Let us use the equality

$$\int_0^t \exp\left\{\int_s^t f(x+\xi(u))du\right\} f(x+\xi(s))ds = -\int_0^t d_s \exp\left\{\int_s^t f(x+\xi(u))du\right\}$$
$$= -1 + \exp\left\{\int_0^t f(x+\xi(u))du\right\}.$$

Therefore,

$$\begin{aligned}
v_t(x) &= 1 + \mathbf{M}\int_0^t \exp\left\{\int_s^t f(x+\xi(u))du\right\} f(x+\xi(s))ds \\
&= 1 + \int_0^t \mathbf{M} f(x+\xi(s))\mathbf{M}(\exp\left\{\int_s^t f(x+\xi(u))du\right\}/\xi(s))ds \\
&= 1 + \int_0^t \mathbf{M} f(x+\xi(s))\mathbf{M}(\exp\left\{\int_s^t f(x+\xi(s)+\xi(u)-\xi(s))du\right\}/\xi(s))ds \\
&= 1 + \int_0^t \mathbf{M}[f(x+\xi(s))(\mathbf{M}\exp\left\{\int_s^t f(y+\xi(u)-\xi(s))du\right\}|_{y=x+\xi(s)}]ds \\
&= 1 + \int_0^t \mathbf{M}[f(x+\xi(s))v_{t-s}(y)|_{y=s+\xi(s)}]ds \\
&= 1 + \int_0^t T_s[fv_{t-s}](x)ds.
\end{aligned}$$

□

Remark. By changing the variables $s = t - u$, we can rewrite Eq. (4.19) in the form

$$v_t(x) = 1 + \int_0^t T_{t-s}[fv_s](x)ds. \tag{4.20}$$

Suppose that $v_t(\cdot) \in C_x^{(2)}$ (this is valid, e.g., for $f \in C_x^{(2)}$) and differentiate (4.20) with respect to t.

Using 3°, we get

$$\frac{\partial}{\partial t}v_t(x) = f(x)v_t(x) + \int_0^t AT_{t-s}[fv_s](x)ds$$
$$= f(x)v_t(x) + A\int_0^t T_{t-s}[fv_s](x)ds.$$

(A can be carried out of the integral sign, because $fv_s \in C_x^{(2)}$ and the derivatives of fv_s are bounded uniformly in s). Taking into account that $A1 = 0$, we obtain the equation

$$\frac{\partial}{\partial t}v_t(x) = Av_t(x) + f(x)v_t(x), \tag{4.21}$$

which is an analogue of Eq. (3.63).

Strong Markov Property

Let us denote by $\mathcal{F}_t$ a σ-algebra generated by the variables $\xi(s)$, $s \leq t$. The process $\xi_h(t) = \xi(t+h) - \xi(h)$ is independent of the σ-algebra $\mathcal{F}_h$ and has the same distribution as $\xi(t)$. This property is preserved for some random moments h.

Let us consider *stopping moments (s.m.)* with respect to a flow of σ-algebras $(\mathcal{F}_t)_{t\geq 0}$. We put $\mathcal{F}_\infty = \bigvee_t \mathcal{F}_t$.

Definition. A nonnegative random variable τ (which can take the value $+\infty$) is called a *s.m. (with respect to the flow $(\mathcal{F}_t)_{t\geq 0}$), if the event $\{\tau \leq t\} \in \mathcal{F}_t$ for all $t \geq 0$.*

If τ is a s.m. , we denote by $\mathcal{F}_\tau$ a σ-algebra of the events $A \in \mathcal{F}_\infty$ such that $A \cap \{\tau \leq t\} \in \mathcal{F}_t$ for all $t \geq 0$.

$\mathcal{F}_\tau$ are events that can happen to the process $\xi(\cdot)$ up to the time τ (every fixed time t is a s.m., and $\mathcal{F}_t$ is a corresponding σ-algebra).

7°. Theorem. *If τ is a finite s.m. and $\mathcal{F}_\tau$ is a corresponding σ-algebra, then the process $\xi_\tau(t) = \xi(t+\tau) - \xi(\tau)$ is independent of $\mathcal{F}_\tau$ and its distributions coincide with the distribution of $\xi(t)$ (i.e., $\xi_\tau(t)$ is also a homogeneous process with independent increments with the same cumulant as $\xi(t)$).*

Proof. To prove the theorem, it suffices to establish the formula

$$\begin{aligned} &\mathbf{M} I_A g(\xi(\tau+s_1) - \xi(\tau), \ldots, \xi(\tau+s_k) - \xi(\tau)) \\ &\quad = \mathbf{P}(A)\mathbf{M} g(\xi(s_1), \ldots, \xi(s_k)), \end{aligned} \tag{4.22}$$

whatever $\mathcal{F}_\tau$-measurable set $A, s_1, \ldots, s_k \in R_+$ and continuous bounded function $g(x_1, x_2, \ldots, x_k)$ we take. First assume that τ possesses only a countable number of values $t_1, t_2, \ldots, t_n, \ldots$. Then

$$\begin{aligned} &\mathbf{M} I_A g(\xi(\tau+s_1) - \xi(\tau), \ldots, \xi(\tau+s_k - \xi(\tau)) \\ &\quad = \sum_n \mathbf{M} I_A I_{\{\tau=t_n\}} g\xi(\tau+s_1) - \xi(\tau), \ldots, \xi(\tau+s_k) - \xi(\tau)) \\ &\quad = \sum_n \mathbf{M} I_{A\cap\{\tau=t_n\}} g(\xi(t_n+s_1) - \xi(t_n), \ldots, \xi(t_n+s_n) - \xi(t_n)). \end{aligned}$$

Note that

$$A \cap \{\tau = t_n\} = A \cap \{\tau \leq t_n\} \setminus \cup_{t_j < t_n} \{\tau \leq t_j\}.$$

All events on the r.h.s. belong to $\mathcal{F}_{t_n}(\mathcal{F}_{t_j} \subset \mathcal{F}_{t_n}$ for $t_j < t_n$; hence, $A \cap \{\tau = t_n\} \in \mathcal{F}_{t_n}$. The process $\xi(s+t_n) - \xi(t_n)$ does not depend on $\mathcal{F}_{t_n}$ and has the same distributions as $\xi(s)$. Thus,

$$\begin{aligned} &\mathbf{M} I_{A\cap\{\tau=t_n\}} g(\xi(t_n+s_1) - \xi(t_n), \ldots, \xi(t_n+s_k) - \xi(t_n)) \\ &\quad = \mathbf{P}(A \cap \{\tau = t_n\})\mathbf{M} g(\xi(s_1), \ldots, \xi(s_k)). \end{aligned}$$

Summing these equalities over n, we get (4.22).

Now let τ be an arbitrary s.m. $A \in \mathcal{F}_\tau$. Denote by τ_n a variable defined by

$$\tau_n = \sum_{k=0}^{\infty} \frac{k}{n} I_{\{\frac{k-1}{n} < \tau \le \frac{k}{n}\}}.$$

τ_n is a s.m., since

$$\{\tau_n \le t\} = \{\tau \le t^{(n)}\},$$

where $t^{(n)} = \frac{k}{n}$ for $\frac{k}{n} < t \le (k+1)/n$ and $\{\tau_n \le t\} \in \mathcal{F}_{t_n} \in \mathcal{F}_t$. Let us show that $A \in \mathcal{F}_{\tau_n}$. We have

$$A \cap \{\tau_n \le t\} = A \cap \{\tau \le t^{(n)}\} \in \mathcal{F}_{t^{(n)}} \subset \mathcal{F}_t;$$

τ_n takes a countable number of values; thus,

$$\begin{aligned} &\mathbf{M} I_A g(\xi(\tau_n + s_1) - \xi(\tau_n), \dots, \xi(\tau_n + s_k) - \xi(\tau_n)) \\ &\quad = \mathbf{P}(A)\mathbf{M} g(\xi(s_1), \dots, \xi(s_k)). \end{aligned} \tag{4.23}$$

Note that $\tau_n \ge \tau$ and $\tau_n \to \tau$ as $n \to \infty$, and that $\xi(s)$ is a continuous from the right process. Hence, $\xi(\tau_n + s_i) \to \xi(\tau + s_i)$ as $n \to \infty$; and we can take the limit (as $n \to \infty$) in (4.23) (g is continuous and bounded). So we get (4.22).

□

Remark. Suppose that τ can take the value $+\infty$. Then the process $\xi_\tau(t)$ is defined only for $\tau < \infty$. Equation (4.22) remains valid if the set A is replaced by $A \cap \{\tau < \infty\}$. The proof of this statement is contained in the aforesaid proof. Therefore,

$$\begin{aligned} &\mathbf{M} I_A I_{\{\tau<\infty\}} g(\xi(\tau + s_1) - \xi(\tau), \dots, \xi(\tau + s_k) - \xi(\tau)) \\ &\quad = \mathbf{P}(A \cap \{\tau < \infty\})\mathbf{M} g(\xi(s_1), \dots, \xi(s_k)) \end{aligned} \tag{4.24}$$

for $A \subset \mathcal{F}_\tau, s_1, \dots, s_k \in R_+$, $g \in C_{X^k}$. This means that the conditional distributions of the process $\xi_\tau(s)$ with respect to $\mathcal{F}_\tau$ on the set $\{\tau < \infty\} \subset \mathcal{F}_\tau$ coincide with the distributions of the process $\xi(s)$.

Let us point out some classes of stopping moments.

Let us consider that the σ-algebra $\mathcal{F}$ of the probability space $\{\Omega, \mathcal{F}, \mathbf{P}\}$ is complete with respect to $\mathbf{P}$ and that the σ-algebras $\mathcal{F}_t$ contain all subsets of $\mathcal{F}$ of measure zero. Under this assumption, a variable coinciding with an $\mathcal{F}_t$-measurable one with probability 1 is also $\mathcal{F}_t$-measurable. If $\tau' = \tau$ with probability 1 and τ is a s.m., then τ' is also a s.m.

The next statement is another important consequence of completeness of the σ-algebras $\mathcal{F}_t$.

Let the real-valued process $\zeta(s, \omega)$ for $s \in [0, t]$ be measurable with respect to $\mathcal{B}_{[0,t]} \otimes \mathcal{F}_t(\mathcal{B}_{[0,t]}$ is a σ-algebra of Borel subsets of $[0, t]$). (A process is called *progressively measurable* if this condition is satisfied for all $t \in R_+$.) Then $\sup_{s \le t} \zeta(s, \omega)$ is also $\mathcal{F}_t$-measurable (see 9°, Chapter 0, the theorem on measurability of projections).

8°. Lemma. *A process $\eta(t,\omega)$ continuous from the right is progressively measurable, provided that $\eta(t,\omega)$ is $\mathcal{F}_t$-measurable for all $t \in R_+$.*

Proof. Given t, we put $\eta_n(s) = \eta(\frac{k}{n}t,\omega)$ for $\frac{k-1}{n}t < s \le \frac{k}{n}t$, $k = 1,2,\dots,n$, $\eta_n(0) = \eta(0)$. Then $\eta_n(s,\omega) \to \eta(s,\omega)$ as $n \to \infty$, since $\eta(s,\omega)$ is continuous from the right. On the other hand, $\eta_n(s,\omega)$ is measurable with respect to $\mathcal{B}_{[0,t]} \otimes \mathcal{F}_t$.

□

Remark. If the process $\eta(t,\omega)$ is progressively measurable and τ is a s.m., then $\eta(\tau,\omega)$ is a $\mathcal{F}_\tau$- measurable variable. In fact, $\{\eta(t,\omega) \in A\} \cap \{\tau \le t\} \in \mathcal{F}_t$, since for $s \le t$, $\eta(s,\omega)$ is measurable with respect to $\mathcal{B}_{[0,t]} \otimes \mathcal{F}_t$, and therefore, we have the superposition of the measurable mappings

$$(\{\tau \le t\}, \mathcal{F}_t) \xrightarrow{(\tau,\omega)} ([0,t] \times \Omega, \mathcal{B}_{[0,t]} \otimes \mathcal{F}_t) \xrightarrow{\eta(\tau(\omega),\omega)} (X, \mathcal{B}_X).$$

on the set $\{\tau \le t\} \in \mathcal{F}_t$.

9°. Theorem. *Let V be some Borel set in $R_+ \times X$, and $\eta(t)$ be a homogeneous process continuous from the right with independent increments. Let us write $\tau_V = \inf\{s : (s,\eta(s)) \in V\}$ (τ_V is called the first time of hitting the set V for the process $\eta(t)$). Then τ_V is a s.m. with respect to the filtration of σ-algebras $(\mathcal{F}_t)_{t\ge 0}$, $\mathcal{F}_t = \sigma\{\eta(s), s \le t\}$.*

Proof. First, we prove the next property of the filtration of σ-algebras $\mathcal{F}_t$, namely *the continuity from the right*. We put $\mathcal{F}_{t+} = \cap_{s>t} \mathcal{F}_s$. Then $\mathcal{F}_{t+} = \mathcal{F}$.

Note that $\xi(t+s) - \xi(t)$ is independent of the σ-algebra $\mathcal{F}_{t+}$. Clearly, for any $\varepsilon > 0$, the process $\xi(t+\varepsilon+s) - \xi(t+\varepsilon)$ does not depend on this σ-algebra. Hence, its limit as $\varepsilon \to 0$, which coincides with $\xi(t+s) - \xi(t)$, is also independent of it. Whatever function $g \in C_{X^{r+l}}$ and $0 < t_1 < \dots < t_r \le t < t_{r+1} < t_{r+l}$ are taken, we have

$$\begin{aligned}
&\mathbf{M}(g(\eta(t_1),\dots,\eta(t_r),\eta(t_{r+1}),\dots,\eta(t_{r+l}))/\mathcal{F}_{t+})\\
&\quad = \mathbf{M}(g(\eta(t_1),\dots,\eta(t_r),\eta(t_r)+\eta(t_{r+1})-\eta(t_r),\dots,\\
&\qquad \eta(t_r)+\eta(t_{r+l})-\eta(t_r))/\mathcal{F}_{t+})\\
&\quad = \mathbf{M}g(x_1,\dots,x_r,x_r+\eta(t_{r+1})-\eta(t_r),\dots,x_r+\eta(t_{r+l})-\eta(t_r))\Big|_{\substack{x_1=\eta(t_1),\\ \dots\dots\dots,\\ x_r=\eta(t_r)}}\\
&\quad = \mathbf{M}(g(\eta(t_1),\dots,\eta(t_{r+l}))/\mathcal{F}_t).
\end{aligned}$$

Since one can approximate (in the sense of the convergence in probability) any bounded $\mathcal{F}_{t+h}$-measurable variable ξ by the variables $g(\eta(t_1),\dots,\eta(t_{r+l}))$, we have $M(\xi|\mathcal{F}_{t+}) = M(\xi|\mathcal{F}_t)$ for all bounded $\mathcal{F}_{t+h}$-measurable variables with probability 1. In particular, if ξ is $\mathcal{F}_{t+}$-measurable, then $\xi = M(\xi|\mathcal{F}_t)$

with probability 1; i.e., every $\mathcal{F}_{t+}$-measurable variable coincides with a $\mathcal{F}_t$-measurable one almost everywhere. This means that $\mathcal{F}_t$ and $\mathcal{F}_{t+}$ coincide.

The event $\{\tau_V < t\}$ coincides with the event $\sup_{s\le t} I_{\{s,\xi(s))\in V\}} = 1$, which belongs to the σ-algebra $\mathcal{F}_t$, because $\zeta(t) = I_{\{(t,\eta(t))\in V\}}$ is a progressively measurable function ($\eta(t)$ is such a function by virtue of Lemma 8°; and $\xi(t) = \psi(t, \eta(t))$, where $\psi(t, x)$ is a Borel function). Thus, $\{\tau_V < t+h\} \in \mathcal{F}_{t+\delta}$ for all $h > 0$ and $\delta \geq h$.

$$\{\tau \leq t\} = \bigcap_h \{\tau_V < t + h\} \in \mathcal{F}_{t+\delta}, \{\tau_V \leq t\} \in \bigcap_{\delta>0} \mathcal{F}_{t+\delta} = \mathcal{F}_{t_+} = \mathcal{F}_t.$$

□

Remark. If V is an open or closed set or a set representable as a countable union of closed ones, then $\mathcal{F}_t$-measurability of $\{\tau_V < t\}$ can be established directly, with no use of the theorem on measurability of projections.

Corollary 1. *Let Γ be some surface in $R_+ \times X$ and G be a region bounded by this surface. If τ is the first time of hitting G, and $\xi(t)$ is a homogeneous process continuous from the right with independent increments, then the process $\xi(t + \tau) - \xi(\tau)$ does not depend on τ and $\xi(\tau)$and has the same distributions as $\xi(t)$.*

Corollary 2. *The first time τ_1 is a s.m. when $\xi(\tau_1) - \xi(\tau_1-) \in A$ (A is a Borel set in X, lying at a positive distance from the point 0), $\xi(t + \tau_1) - \xi(\tau_1)$ is independent of τ_1 and $\xi(\tau_1) - \xi(t_1-)$ and has the same distributions as $\xi(t)$.*

To show this, one can examine a composed process $(\xi(t), \nu_t(A))$ in $X \times R$, which is also a homogeneous process with independent increments. One should take V to be the set $\{(s, x, \alpha), s \in R_+, \alpha \in R : |\alpha| > 0\}$.

Another class of s.m. is connected with integrals.

10°. Theorem. *Let $\xi(t)$ be a homogeneous process continuous from the right with independent increments, and let $g(t, x)$ be a bounded Borel function from $R_+ \times X$ into R. Let us put*

$$\eta_t = \int_0^t g(s, \xi(s))ds.$$

If V is a Borel set in $R_+ \tau X$, $\tau = \{\inf t : (s, \eta_s) \in V\}$, then τ is a stopping moment.

Proof. Note that η_t is an $\mathcal{F}_t$-measurable variable. This is valid for the continuous function $g(s, x)$, because in this case, the integral is a limit of Riemannian integral sums, which are evidently $\mathcal{F}_t$-measurable. If the g_n are such that $\int_0^t g_n(s, \xi(s))ds$ are $\mathcal{F}_t$-measurable, $g_n(s, x) \to g(s, x)$ for all s and x, and if the g_n are uniformly bounded, then $\int_0^t g(x, \xi(s))ds$ is also $\mathcal{F}_t$-measurable.

Since bounded Borel functions are obtained from continuous ones by the limit transition (which is applied at most a countable number of times), η_t is $\mathcal{F}_t$-measurable. By virtue of Lemma 8°, η_t is progressively measurable. Further, the proof repeats the proof of Theorem 9°.

□

§4.2 Additive Functionals

The simplest example of an additive functional of the homogeneous process with independent increments $\xi(t)$ is an integral of some function along the trajectory of the process, namely

$$\alpha_t^s = \int_s^t g(\xi(u))du, \qquad s < t \tag{4.25}$$

(g is a Borel function for which this integral exists). It has the following properties:

(I) $\alpha_u^s = \alpha_t^s + \alpha_u^t$ for $s < t < u$ (with probability 1);

(II) α_t^s is continuous with respect to its arguments $s < t$;

(III) α_t^s is $\mathcal{F}_t$-measurable, where $\mathcal{F}_t$ is a σ-algebra generated by $\xi(s)$, $s \le t$;

(IV) a conditional distribution of α_t^s with respect to the σ-algebra $\mathcal{F}_s$ depends only on $\xi(s)$ and $t - s$.

The last property follows from the fact that the conditional distribution of (4.25) with respect to the σ-algebra $\mathcal{F}_s$ coincides with the distribution of $G(t-s, \xi(s), \alpha)$, where

$$G(t, x, \alpha) = \mathbf{P}\Big\{\int_0^t g(\xi(u) + x)du < \alpha\Big\}. \tag{4.26}$$

It is convenient to consider integrals of the form $\int_s^t g(\xi(u) - \xi(s) + x)du = \widehat{\alpha}_t^s(x)$, $s \le t$. This variable is measurable with respect to the σ-algebra $\mathcal{F}_t^s$ generated by $\{\xi(u) - \xi(s), s \le u \le t\}$; its distribution depends only on x and $t - s$. In particular, $\widehat{\alpha}_u^s$ and $\xi(u) - \xi(s)$ have the same joint distributions as $\widehat{\alpha}_{u-s}^0(x)$ and $\xi(u-s)$, $u \ge s$. Since $\widehat{\alpha}_t^s(x)$ is some measurable function of $\xi(u) - \xi(s)$, $u \in [s, t]$ (this means that $\widehat{\alpha}_t^s(x)$ is $\mathcal{F}_t^s$- measurable), we have $\widehat{\alpha}_t^s(x) = \widehat{\alpha}_{t-s}^0(x, \omega_s')$. Here the variable to the right can be expressed in terms of the process $\xi(u) - \xi(s)$, $s \le u \le t$, in exactly the same way as $\widehat{\alpha}_{t-s}^0(x, \omega)$ is expressed in terms of $\xi(u)$, $0 \le u \le t - s$. Clearly, $\widehat{\alpha}_t^s(x)$ is also continuous in s and t.

For the variables $\alpha_t^s(x)$, property (I) can be transformed as follows:

$$\alpha_t^s(x) + \alpha_u^t(x + \xi(t) - \xi(s)) = \alpha_u^s(x) \tag{4.27}$$

for $s < t < u$.

Definition. A family of random variables $\alpha_t(x)$, $t \in R_+$, $x \in X$, is called *an additive functional of the homogeneous process with independent increments* $\xi(t)$ if it satisfies the following conditions:

(1) $\alpha_t(x)$ is $\mathcal{F}_t$-measurable for every $x \in X$;

(2) $\alpha_s(x) = \alpha_s(x, \omega)$ is measurable with respect to $\mathcal{B}_{[0,t]} \otimes \mathcal{B}_X \otimes \mathcal{F}_{[0,t]}$ for $s \le t$, $x \in X$;

(3) $\alpha_t(x)$ is continuous in t, and $\alpha_0(x) = 0$;

(4) for $0 < s < t$,

$$\alpha_t(x) = \alpha_s(x) + \theta_s \alpha_{t-s}(x + \xi(s)), \tag{4.28}$$

where $\theta_u \alpha_v(x)$ denotes the substitution of the process $\xi(u+\tau) - \xi(u)$, $\tau \le v$, for the process $\xi(\tau)$, $\tau \le v$ in $\alpha_v(x)$. Here we shall consider nonnegative additive functionals, which can be obtained, e.g., by taking $g \ge 0$ in (4.25) or (4.26). Nonnegative additive functionals satisfy the following condition, in addition to properties (1)–(4):

(5) $\mathbf{P}\{\alpha_t(x) \ge 0\} = 1$ for all $t \in R_+$, $x \in X$.

11°. *Theorem Let $\alpha_t(x)$ be a nonnegative additive functional such that $\sup_{x \in X} \mathbf{M}\alpha_t(x) < \infty$ for some $t > 0$. We put $g_h(x) = \frac{1}{h}\mathbf{M}\alpha_h(x)$. Then*

$$\alpha_t(x) = \lim_{h \to 0} \int_0^t g_h(x + \xi(s))ds. \tag{4.29}$$

Proof. (4.28) implies

$$\mathbf{M}(\alpha_{s+h}(x) - \alpha_s(x)|\mathcal{F}_s) = \mathbf{M}(\theta_s \alpha_h(\xi(s) + x)|\mathcal{F}_s) = hg_h(x + \xi(s))$$

(we have used the fact that $\theta_s \alpha_h(y)$ does not depend on $\mathcal{F}_s$). Thus,

$$\begin{aligned}\int_0^t & g_h(x+\xi(s))ds \\ &= \frac{1}{h}\int_0^t \mathbf{M}(\alpha_{s+h}(x)|\mathcal{F}_s)ds - \frac{1}{h}\int_0^t \alpha_s(x)ds \\ &= \frac{1}{h}\int_0^t [\mathbf{M}(\alpha_{s+h}(x)|\mathcal{F}_s) - \alpha(s+h)]ds + \frac{1}{h}\int_0^t [\alpha_{s+h}(x) - \alpha_s(x)]ds.\end{aligned}$$

Denote

$$\zeta_h(s) = \mathbf{M}(\alpha_{s+h}(x)|\mathcal{F}_s) - \alpha(s+h).$$

Then

$$\int_0^t g_h(x+\xi(s))ds = \frac{1}{h}\int_t^{t+h} \alpha_s(x)\,ds - \frac{1}{h}\int_0^h \alpha_s(x)ds + \frac{1}{h}\int_0^t \zeta_h(s)ds.$$

By virtue of the continuity of $\alpha_s(x)$, we have

$$\frac{1}{h}\int_t^{t+h} \alpha_s(x)ds \to \alpha_t(x), \qquad \frac{1}{h}\int_0^h \alpha_s(x)\,ds \to \alpha_0(x) = 0$$

as $h \to 0$ with probability 1. Hence, to prove the theorem it suffices to show that $\frac{1}{h}\int_0^t \zeta_h(s)ds \to 0$ as $h \to 0$ in probability.

The following lemma is necessary to complete the proof.

12°. Lemma. *If* $\sup_x \mathbf{M}\alpha_{t_0}(x) < \infty$ *for some* t_0, *then* $\sup_x \mathbf{M}\alpha_t(x)$ *and* $\sup_x \mathbf{M}\alpha_t^2(x)$ *are locally bounded.*

Proof. Let $\sup_x \mathbf{M}\alpha_t(x) = c_t$ (maybe $c_t = +\infty$). Then $c_t \le c_s$ for $t < s$, and thus, $c_t \le c_{t_0}$ for $t \le t_0$. Further, property (4) implies for $s < t$,

$$\mathbf{M}\alpha_t(x) = \mathbf{M}\alpha_s(x) + \mathbf{M}\mathbf{M}(\theta_s\alpha_{t-s}(x + \xi(s))|\mathcal{F}_s) \le \mathbf{M}\alpha_s(x) + c_{t-s}.$$

Hence, $c_t \le c_s + c_{t-s}$. Therefore, $c_t \le nc_{t_0} < \infty$ for $t \le nt_0$. Further, since $\alpha_t(x)$ is a continuous increasing function, we have

$$\begin{aligned}&\lim_{n\to\infty} \sum_{k=0}^{n-1} |\alpha_{(k+1)t/n}(x) - \alpha_{kt/n}(x)|^2 \\ &\quad\le \lim_{n\to\infty} \alpha_t(x) \max_k[\alpha_{(k+1)t/n}(x) - \alpha_{kt/n}(x)] = 0.\end{aligned}$$

Hence,

$$\begin{aligned}\alpha_t^2(x) &= 2\lim_{n\to\infty} \sum_{0\le k<j<n-1} [\alpha_{(k+1)t/n}(x) - \alpha_{kt/n}(x)][\alpha_{(j+1)t/n}(x) - \alpha_{jt/n}(x)] \\ &= 2\lim_{n\to 0} \sum_{k=0}^{n-1} [\alpha_{(k+1)t/n}(x) - \alpha_{kt/n}(x)][\alpha_t - \alpha_{(k+1)t/n}],\end{aligned}$$

$$\begin{aligned}\mathbf{M}\alpha_t^2(x) \le 2\varliminf_{n\to\infty} \sum_{k=0}^{n-1} &\mathbf{M}[\alpha_{(k+1)t/n}(x) - \alpha_{kt/n}(x)] \\ &\times \mathbf{M}([\alpha_t(x) - \alpha_{(k+1)t/n}(x)]|\mathcal{F}_{(k+1)t/n}).\end{aligned}$$

But for $s < t$,

$$\mathbf{M}([\alpha_t(x) - \alpha_s(x)]|\mathcal{F}_s) = \mathbf{M}(\theta_s\alpha_{t-s}(x + \xi(s))|\mathcal{F}_s) \le c_{t-s} \le c_t;$$

and thus,

$$\mathbf{M}\alpha_t^2(x) \le 2\varliminf_{n\to\infty} c_t \sum_{k=0}^{n-1} \mathbf{M}[\alpha_{(k+1)t/n}(x) - \alpha_{kt/n}(x)] = 2c_t\mathbf{M}\alpha_t(x) \le 2c_t^2.$$

□

We now return to the proof of the theorem. Let us estimate

$$\mathbf{M}\Big(\frac{1}{h}\int_0^t \zeta_h(s)ds\Big)^2 = \frac{1}{h^2}\int_0^t\int_0^t \mathbf{M}\zeta_h(s)\zeta_h(u)ds\,du.$$

Note that $\mathbf{M}\zeta_h(s)\zeta_h(u) = \mathbf{M}\zeta_h(s)\mathbf{M}(\zeta_h(u)/\mathcal{F}_u) = 0$ for $s+h>u$. We have used that $\zeta_h(s)$ is $\mathcal{F}_{s+h}$-measurable, and hence, $\mathcal{F}_u$-measurable, and that $\mathbf{M}(\zeta_h(u)/\mathcal{F}_u)=0$. Thus,

$$\begin{aligned}
\mathbf{M}\Big(\frac{1}{h}\int_0^t \zeta_h(s)ds\Big)^2 &= \frac{1}{h^2}\iint\limits_{\substack{|s-u|\le h\\ 0\le s\le t\\ 0\le u\le t}} \mathbf{M}\zeta_h(s)\zeta_h(u)ds\,du \\
&\le \frac{1}{2h^2}\iint\limits_{\substack{|s-u|\le h\\ 0\le s\le t\\ 0\le u\le t}} [\mathbf{M}\zeta_h^2(s)+\mathbf{M}\zeta_h^2(u)]ds\,du \\
&\le \frac{2}{h}\int_0^t \mathbf{M}\zeta_h^2(s)ds.
\end{aligned}$$

But

$$\begin{aligned}
\mathbf{M}(\zeta_h(s))^2 &= \mathbf{M}(\alpha_{s+h}(x)-\alpha_s(x)-\mathbf{M}(\alpha_{s+h}(x)-\alpha_s(x)|\mathcal{F}_s))^2 \\
&\le \mathbf{M}(\alpha_{s+h}(x)-\alpha_s(x))^2.
\end{aligned}$$

Hence, Theorem 11° follows from the equality

$$\lim_{h\to 0}\mathbf{M}\frac{1}{h}\int_0^t(\alpha_{s+h}(x)-\alpha_s(x))^2ds = 0. \tag{4.30}$$

Note that by virtue of the continuity of α_s,

$$\begin{aligned}
&\frac{1}{h}\int_0^t(\alpha_{s+h}(x)-\alpha_s(x))^2ds \\
&\qquad\le \sup_{u\le t}(\alpha_{u+h}(x)-\alpha_u(x))\frac{1}{h}\int_0^t[\alpha_{s+h}(x)-\alpha_s(x)]ds \\
&\qquad= \sup_{u\le t}(\alpha_{u+h}(x)-\alpha_u(x))\frac{1}{h}\int_t^{t+h}\alpha_s(x)ds \\
&\qquad\le \sup_{u\le t}(\alpha_{u+h}(x)-\alpha_u(x))\alpha_{t+h}\to 0
\end{aligned}$$

as $h\to 0$. Moreover,

$$\begin{aligned}
\frac{1}{h}\int_0^t(\alpha_{s+h}(x)-\alpha_s(x))^2ds &\le \frac{1}{h}\int_0^t[a_{s+h}^2(x)-\alpha_s^2(x)]ds \\
&\le \frac{1}{h}\int_t^{t+h}\alpha_s^2(x)ds \le \alpha_{t+h}^2(x).
\end{aligned}$$

Therefore, $\frac{1}{h}\int_0^t(\alpha_{s+h}(x)-\alpha_s(x))^2ds$ tends to zero and is bounded for $h<h_1$ by the variable $\alpha^2_{t+h_1}(x)$, for which $\mathbf{M}\alpha^2_{t+h_1}(x)<\infty$. Consequently, (4.30) follows from the possibility of limit transition under the sign of mathematical expectation.

□

Now consider some transformations of additive functionals.

13°. Lemma. *Let $\alpha_t(x)$ be a nonnegative additive functional, and let $f(x)$ be a nonnegative locally bounded Borel function on X. Then*

$$\beta_t(x)=\int_0^t f(x+\xi(s))d_s\alpha_s(x)$$

(the Stieltjes integral with respect to a continuous increasing function) is also a nonnegative additive functional.

Proof. Properties (1) and (2) are evident for a continuous function f. And since they remain valid when we take the limit with respect to f, they hold for a Borel function f, too. Property (3) follows from the continuity of $\alpha_t(x)$, whereas (5) arises from the nonnegativeness of f. Property (4) follows from the next equality: for $t>0$ and $h>0$,

$$\begin{aligned}\beta_{t+h}(x)-\beta_t(x)&=\int_t^{t+h} f(x+\xi(s))d_s\alpha_s(x)\\&=\int_t^{t+h} f(x+\xi(t)+\xi(s)-\xi(t))\,d\alpha_s(x)\\&=\int_0^h f(x+\xi(t)+\theta_t(\xi(u)))d[\alpha_{t+u}(x)-\alpha_t(x)]\\&=\int_0^h f(x+\xi(t)+\theta_t\xi(u))d_u\theta_t d_u(x+\xi(t))\\&=\theta_t\beta_h(x+\xi(t)).\end{aligned}\tag{4.31}$$

□

14°. Lemma. *Let $\alpha_t(x)$ be a nonnegative additive functional for which* $\mathbf{P}\{\alpha_t(x)>\delta\}\le q<1$ *for some $\delta>0$ (q is independent of x). Then*

$$\mathbf{M}\alpha_t(x)\le\frac{\delta}{1-q}.$$

Proof. Let us estimate the probability of the event $\{\alpha_t(x)>2\delta\}$. Denote by τ the maximal time for which $\alpha_\tau(x)=\delta$. Let us show that τ is a s.m. Let $\tau_n=k/n$ if $\alpha_{i/n}(x)\le\delta$ for $i<k$, and $\alpha_{k/n}(x)>\delta$. Then

$$\{\tau_n\le k/n\}=\{\alpha_{k/n}(x)>\delta\}\in\mathcal{F}_{k/n},$$

and hence, τ_n is a s.m. Furthermore, $\tau = \lim \tau_n$, and thus,

$$\{\tau < s\} = \bigcup_n \bigcap_{m=n}^{\infty} \{\tau_m < s\} \in \mathcal{F}_s$$

$$\{\tau \le s\} = \bigcap_{h>0} \{\tau < s + h\} \in \bigcap_{h>0} \mathcal{F}_{s+h} = \mathcal{F}_{s+} = \mathcal{F}_s.$$

For the s.m. τ, we can write (4.28): for $\tau < \delta$,

$$\alpha_t(x) = \alpha_\tau(x) + \theta_\tau \alpha_{t-\tau}(x + \xi(\tau)) = \delta + \theta_\tau \alpha_{t-\tau}(x + \xi(\tau)).$$

If $\alpha_t(x) > 2\delta$, then $\tau < t$ and $\theta_\tau \alpha_{t-\tau}(x+\xi(\tau)) > \delta$; moreover, $\theta_\tau \alpha_t(x+\xi(\tau)) > \delta$. Therefore,

$$\mathbf{P}\{\alpha_t(x) > 2\delta\} \le \mathbf{M} I_{\{\tau<\delta\}} \mathbf{P}\{\theta_\tau \alpha_t(x + \xi(\tau)) > \delta / \mathcal{F}_\tau\},$$

$\theta_\tau \alpha_t(y)$ results from substituting the process $\xi_\tau(s) = \xi(\tau + s) - \xi(\tau)$, $s \le t$, for $\xi(s)$, $s \le t$, in $\alpha_t(y)$. This process does not depend on $\mathcal{F}_\tau$ (ξ_τ is $\mathcal{F}_\tau$-measurable). Hence,

$$\mathbf{P}\{\theta_\tau \alpha_t(x + \xi(\tau)) > \delta / \mathcal{F}_\tau\} = \mathbf{P}\{\alpha_t(y) > \delta\}|_{y=\xi(\tau)+x} \le q,$$

$$\mathbf{P}\{\alpha_t(x) > 2\delta\} \le \mathbf{M} I_{\{\tau \le \delta\}} q = q\mathbf{P}\{\alpha_t(x) > \delta\} \le q^2.$$

Similarly

$$\begin{aligned} \mathbf{P}\{\alpha_t(x) > m\delta\} &\le \mathbf{M} I_{\{\tau<\delta\}} \mathbf{P}\{\theta_\tau \alpha_t(x + \xi(\tau)) > \delta / \mathcal{F}_\tau\} \\ &\le \mathbf{M} I_{\{\tau \le \delta\}} \mathbf{P}\{\alpha_t(y) > (m-1)\delta\}|_{y=\xi(\tau)+x} \\ &\le q \sup_y \mathbf{P}\{\alpha_t(y) > (m-1)\delta\}. \end{aligned}$$

By induction we obtain

$$\mathbf{P}\{\alpha_t(x) > m\delta\} \le q^m.$$

Consequently,

$$\mathbf{M}\alpha_t(x) \le \delta \sum_{m=0}^{\infty} \mathbf{P}\{\alpha_t(x) > m\delta\} = \frac{\delta}{1-q}.$$

□

Remark. *Provided that conditions of Lemma 14° hold, $\alpha_t(x)$ has all the moments, which are uniformly bounded on every bounded interval of time.*

It has been proved in Lemma 12° that $\sup_{s \le t, x \in X} \mathbf{M}\alpha_s(x) < \infty$, and $\sup_{s \le t, x \in X} \mathbf{M}\alpha_s^2(x) < \infty$ for all t. Furthermore,

$$\begin{aligned} \alpha_t^m(x) &= m \int_0^t (\alpha_t(x) - \alpha_s(x))^{m-1} d\alpha_s(x) \\ &= \lim_{n \to \infty} m \sum_{k=0}^{n-1} (\alpha_t(x) - \alpha_{(k+1)t/n}(x))^{m-1} (\alpha_{(k+1)t/n}(x) - \alpha_{kt/n}(x)). \end{aligned}$$

Hence,

$$\mathbf{M}\alpha_t^m(x) \le m \sup_y \mathbf{M}\alpha_t^{m-1}(y)\mathbf{M}\alpha_t(x);$$
$$\sup_x \mathbf{M}\alpha_t^m(x) \le m \sup_x \mathbf{M}\alpha_t^{m-1}(x) \sup_x \mathbf{M}\alpha_t(x). \tag{4.32}$$

The following inequality can be obtained from (4.32) by induction:

$$\sup_x \mathbf{M}\alpha_t^m(x) \le m!(\sup_x \mathbf{M}\alpha_t(x))^m.$$

15°. Lemma. *There exists a function $f(x)$ positive everywhere such that*

$$\sup_x \mathbf{M} \int_0^t f(x+\xi(s))d_s\alpha_s(x) < \infty.$$

Proof. We fix t_0. Then

$$\int_0^{t_0} \exp\{-\alpha_{t_0}(x) + \alpha_s(x)\}d\alpha_s(x) = 1 - \exp\{-\alpha_{t_0}(x)\}.$$

By virtue of property (4),

$$1 - \exp\{-\alpha_{t_0}(x)\} = \int_0^{t_0} \exp\{-\theta_s\alpha_{t_0-s}(x+\xi(s))\}d\alpha_s(x),$$
$$\int_0^{t_0/2} \exp\{-\theta_s\alpha_{t_0/2}(x+\xi(s))\}d\alpha_s(x) \le 1.$$

Therefore,

$$\begin{aligned} 1 &\ge \mathbf{M}\int_0^{t_0/2} \exp\{-\theta_s\alpha_{t_0/2}(x+\xi(s))\}d\alpha_s(x) \\ &= \mathbf{M}\int_0^{t_0/2} \mathbf{M}(\exp\{-\theta_s\alpha_{t_0/2}(x+\xi(s))\}|\mathcal{F}_s)d\alpha_s(x). \end{aligned} \tag{4.33}$$

We have used the following equality: if $\varphi(s,\omega)$ is some nonnegative measurable process, then

$$\mathbf{M}\int_0^t \varphi(s,\omega)\, d_s\alpha_s(x) = \mathbf{M}\int_0^t \mathbf{M}(\varphi(s,\omega)|\mathcal{F}_s)d_s\alpha_s(x). \tag{4.34}$$

It suffices to establish this equality for bounded φ and increasing continuous $\mathcal{F}_s$-measurable processes α_s, because (4.34) can be obtained by the limit transition in

$$\mathbf{M}\int_0^t (\varphi(s,\omega)\wedge m)d_s(\alpha_s \wedge n) = \mathbf{M}\int_0^t \mathbf{M}(\varphi(s,\omega)\wedge m|\mathcal{F}_s)d_s(\alpha_s\wedge n) \tag{4.35}$$

If $\varphi(s,\omega)$ is a bounded continuous function and if α_s is also bounded, we have

$$\begin{aligned}\mathbf{M}\int_0^t \varphi(s,\omega)d\alpha_s &= \mathbf{M}\lim_{n\to\infty}\sum_{k=1}^n \varphi(\frac{k}{n}t,\omega)[\alpha_{kt/n}-\alpha_{(k-1)t/n}]\\ &= \mathbf{M}\lim_{n\to\infty}\sum_{k=1}^n \mathbf{M}(\varphi(\frac{k}{n}t,\omega)|\mathcal{F}_{kt/n})[\alpha_{kt/n}-\alpha_{(k-1)t/n}]\\ &= \mathbf{M}\int_0^t \mathbf{M}(\varphi(s)|\mathcal{F}_s)d\alpha_s.\end{aligned}$$

We can get (4.34) in the general case by the limit transition from continuous $\varphi(s,\omega)$.

Denote

$$f(x) = \mathbf{M}\exp\{-\alpha_{t_0/2}(x)\}; \tag{4.36}$$

$0 < f(x) \leq 1$; $f(x)$ is a measurable function. (4.33) implies

$$\sup_x \mathbf{M}\int_0^{t_0/2} f(x+\xi(s))d\alpha_s(x) \leq 1.$$

It remains to use Lemma 12° and the fact that $\int_0^t f(x+\xi(s))d\alpha_s(x)$ is an additive functional.

□

Corollary. *Let $\alpha_t(x)$ be a nonnegative additive functional. Then there exist a nonnegative additive functional $\widetilde{\alpha}_t(x)$ for which $\sup_x \mathbf{M}\widetilde{\alpha}_t(x) < \infty$, and a positive Borel function $g(x)$ such that*

$$\alpha_t(x) = \int_0^t g(x+\xi(s))d\widetilde{\alpha}_s(x) \tag{4.37}$$

(we take $\int_0^t f(x+\xi(s))d\alpha_s(x)$ as $\widetilde{\alpha}_t(x)$, and f is the same as in Lemma 15°).

16°. Theorem. *For every nonnegative additive functional $\alpha_t(x)$, there exists a sequence of nonnegative functions $g_n(x)$ such that*

$$\alpha_t(x) = \lim_{n\to\infty}\int_0^t g_n(x+\xi(s))ds. \tag{4.38}$$

in the sense of the convergence in probability.

Proof. We put

$$g_n(x) = n\mathbf{M}(1-e^{-\alpha_{1/n}(x)}). \tag{4.39}$$

Then

$$\begin{aligned}\int_0^t & g_n(x+\xi(s))ds \\ &= \int_0^t n\mathbf{M}(1-\exp\{-\theta_s\alpha_{1/n}(x+\xi(s))\}|\mathcal{F}_s)ds \\ &= \int_0^t n\mathbf{M}(1-\exp\{-\alpha_{s+1/n}(x)+\alpha_s(x)\}|\mathcal{F}_s)ds \\ &= \int_0^t n(1-e^{-\alpha_{s+1/n}(x)+\alpha_s(x)})ds \\ &\quad + \int_0^t n(e^{-\alpha_{s+1/n}(x)+\alpha_s(x)} - \mathbf{M}(\exp\{-\alpha_{s+1/n}(x)+\alpha_s(x)\}|\mathcal{F}_s)ds.\end{aligned}$$

The continuity of $\alpha_s(x)$ and the inequalities

$$\alpha - \alpha^2/2 < 1 - e^{-\alpha} < \alpha$$

yield

$$\begin{aligned}&\left|\int_0^t n(1-\exp\{-\alpha_{s+1/n}(x)+\alpha_s(x)\})ds - \int_0^t n(\alpha_{s+1/n}(x)-\alpha_s(x))ds\right| \\ &= \left|\int_0^t n(1-\exp\{-\alpha_{s+1/n}(x)+\alpha_s(x)\})ds - n\int_t^{t+1/n}\alpha_s(x)ds - n\int_0^{1/n}\alpha_s(x)ds\right| \\ &\leq \frac{1}{2}\int_0^t n(\alpha_{s+1/n}(x)-\alpha_s(x))^2 ds \leq \frac{1}{2}\max_n |\alpha_{u+1/n}(x)-\alpha_u(x)|\alpha_{t+1/n}(x).\end{aligned}$$

Hence,

$$\lim_{n\to\infty}\int_0^t n(1-\exp\{-\alpha_{s+1/n}(x)+\alpha_s(x)\})ds = \alpha_t(x).$$

with probability 1. To prove the theorem it suffices to show that

$$\begin{aligned}\int_0^t & n(\exp\{-\alpha_{s+1/n}(x)+\alpha_s(x)\} \\ &- \mathbf{M}(\exp\{-\alpha_{s+1/n}(x)+\alpha_s(x)\}|\mathcal{F}_s))ds \to 0\end{aligned}$$

in probability as $n \to \infty$. Denote $\alpha_s(x) \wedge c = \alpha_s^c(x)$. Then

$$\mathbf{P}\Big\{\Big|\int_0^t n(\exp\{-\alpha_{s+1/n}(x) + \alpha_s(x)\}$$

$$- \mathbf{M}(\exp\{-\alpha_{s+1/n}(x) + \alpha_s(x)\}/\mathcal{F}_s))ds\Big| > \varepsilon\Big\}$$

$$\leq \mathbf{P}\{\alpha_{t+1/n}(x) > c\} + \mathbf{P}\Big\{\Big|\int_0^t n(\exp\{-\alpha^c_{s+1/n}(x) + \alpha^c_s(x)\}$$

$$- \mathbf{M}(\exp\{-\alpha^c_{s+1/n} + \alpha^c_s\}|\mathcal{F}_s)ds\Big| > \varepsilon\Big\}$$

$$\leq \mathbf{P}\{\alpha_s(x) > c\} + \frac{1}{\varepsilon^2}\mathbf{M}\Big(\int_0^t \zeta_s^{(n)} ds\Big)^2,$$

where

$$\zeta_s^{(n)} = n(\exp\{-\alpha^c_{s+1/n} + \alpha^c_s\} - \mathbf{M}(\exp\{-\alpha^c_{s+1/n} + \alpha^c_s\}|\mathcal{F}_s)).$$

Just as in the proof of Theorem 11°, we get

$$\mathbf{M}\Big(\int_0^t \zeta_s^{(n)} ds\Big)^2 \leq 2n \int_0^t \mathbf{M}(\exp\{-\alpha^c_{s+1/n} + \alpha^c_s\} - 1)^2 ds$$

$$\leq 2n\mathbf{M}\int_0^t (\alpha^c_{s+1/n} - \alpha^c_s)^2 ds$$

and find that the last expression tends to zero. $\mathbf{P}\{\alpha_{t+1/n}(x) > c\}$ can be made as small as desired.

□

17°. Lemma. *Let $\alpha_t(x)$ be a nonnegative additive functional. If $\sup_x \mathbf{M}\alpha_t(x) < \infty$, and $\psi(x)$ is a nonnegative function integrable on X, then*

$$\gamma_t(x) = \int \alpha_t(y)\psi(x+y)dy \tag{4.40}$$

is a nonnegative additive functional. If $\lim_{x\to 0}\int |\psi(x+y) - \psi(y)|dy = 0$, then $\gamma_t(x)$ is continuous in mean with respect to x. (4.40) can be differentiated term-by-term (in mean) as many times as there exist absolutely integrable derivatives of the function ψ.

Proof. Conditions of the continuity and differentiability of $\gamma_t(x)$ are evident. Show that $\gamma_t(x)$ is a nonnegative additive functional. The existence of the integral on the r.h.s. of (4.40) follows from the measurability of the integrand

and from the existence of $\int \mathbf{M}\alpha_t(y)\psi(x+y)dy$. Conditions (1), (2), (3), and (5) are obviously satisfied. Let us check condition (4). Substituting y for x in (4.28), multiplying by $\psi(x+y)$, and integrating with respect to y, we get

$$\begin{aligned}\gamma_t(x) &= \gamma_s(x) + \int [\theta_s \alpha_{t-s}(y+\xi(s))]\psi(x+y)dy \\ &= \gamma_s(x) + \theta_s \gamma_{t-s}(x+\xi(s)),\end{aligned}$$

because

$$\int \theta_s \alpha_{t-s}(y+z)\psi(x+y)dy = \theta_s \int \alpha_{t-s}(y+z)\psi(x+y)dy.$$

□

18°. Lemma. *Let $\alpha_t(x)$ be a nonnegative additive functional for which $\sup_x \mathbf{M}\alpha_t(x) < \infty$. If $g_n(y)$ is a sequence of nonnegative Borel functions, for which* (4.38) *holds, then the following limit*

$$\lim_{n\to\infty} \int g_n(y)\psi(y)dy = \lim_{t\to 0} \frac{1}{t} \int \alpha_t(y)\psi(y)dy. \tag{4.41}$$

exists, whatever finite continuous function $\psi(x)$ on X we take.

Proof. Suppose that $\gamma_t(x)$ is defined by (4.40). Then

$$\begin{aligned}\gamma_t(x) &= \lim_{n\to\infty} \int \left[\int_0^t g_n(y+\xi(s))ds\right] \psi(x+y)dy \\ &= \lim_{n\to\infty} \int_0^t \left[\int g_n(y+\xi(s))\psi(x+y)dy\right] ds \\ &= \lim_{n\to\infty} \int_0^t \left[\int g_n(y)\psi(x+y-\xi(s))dy\right] ds \\ &= \lim_{n\to\infty} \int g_n(y) \left(\int_0^t \psi(x+y-\xi(s))ds\right) dy.\end{aligned} \tag{4.42}$$

Here, by taking $\psi(y) = 1$ for $|y| \le 2c$, we obtain

$$t \overline{\lim_{n\to\infty}} \int_{|y|\le c} g_n(y)dy \cdot I_{\{\sup_{s\le t} |\xi(s)|\le c\}} \le \gamma_t(0).$$

Hence,

$$\overline{\lim_{n\to\infty}} \int_{|y|\le c} g_n(y)dy \le \mathbf{M}\gamma_t(0)(t\mathbf{P}\{\sup_{s\le t} |\xi(s)| \le c\})^{-1}.$$

The sequence of measures $m_n(A) = \int_A g_n(y)dy$ is thus compact with respect to the convergence on finite functions. If $\overline{m}(A)$ is some limit measure of the sequence m_n (it is locally bounded), then it follows from (4.41) that

$$\gamma_t(x) = \iint_0^t \psi(x + y - \xi(s))\overline{m}(dy)ds. \tag{4.43}$$

Hence, if $\overline{m}_1$ is another limit measure of the sequence m_n, then

$$\iint_0^t \psi(y - \xi(s)\overline{m}(dy)ds = \iint_0^t \psi(y - \xi(s))\overline{m}_1(dy)ds$$

with probability 1 for every finite continuous function ψ. Dividing both parts of this equality by t and proceeding to the limit as $t \to 0$, we obtain

$$\int \psi(y)\overline{m}(dy) = \int \psi(y)\overline{m}_1(dy).$$

The sequence of measures m_n is thus compact with respect to the convergence on finite functions; moreover, it has the unique limit point and is, therefore, convergent on finite functions. This implies the existence of the limit on the l.h.s. of (4.41). Denoting the limit measure by $\overline{m}$, we obtain from (4.43)

$$\int \psi(y)\overline{m}(dy) = \lim_{t\to 0} \frac{1}{t} \iint_0^t \psi(y - \xi(s))\overline{m}(dy).$$

This proves (4.41).

□

Remark. Examining the proof of this lemma, we find that it remains valid if $\mathsf{M}\alpha_t(x)$ is a locally bounded function.

19°. Theorem. *Let $\alpha_t(x)$ be a nonnegative additive functional for which $\sup_x \mathsf{M}\alpha_t(x) < \infty$. It is determined uniquely by the finite function*

$$v_\alpha(\lambda, x) = \mathsf{M} \int_0^\infty e^{-\lambda t} d\alpha_t(x) \tag{4.44}$$

or by the measure $m_\alpha(dy)$ such that for any continuous finite function $\psi(x)$,

$$\int v_\alpha(\lambda, x)\psi(x)dx = \int R_\lambda \psi(x) m_\alpha(dx) \tag{4.45}$$

(R_λ is a resolvent of the process). In order that some nonnegative additive functional correspond to a locally finite measure $m(dx)$, it is necessary that for all $t > 0$, the measure $\int_0^t F_s(A - y)m(dy)ds = q_t(A)$ be absolutely continuous

with respect to the Lebesgue measure (here $\mathcal{F}_s(A-y) = \mathbf{P}\{\xi(s)+y \in A\}$), and have a bounded density.

Suppose now that the measure $m(dx)$ is such that for all $t > 0$, the function $q_t(A)$ has the bounded density with respect to the Lebesgue measure, and that this density is uniformly continuous in x in the whole space (uniformly in t on every bounded domain) . Then there exists a nonnegative additive functional $\alpha_t(x)$ for which $m = m_\alpha$ (m_α is connected with the functional α by (4.44) and (4.45)).

Proof. The function (4.44) can be represented as follows:

$$v_\alpha(\lambda, x) = \int_0^\infty e^{-\lambda t} d_t \Phi_\alpha(t, x),$$

where $\Phi_\alpha(t,x) = \mathbf{M}\alpha_t(x)$. $\Phi_\alpha(t,x)$ is a continuous nondecreasing function of t; moreover, as follows from property (4), if $\sup_x \Phi_\alpha(t,x) \le c_t$, then

$$\sup_x [\Phi_\alpha(s+t,x) - \Phi_\alpha(s,x)] = \sup_x \mathbf{M}\theta_s \alpha_t(x+\xi(t)) \le c_t.$$

Thus,

$$\begin{aligned}\int_0^\infty e^{-\lambda t} d_t \Phi_\alpha(t,x) &\le \sum_{k=0}^\infty e^{-k\lambda t_0}[\Phi_\alpha((k+1)t_0, x) - \Phi_\alpha(kt_0, x)] \\ &\le c_{t_0} \cdot \frac{1}{1-e^{-\lambda t_0}} < \infty\end{aligned}$$

for all $\lambda > 0$. Hence, the function $v_\alpha(\lambda, x)$ is defined and bounded for all $\lambda > 0$. It determines the function $\Phi_\alpha(t,x)$ uniquely, and Theorem 11° implies that

$$\alpha_t(x) = \lim_{h \to 0} \int_0^t \frac{1}{h} \Phi_\alpha(h, \xi(s)+x) ds.$$

The existence of the measure $m_\alpha(dx)$, satisfying equation (4.45), follows from Lemma 18°. Let $m_a(dx)$ be a measure such that for every continuous finite function $\psi(x)$, we have

$$\int \alpha_t(x)\psi(x)dx = \iint_0^t \psi(x+\xi(s)) m_\alpha(dx) ds. \tag{4.46}$$

Assume that $\psi \ge 0$. Then

$$\int e^{-\lambda t} dt \int \alpha_t(x)\psi(x)dx = \int_0^\infty \left[\int e^{-\lambda t}\psi(x+\xi(t))\right] m_\alpha(dx).$$

Here we take the mathematical expectation and change the order of integration, we find

$$\iint_0^\infty e^{-\lambda t}\mathbf{M}\psi(x+\xi(t))dt\, m(dx) = \mathbf{M}\int_0^\infty e^{-\lambda t}dt\int \alpha_t(x)\psi(x)dx$$
$$= -\frac{1}{\lambda}\mathbf{M}\int_0^\infty \Big(\int \alpha_t(x)\psi(x)dx\Big)de^{-\lambda t}$$
$$= -\frac{1}{\lambda}\iint_0^\infty \Phi_\alpha(t,x)\psi(x)dx\, de^{-\lambda t}$$
$$= \int e^{-\lambda t}d_t\Phi_\alpha(t,x)m(dx).$$

This yields (4.45). Using the fact that

$$\lim_{\lambda\to\infty}\lambda R_\lambda\psi(x) = \lim_{\lambda\to\infty}\lambda\int_0^\infty e^{-\lambda t}\mathbf{M}\psi(x+\xi(t))dt$$
$$= \lim_{\lambda\to\infty}\int_0^\infty e^{-s}\mathbf{M}\psi(x+\xi(\tfrac{s}{\lambda}))ds = \psi(x)$$

uniformly in x, we find that $\int\psi(x)m_\alpha(dx) = \lim_{\lambda\to\infty}\int\lambda v_\alpha(\lambda,x)\psi(x)dx$, so m_α is the unique measure satisfying (4.45). Let us take the mathematical expectation in (4.46) and substitute the function I_A (A is a bounded Borel set) for ψ. We get

$$\int_A \Phi_\alpha(t,x)dx = \int_0^t F_s(A-y)m_\alpha(dy)ds = q_t(A),$$

and this yields that the measure $q_t(A)$ is absolutely continuous with respect to the Lebesgue measure and its density $\Phi_\alpha(t,x)$ is bounded.

Whatever locally finite measure m is taken, we can define a family of non-negative additive functionals

$$\gamma_t^{(n)}(x) = \frac{1}{l(U_n)}\iint_0^t I_{U_n}(x+y+\xi(s))ds\, m(dy),$$

where $l(U_n)$ is a Lebesgue measure of the sphere U_n of radius $\frac{1}{n}$ centered at the point 0. We shall show that under accepted assumptions on the measure m the sequences $\gamma_t^{(n)}(x)$ converge for all t and x to a certain limit, which is just the desired additive functional $\alpha_t(x)$. Denote by $\Phi(t,x)$ a density of the measure $q_t(A)$ with respect to the Lebesgue measure. It is bounded and uniformly continuous. We put

$$\Psi_n(t,x) = \mathbf{M}\gamma_t^{(n)}(x) = \frac{1}{l(U_n)}\int_0^t F_s(U_n-x-y)m(dy)$$
$$= \frac{1}{l(U_n)}\int_{U_n}\Phi(t,x+y)dy.$$

Properties of $\Phi(t,x)$ imply that for all $c>0$,

$$\lim_{n\to\infty}\sup_{t\le c}\sup_{x}|\Phi_n(t,x)-\Phi(t,x)|=0.$$

We put

$$g_n(x)=\int I_{U_n}(x+y)m(dy).$$

Then $\gamma_t^{(n)}=\int_0^t g_n(x+\xi(s))ds$. Let us estimate

$$\begin{aligned}
\mathbf{M}(\gamma_t^{(n)}(x)-\gamma_t^{(m)}(x))^2 &= \mathbf{M}\Big(\int_0^t (g_n(x+\xi(s))-g_m(x+\xi(s)))ds\Big)^2\\
&= 2\mathbf{M}\iint_{0<s<u<t}(g_n(x+\xi(s))-g_m(x+\xi(s)))\\
&\quad\times(g_n(x+\xi(u))-g_m(x+\xi(u)))ds\,du\\
&= 2\mathbf{M}\int_0^t (g_n(x+\xi(s))-g_m(x+\xi(s)))\\
&\quad\times\mathbf{M}\left[\int_s^t (g_n(x+\xi(u))-g_m(x+\xi(u)))du/\xi(s)\right]ds\\
&= 2\mathbf{M}\int_0^t (g_n(x+\xi(s))-g_m(x+\xi(s)))\\
&\quad\times[\Phi_n(t-s,x+\xi(s))-\Phi_m(t-s,x+\xi(s))]ds\\
&\le 2\sup_{u\le t}\sup_{y}|\Phi_n(u,y)-\Phi_m(u,y)|\\
&\quad\times\mathbf{M}\int_0^t [g_n(x+\xi(s))+g_m(x+\xi(s))]ds\\
&= 2\sup_{u\le t}\sup_{y}|\Phi_n(u,y)-\Phi_m(u,y)|[\Phi_n(t,x)+\Phi_m(t,x)].
\end{aligned}$$

This yields the existence (in mean square) of the limit $\lim_{n\to\infty}\gamma_t^{(n)}(x)$. As a limit of nonnegative additive functionals it will also be a nonnegative additive functional. If we denote this limit by $\alpha_t(x)$, we obtain

$$\mathbf{M}\alpha_t(x)=\lim_{n\to\infty}\Phi_n(t,x)=\Phi(t,x),$$

$$v_\alpha(\lambda,x)=\int_0^\infty e^{-\lambda t}d_t\Phi(t,x),$$

$$\begin{aligned}
\int v_\alpha(\lambda,x)\psi(x)dx &= \int e^{-\lambda t}d_t\int\Phi(t,x)\psi(x)dx\\
&= \int_0^\infty e^{-\lambda t}d_t\int q_t(dx)\psi(x)\\
&= \int_0^\infty e^{-\lambda t}d_t\int_0^t T_s\psi(x)m(dx)\\
&= \int R_\lambda\psi(x)m(dx).
\end{aligned}$$

□

Remark. If measures m_1 and m_2 are such that $m_1 < m_2$ and if nonnegative additive functionals $\alpha_t^{(1)}(x)$ and $\alpha_t^{(2)}(x)$ correspond to m_1 and m_2, respectively, then $\alpha_t^{(2)}(x) - \alpha_t^{(1)}(x)$ is a nonnegative additive functional that corresponds to $m_2 - m_1$. Using the fact that the limit of a monotone sequence of nonnegative additive functionals $\alpha_t^{(n)}(x)$ is also a nonnegative additive functional, one can show that a nonnegative additive functional corresponds to every measure m for which there exists (for all $t > 0$) a bounded density of the measure $q_t(A)$ with respect to the Lebesgue measure.

§4.3 Composed Poisson Process

A composed Poisson process in a finite-dimensional Euclidean space is a homogeneous process $\xi(t)$ with independent increments, which has the characteristic function

$$\mathbf{M}\exp\{i(z,\xi(t))\} = \exp\{t\int(e^{i(z,x)} - 1)\Pi(dx)\}, \tag{4.47}$$

where Π is a finite measure on X. Denote the time of the first jump of the process by τ_1 and its value by η_1, the time of the second jump by $\tau_2 + \tau_1$ and its value by η_2, the time of the kth jump by $\tau_k + \tau_{k-1} + \cdots + \tau_1$ and its value by η_k, etc.

18°. Theorem. *The pairs of random variables $\{(\tau_k, \eta_k), k = 1, 2, \ldots\}$ are independent and equally distributed; moreover,*

$$\mathbf{P}\{\tau_1 > t, \eta_1 \in A\} = e^{-t\Pi(X)}\Pi(A)(\Pi(X))^{-1}. \tag{4.48}$$

Proof. By virtue of Theorem 19°, Chapter 3, we can write

$$\mathbf{P}\{\tau > t\} = \exp\{-t\Pi(X)\},$$
$$\mathbf{P}\{\eta_1 \in A|\tau_1\} = \frac{\tau_1\Pi(A)}{\tau_1\Pi(X)} = \frac{\Pi(A)}{\Pi(X)}.$$

These two formulas imply (4.48). Note now that $\zeta_k = \tau_1 + \cdots + \tau_k$ is a s.m. with respect to the filtration of σ-algebras $\mathcal{F}_t = \sigma(\{\xi(s), s \leq t\})$, $t \geq 0$. Thus, it follows from Theorem 7° that the process $\xi_{\zeta_k}(t) = \xi(\zeta_k + t) - \xi(\zeta_k)$ has the same distributions as $\xi(t)$ and $\xi_{\zeta_k}(t)$ and is independent of the σ-algebra $\mathcal{F}_{\zeta_k}$. Consequently, if τ' and η' are, respectively, the time of the first jump and its value, then the conditional distribution of the pair of variables (τ', η') with respect to the σ-algebra $\mathcal{F}_{\zeta_k}$ is the same as the distribution of the pair (τ_1, η_1). But $\tau' = \tau_{k+1}$, $\eta' = \eta_{k+1}$. Hence,

$$\mathbf{P}\{\tau_{k+1} > t, \eta_{k+1} \in A|\mathcal{F}_{\zeta_k}\} = \mathbf{P}\{\tau_1 > t, \eta_1 \in A\}. \tag{4.49}$$

Now show that $\tau_1, \eta_1, \dots, \tau_k, \eta_k$ are measurable with respect to $\mathcal{F}_{\zeta_k}$. It suffices to prove that the variables $\zeta_1, \xi(\zeta_1), \dots, \zeta_k, \xi(\zeta_k)$ are measurable with respect to $\mathcal{F}_{\zeta_k}$. Remark 8° implies that $\xi(\zeta_i)$ is $\mathcal{F}_{\zeta_i}$-measurable. ζ_i is also $\mathcal{F}_{\zeta_i}$-measurable, since

$$\{\zeta_i \in B\} \cap \{\zeta_i \le t\} = \{\zeta_i \in B \cap [0,t]\} \in \mathcal{F}_t \tag{4.50}$$

(the collection of those Borel $B \in R_+$, for which the last relation (4.50) holds, is clearly a σ-algebra containing intervals $[0,s]$ for all s; therefore, it contains all Borel sets). In order to show that $\zeta_i, \xi(\zeta_i)$ are $\mathcal{F}_{\zeta_k}$-measurable it suffices to prove that $\mathcal{F}_{\zeta_i} \subset \mathcal{F}_{\zeta_k}$ for $i < k$. If $A \in \mathcal{F}_{\zeta_i}$ then $A \cap \{\zeta_i \le t\} \in \mathcal{F}_t$, and so

$$\begin{aligned} A \cap \{\zeta_k \le t\} &= A \cap \{\zeta_i \le t\} \cap \{\zeta_i \le \zeta_k\} \cap \{\zeta_k \le t\} \\ &= (A \cap \{\zeta_i \le t\}) \cap \{\zeta_k \le t\} \in \mathcal{F}_t \end{aligned}$$

(we have used the fact that $\{\zeta_i \le \zeta_k\}$ is a certain event and that $\{\zeta_k \le t\} \in \mathcal{F}_t$, because ζ_k is a s.m.). Thus, we have proved that (τ_{k+1}, η_{k+1}) is independent of $\tau_1, \dots, \tau_k, \eta_1, \dots, \eta_k$. This property and (4.49) together imply the proof of the theorem.

□

Corollary. *Let ν_t be a number of jumps of the process $\xi(t)$. Then ν_t does not depend on the sequence of independent variables $\{\eta_k, k = 1, 2, \dots\}$; ν_t is a homogeneous Poisson process,*

$$\xi(t) = \sum_{k=1}^{\nu_t} \eta_k \qquad (\sum_1^0 \eta_k = 0). \tag{4.51}$$

In fact, ν_t is determined by the sequence of variables $\{\tau_k, k = 1, 2, \dots\}$, which by virtue of Theorem 18° does not depend on the sequence $\{\eta_k, k = 1, 2, \dots\}$. ν_t is a number of jumps of the process during $[0, t]$; hence, it is a Poisson process, its parameter is $t\Pi(X)$, and thus this process is homogeneous. Formula (4.51) is evident (a process is a sum of its own jumps, which have happened up to the given time).

19°. Theorem. *If $\xi(t)$ is a composed Poisson process and $f(x)$ is a bounded Borel function, then*

$$\lim_{t \to 0} \frac{1}{t}(\mathbf{M} f(\xi(t) + x) - f(x)) = \int (f(x+y) - f(x))\Pi(dy). \tag{4.52}$$

The function $T_s f(x) = \mathbf{M} f(x + \xi(s))$ satisfies the equation

$$\frac{\partial}{\partial s} T_s f(x) = \int [T_s f(x+y) - T_s f(x)]\Pi(dy); \tag{4.53}$$

The resolvent of the process $R_\lambda f(x) = \int_0^\infty e^{-\lambda t} T_t f(x)dt$ *has the form,*

$$R_\lambda f(x) = \frac{1}{\lambda + a}\sum_{n=0}^{\infty} \Pi_\lambda^n f(x), \tag{4.54}$$

where $a = \Pi(X)$, $\Pi_\lambda f(x) = \frac{1}{a+\lambda}\int f(x+y)\Pi(dy)$, *and* Π_λ^n *is the nth power of the operator* Π_λ, *which is defined on bounded Borel functions.*

Proof. Using (4.51) we find

$$\begin{aligned}\mathbf{M}f(\xi(t)+x) &= \mathbf{M}f\left(\sum_{k=1}^{\nu_t}\eta_k + x\right) = f(x)\mathbf{P}\{\nu_t = 0\}\\ &\quad + \mathbf{M}f(x+\eta_1)\mathbf{P}\{\nu_t = 1\} + O(\mathbf{P}\{\nu_t > t\})\\ &= f(x)e^{-at} + \int f(x+y)\frac{1}{a}\Pi(dy)(at)e^{-at} + O(t^2)\end{aligned}$$

(we have used the fact that ν_t has the Poisson distribution with the parameter ta and $\mathbf{P}\{\eta_1 \in A\} = \frac{1}{a}\Pi(A)$). Therefore,

$$\frac{1}{t}[\mathbf{M}f(\xi(t)+x) - f(x)] = e^{-at}\int f(x+y)\Pi(dy) + f(x)\frac{e^{-at}-1}{t} + O(t).$$

Proceeding to the limit as $t \to 0$, we obtain (4.52). Applying (4.52) to the function $T_s f(x)$, we get (4.53). Further,

$$\begin{aligned}R_\lambda f(x) &= \mathbf{M}\int_0^\infty e^{-\lambda t} f(x+\xi(t))dt\\ &= \mathbf{M}\sum_{n=0}^{\infty}\int_{\zeta_n}^{\zeta_{n+1}} e^{-\lambda t} f(x+\xi(t))dt,\end{aligned} \tag{4.55}$$

Here $\zeta_0 = 0$, $\zeta_n = \tau_1 + \cdots + \tau_n$. For $\zeta_n < t < \zeta_{n+1}$, we have $\xi(t) = \sum_{k=1}^n \eta_k$; thus,

$$\begin{aligned}&\mathbf{M}\int_{\zeta_n}^{\zeta_{n+1}} e^{-\lambda t} f(x+\xi(t))dt\\ &= \mathbf{M}e^{-\lambda\zeta_n}\int_0^{\tau_{n+1}} e^{-\lambda t}dt\, f\left(x+\sum_{k=1}^n \eta_k\right)\\ &= \mathbf{M}e^{-\lambda\zeta_n} f\left(x+\sum_{k=1}^n\eta_k\right)\cdot\frac{1-e^{-\lambda\tau_{n+1}}}{\lambda}\\ &= \mathbf{M}e^{-\lambda\zeta_n} f\left(x+\sum_{k=1}^n\eta_k\right)\mathbf{M}\left(\frac{1-e^{-\lambda\tau_{n+1}}}{\lambda}/\mathcal{F}_{\zeta_k}\right)\\ &= \frac{1-a\int_0^\infty e^{-\lambda t - at}dt}{\lambda}\mathbf{M}e^{-\lambda\zeta_n} f\left(x+\sum_{k=1}^n\eta_k\right)\\ &= \frac{1}{a+\lambda}\mathbf{M}e^{-\lambda\zeta_n} f\left(x+\sum_{k=1}^n\eta_k\right).\end{aligned}$$

If

$$\Pi_\lambda f(x) = \mathbf{M}e^{-\lambda\tau_1} f(x+\eta_1)$$
$$= \mathbf{M}e^{-\lambda\tau_1}\mathbf{M}f(x+\eta_1) = \frac{a}{a+\lambda}\int f(x+y)\frac{1}{a}\Pi(dy),$$

then

$$\mathbf{M}e^{-\lambda\zeta_n} f\left(x+\sum_{k=1}^{n}\eta_k\right)$$
$$= \mathbf{M}e^{-\lambda\zeta_{n-1}}\mathbf{M}\left(e^{-\lambda\tau_n} f\left(x+\sum_{k=1}^{n-1}\eta_k+\eta_n\right)\,\Big|\,\mathcal{F}_{\zeta_{n-1}}\right)$$
$$= \mathbf{M}e^{-\lambda\zeta_{n-1}}\Pi_\lambda f(x+\sum_{k=1}^{n-1}\eta_k).$$

Consequently,

$$\mathbf{M}\int_{\zeta_n}^{\zeta_{n+1}} e^{-\lambda t} f(x+\xi(t))dt = \frac{1}{a+\lambda}\Pi_\lambda^n f(x),$$

and (4.54) follows from (4.55).

□

Real-Valued Composed Poisson Processes. Distribution of Ladder Functionals

Consider a composed Poisson process $\xi(t)$ in R. Assume that its characteristic function is

$$\mathbf{M}e^{iz\xi(t)} = \exp\left\{a\int(e^{i(z,x)}-1)d\Phi(x)\right\}, \tag{4.56}$$

where Φ is a distribution function in R (it is a distribution of the jump η_1), and a is a parameter of the exponential distribution of the time of the first jump $\tau_1 : \mathbf{P}\{\tau_1 > t\} = e^{-at}$. Consider distributions of some characteristics of this process. We put

$$Q_+(t,x) = \mathbf{P}\{\sup_{s\le t}\xi(s) < x\}. \tag{4.57}$$

20°. Lemma. *For $x > 0$, the function $Q_+(t,x)$ satisfies the equation*

$$Q_+(t,x) = e^{-at} + a\int_0^t e^{-au}du\int_{-\infty}^{x} Q_+(t-u,x-y)d\Phi(y). \tag{4.58}$$

If

$$q_+(\lambda, x) = \int_0^\infty e^{-\lambda t} Q_+(t, x) dt,$$

then for $x > 0$,

$$q_+(\lambda, x) = \frac{1}{a+\lambda} + \frac{a}{a+\lambda} \int_{-\infty}^{x} q_+(\lambda, x-y) d\Phi(y). \qquad (4.59)$$

Proof. We have

$$\begin{aligned}
&\mathbf{P}\{\sup_{s \le t} \xi(s) < x\} \\
&= \mathbf{P}\{\tau_1 > t\} + \mathbf{P}(\{\tau_1 \le t\} \cap \{\eta_1 < x\} \cap \{\sup_{\tau_1 \le s < t} \xi(s) < x\}) \\
&= \mathbf{P}\{\tau_1 > t\} + \mathbf{P}(\{\tau_1 \le t\} \cap \{\eta_1 < x\} \\
&\quad \cap \{\sup_{0 \le s \le t - \tau_1} \xi(s+\tau_1) - \xi(\tau_1) < x - \eta_1\}) \\
&= \mathbf{P}\{\tau_1 > t\} + \mathbf{M} I_{\{\tau_1 \le t\} \cap \{\eta_1 < x\}} \\
&\quad \times \mathbf{M}(I_{\{\sup_{0 \le s \le t-\tau_1} \xi(s+\tau_1) - \xi(\tau_1) < x - \eta_1\}} | \mathcal{F}_{\tau_1}).
\end{aligned}$$

We now use $\mathcal{F}_{\tau_1}$-measurability of τ_1 and η_1 and the fact that $\xi(s+\tau_1) - \xi(\tau_1)$ does not depend on $\mathcal{F}_{\tau_1}$. We obtain

$$\mathbf{M}(I_{\{\sup_{0 \le s \le t-\tau_1} \xi(s+\tau_1) - \xi(\tau_1) < x - \eta_1\}} | \mathcal{F}_{\tau_1}) = Q_+(t - \tau_1, x - \eta_1).$$

Hence,

$$\begin{aligned}
Q(t, x) &= e^{-at} + \mathbf{M} I_{\{\tau_1 \le t\} \cap \{\eta_1 < x\}} Q_+(t - \tau_1, x - \eta_1) \\
&= e^{-at} + a \int_0^t e^{-au} du \int_{-\infty}^{x} d\Phi(y) Q_+(t - u, x - y).
\end{aligned}$$

We have obtained Eq. (4.58). Multiplying both sides of this equality by $e^{-\lambda t}$ and integrating them, we get (4.59).

□

21°. Theorem. *Suppose that* $q_+(\lambda, x) = 0$ *for* $x \le 0$. *Then the function*

$$\tilde{q}(\lambda, z) = \int e^{izx} d_x q_+(\lambda, x) \qquad (4.60)$$

is defined by the formula

$$\tilde{q}(\lambda, z) = \frac{1}{\lambda} \exp\left\{\int_0^\infty \frac{e^{-\lambda t}}{t} \int_0^\infty (e^{izx} - 1) dF_t(x)\right\}, \qquad (4.61)$$

where

$$F_t(x) = \mathbf{P}\{\xi(t) < x\}.$$

Proof. The convolution-type equation on the semi-axis has already been considered in Section 1.4, where Wiener's method was used to solve it. Eq. (4.59) has just the same form. Suppose that $\varepsilon(x) = 1$ for $x > 0$ and that $\varepsilon(x) = 0$ for $x \le 0$. Then (4.59) can be rewritten as follows:

$$\frac{\varepsilon(x)}{a+\lambda} = \varepsilon(x) \int q(\lambda, x-y)\, d[\varepsilon(y) - \frac{a}{a+\lambda}\Phi(y)] \tag{4.63}$$

and the following relation can be obtained:

$$\varepsilon(x) \int \frac{\varepsilon(x-y)}{a+\lambda}\, dv_1(y) = \int q(\lambda, x-y) dv_2(y), \tag{4.64}$$

where v_1 and v_2 are functions with bounded variation, $v_2(x) = 0$ for $x \le 0$, $v_2(0+) = 0$, $v_1(x)$ is constant on R_+. The functions v_1 and v_2 are defined by the following relation: if

$$\tilde{v}_k(z) = \int e^{izx} dv_k(x), \quad k = 1, 2, \qquad \varphi(z) = \int e^{izy} d\Phi(y),$$

then

$$\tilde{v}_2(z) = \tilde{v}_1(z)\left(1 - \frac{a}{a+\lambda}\varphi(z)\right).$$

We take

$$\tilde{v}_1(z) = \exp\left\{\sum_{n=1}^{\infty} \frac{1}{n}(\frac{a}{a+\lambda})^n \int_{-\infty}^{+0} e^{izx}\, d\Phi_n(x)\right\}, \tag{4.65}$$

$$\tilde{v}_2(z) = \exp\left\{-\sum_{n=1}^{\infty} \frac{1}{n}(\frac{a}{a+\lambda})^n \int_{+0}^{\infty} e^{izx}\, d\Phi_n(x)\right\}. \tag{4.66}$$

Here Φ_n is the nth (iterated n times) convolution of the function Φ. (4.64) implies

$$\tilde{q}(\lambda, z)\tilde{v}_2(z) = \frac{1}{a+\lambda}\int dv_1(y) = \frac{\tilde{v}_1(0)}{a+\lambda};$$

$$\begin{aligned} \tilde{q}(\lambda, z) &= \frac{\tilde{v}_1(0)}{\tilde{v}_2(z)} \cdot \frac{1}{a+\lambda} \\ &= \frac{1}{\lambda}\exp\left\{\sum_{n=1}^{\infty} \frac{1}{n}(\frac{a}{a+\lambda})^n \int_0^{\infty} (e^{izx} - 1) d\Phi_n(x)\right\}. \end{aligned} \tag{4.67}$$

Note now that

$$F_t(x) = \mathbf{P}\left\{\sum_{k=1}^{\nu_t} \eta_k < x\right\} = \sum_{n=0}^{\infty} \mathbf{P}\{\nu_t = n\}\Phi_n(x)$$
$$= \sum_{n=0}^{\infty} \frac{(at)^n}{n!} e^{-at}\Phi_n(x),$$
$$\int_0^{\infty} [F_t(x) - F_t(+0)]\frac{e^{-\lambda t}}{t} dt = \int_0^{\infty} \sum_{n=1}^{\infty} \frac{(at)^n}{tn!} e^{-at-\lambda t}[\Phi_n(x) - \Phi_n(0+)]dt$$
$$= \sum_{n=1}^{\infty} \frac{a^n}{n!}\left(\int_0^{\infty} t^{n-1}e^{-(a+\lambda)t}dt\right)[\Phi_n(x) - \Phi_n(0+)]$$
$$= \sum_{n=1}^{\infty} \frac{1}{n}\left(\frac{a}{a+\lambda}\right)^n [\Phi_n(x) - \Phi_n(0+)].$$

Hence,

$$\sum_{n=1}^{\infty} \frac{1}{n}\left(\frac{a}{a+\lambda}\right)^n \int_0^{\infty} (e^{izx} - 1)d\Phi_n(x) = \int_0^{\infty} \frac{e^{-\lambda t}}{t} \int_0^{\infty} (e^{izx} - 1)dF_t(x)dt.$$

Formula (4.61) follows from (4.67) and from the last equality.

□

Remark. Assume that

$$Q_-(t,x) = \mathbf{P}\{\inf_{s\le t} \xi(s) < x\}, \qquad x < 0;$$
$$q_-(\lambda, x) = \int_0^{\infty} e^{-\lambda t}Q_-(t,x)dt, \qquad \tilde{q}_-(\lambda, x) = \int e^{izx}dq_-(\lambda, x). \tag{4.68}$$

Then

$$\tilde{q}_-(\lambda, x) = \frac{1}{\lambda}\exp\left\{\int_0^{\infty} \frac{e^{-\lambda t}}{t}\int_{-\infty}^{0} (e^{izx} - 1)d_xF_t(x)dt\right\}. \tag{4.69}$$

Now define "ladder" variables for the process $\xi(t)$. Let $x > 0$. We put

$$\tau^x = \inf\{t : \xi(t) \ge x\}, \qquad \gamma_x = \xi(\tau^x+) - x$$

(if the set in brackets is empty, we consider that $\tau^x = +\infty$; γ_x is not defined). τ^x is a time of the first jump over the level x, γ_x is a value of the first jump over the level x.

We put

$$N(t, y, x) = \mathbf{P}\{\tau^x < t, \gamma_x \ge y\}. \tag{4.70}$$

22°. Lemma. *For $x \geq 0$ and $y > 0$, the function $N(t, x, y)$ satisfies the equation*

$$N(t, y, x) = (1 - e^{-at})(1 - \Phi(x + y)) + \int_0^t \int_{-\infty}^x ae^{-as} N(t - s, y, x - u) d\Phi(u) ds, \tag{4.71}$$

whereas its Laplace transform with respect to t

$$n(\lambda, y, x) = \int_0^\infty e^{-\lambda t} d_t N(t, y, x) \tag{4.72}$$

satisfies the equation

$$n(\lambda, y, x) = \frac{a}{a + \lambda}(1 - \Phi(x + y)) + \frac{a}{a + \lambda} \int_{-\infty}^x n(\lambda, y, x - u)\, d\Phi(u) \qquad (x > 0). \tag{4.73}$$

Proof. (4.73) can be obtained from (4.71) by means of the Laplace transform. Let us deduce Eq. (4.71). If $\eta_1 \geq x$, then $\gamma_x = \eta_1$, $\tau^x = \tau_1$; otherwise, if $\eta_1 < x$, then assuming that $\hat{\xi}(t) = \xi(t + \tau_1) - \xi(\tau_1)$, we get $\tau^x = \tau_1 + \hat{\tau}^{x - \eta_1}$, $\gamma_x = \eta_1 + \hat{\gamma}_{x - \eta_1}$, where $\hat{\tau}^u$ and $\hat{\gamma}_u$ are, respectively, the time and value of the first jump over the level u for the process $\hat{\xi}(t)$. Hence,

$$\tau^x = \tau_1 I_{\{\eta_1 \geq x\}} + \hat{\tau}^{x - \eta_1} I_{\{\eta_1 < x\}},$$
$$x + \gamma_x = \eta_1 + \hat{\gamma}_{x - \eta_1} I_{\{\eta_1 < x\}},$$
$$\{\tau^x < t\} \cap \{\gamma_x \geq y\} = \{\tau_1 < t\} \cap [\{\eta_1 \geq x + y\} \cup (\{\eta_1 < x\} \cap \{\hat{\tau}^{x - \eta_1} < t - \tau_1\} \cap \{\hat{\gamma}_{x - \eta_1} \geq y\})].$$

Since $\hat{\xi}(t)$ is independent of τ_1 and η_1 and has the same distributions as $\xi(t)$, then $\hat{\tau}^u$, $\hat{\gamma}_u$ do not depend on τ_1 and η_1 and have identical distributions with τ^u and γ_u, respectively. Thus,

$$\mathbf{P}\{\tau^x < t, \gamma_x \geq y\} = \mathbf{P}\{\tau_1 < t, \eta_1 \geq x + y\} + \int_0^t \int_{-\infty}^x \mathbf{P}\{\tau_1 \in ds, \eta_1 \in du\} \mathbf{P}\{\hat{\tau}^{x - u} < t, \hat{\gamma}_{x - u} \geq y\}.$$

This implies (4.71).

□

Eq. (4.73) is also a convolution-type equation on the semi-axis. By rewriting it as follows:

$$\varepsilon(x) \frac{a}{a + \lambda}(1 - \Phi(x + y)) = \varepsilon(x) \int n(\lambda, y, x - u) d \left[\varepsilon(u) - \frac{a}{a + \lambda}\Phi(u)\right] \tag{4.74}$$

(we assume that $n(\lambda, y, x) = 0$ for $x \leq 0$), we find that it differs from Eq. (4.63) by the l.h.s. only.

23°. Theorem. *The function $n(\lambda, y, x)$ is defined by the equality*

$$n(\lambda, y, x) = \lambda a \int_0^x \left[\int_{-\infty}^0 (1 - \Phi(x + y - u - z))dq_-(\lambda, z)\right] dq_+(\lambda, u). \quad (4.75)$$

Proof. The relation (4.74) yields

$$\varepsilon(x) \int \varepsilon(x - u) \frac{a}{a + \lambda} [1 - \Phi(x + y - u] dv_1(u) = \int n(\lambda, y, x - u) dv_2(u).$$

Multiplying this relation by $e^{-\mu x}$ and integrating from 0 to ∞, we get

$$\begin{aligned} &\frac{a}{a + \lambda} \int_0^\infty \int [1 - \Phi(x + y - u)] e^{-\mu x} dx\, dv_1(u) \\ &\quad = \int e^{-\mu x} n(\lambda, y, x) dx \int e^{-\mu u}\, dv_2(u), \qquad (4.76) \\ &\int e^{-\mu x} n(\lambda, y, x) dx \\ &\quad = \frac{a}{a + \lambda} \int_0^\infty e^{-\mu x} \int [1 - \Phi(x + y - u)] dv_1(u)] dx \\ &\quad \times \left[\int e^{-\mu u} dv_2(u)\right]^{-1}. \end{aligned}$$

(4.66) implies that

$$\begin{aligned} \left[\int e^{-\mu u} dv_2(u)\right]^{-1} &= \exp\left\{\sum_{n=1}^\infty \frac{1}{n} \left(\frac{a}{a + \lambda}\right)^n \int_{+0}^\infty e^{-\mu u}\, d\Phi_n(u)\right\} \\ &= \frac{a + \lambda}{\tilde{v}_1(0)} \int_0^\infty e^{-\mu x} dq_+(\lambda, x). \end{aligned}$$

The product of the the Laplace transforms is a Laplace transform of a convolution, and thus,

$$n(\lambda, y, x) = \frac{a}{\tilde{v}_1(0)} \int_0^x \left[\int_{-\infty}^0 [1 - \Phi(x + y - z - u)] dv_1(u)\right] dq_+(\lambda, z). \quad (4.77)$$

Using the expressions for $q_-(\lambda, u)$ in terms of $v_1(u)$ (see (4.65) and (4.69)), we obtain (4.75).

□

Distributions of Boundary Functionals in the Multi-Dimensional Case

Let G be some region in X, $\xi(t)$ be a composed Poisson process with the characteristic function (4.47). Denote $\tau(x) = \inf\{t : x + \xi(t)\overline{\in} G\}$ (if the set in brackets is empty, then $\tau(x) = +\infty$). Let us find the joint distribution of $\tau(x)$ and $x + \xi(\tau(x))$.

24°. Theorem. *For $\lambda > 0$ and some bounded continuous function $g(x)$, we put*

$$u(x) = \mathbf{M}e^{-\lambda\tau(x)}g(x + \xi(\tau(x))). \tag{4.78}$$

Then $u(x)$ satisfies the equation

$$\begin{aligned} u(x) = {} & \frac{1}{a+\lambda}\int_{x+y\in X\setminus G} g(x+y)\Pi(dy) \\ & + \frac{1}{a+\lambda}\int_{x+y\in G} u(x+y)\Pi(dy) \end{aligned} \tag{4.79}$$

for $x \in G$.

Proof. If τ_1 is a time of the first jump, η_1 is a value of this jump and $\xi'(t) = \xi(t+\tau) - \eta_1$, then $\xi'(t)$ does not depend on τ_1 and η_1, and distributions of $\xi'(t)$ coincide with distributions of $\xi(t)$. Denote $\tau'(x) = \inf\{t : x + \xi'(t)\overline{\in} G\}$.
Then

$$\tau(x) = \tau_i + I_{\{x+\eta_1\in G\}}\tau'(x+\eta_1). \tag{4.80}$$

Hence,

$$\begin{aligned} u(x) &= \mathbf{M}e^{-\lambda\tau(x)}g(x+\tau(x)) \\ &= \mathbf{M}e^{-\lambda\tau_1}g(x+\eta_1)I_{\{\eta_1+x\overline{\in}G\}} + \mathbf{M}I_{\{\eta_1+x\in G\}}e^{-\lambda\tau_1} \\ &\quad \times \mathbf{M}[e^{-\lambda\tau'(x+\eta_1)}g(\xi'(\tau'(x+\eta_1)) + x + \eta_1)/\eta_1, \tau_1] \\ &= \mathbf{M}e^{-\lambda\tau_1}\mathbf{M}g(x+\eta_1)I_{\{\eta_1+x\overline{\in}G\}} + \mathbf{M}e^{-\lambda\tau_1}\mathbf{M}I_{\{\eta_1+x\in G\}}u(x+\eta_1). \end{aligned}$$

Taking into account that

$$\mathbf{M}e^{-\lambda\tau_1} = \frac{a}{a+\lambda}, \qquad \mathbf{P}\{\eta_1 \in dy\} = \frac{1}{a}\Pi(dy),$$

we obtain (4.79) from the last relation.

□

Remark. The solution of Eq. (4.79) can be represented as a series. Denote by Π_λ^G the operator, which acts according to

$$\Pi_\lambda^G g(x) = \frac{1}{a+\lambda}I_G\int g(x+y)\Pi(dy).$$

This operator acts in the space of bounded Borel functions g with the norm $\|g\| = \sup_x |g(x)|$, therefore

$$\|\Pi_\lambda^G g\| \leq \frac{1}{a+\lambda}\int \|g\|\Pi(dy) = \|g\|\frac{a}{a+\lambda}, \qquad \|\Pi_\lambda^G\| \leq \frac{a}{a+\lambda}.$$

We extend the definition of $u(x)$ for $x \in X\backslash G$ by the equality $u(x) = g(x)$ (this is quite natural if we assume that $\tau(x) = 0$ for $x\overline{\in}G$). Furthermore, suppose that $g(x) = 0$ for $x \in G$. Then Eq. (4.79) can be rewritten as follows:

$$u(x) = g(x) + \Pi_\lambda^G u(x),$$

and thus,

$$u(x) = g(x) + \sum_{k=1}^{\infty}(\Pi_\lambda^G)^k g(x). \tag{4.81}$$

§4.4 Homogeneous Processes in R

Let us consider the homogeneous process with independent increments $\xi(t)$ in R, which has the characteristic function

$$\mathbf{M}e^{iz\xi(t)} = \exp\left\{t\left[iaz - \frac{1}{2}bz^2 + \int(e^{izx} - 1 - izxI_{\{|x|\leq 1\}})\Pi(dx)\right]\right\}. \tag{4.82}$$

The process $\xi(t)$ is assumed to be continuous from the right and to have limits from the left.

Distributions of Ladder and Boundary Functionals

First, we apply the results obtained for the distributions of ladder functionals (Theorems 21° and 23°) to the general case. For this purpose, we shall need the uniform approximation of processes with the characteristic function (4.82) by composed Poisson processes.

25°. *Let $\nu(t)$ be a Poisson process with the parameter 1, i.e., $\nu(t)$ is a homogeneous process with independent increments, for which*

$$\mathbf{P}\{\nu(t) = k\} = \frac{t^k}{k!}e^{-t}.$$

Then

$$\mathbf{P}\left(\bigcap_{T>0}\left\{\lim_{n\to\infty}\sup_{t\leq T}\left|t - \frac{1}{n^2}\nu(tn^2)\right| = 0\right\}\right) = 1. \tag{4.83}$$

Proof. By virtue of the Kolmogorov's inequality (see Lemma 14°, Chapter 3), we have

$$\mathbf{P}\left\{\sup_{t\le T}\left|t-\frac{1}{n^2}\nu(tn^2)\right|>\varepsilon\right\}\le\frac{1}{\varepsilon^2}\mathbf{D}\left(\frac{1}{n^2}\nu(Tn^2)\right)\le\frac{1}{\varepsilon^2}\frac{Tn^2}{n^4}=\frac{T}{\varepsilon^2n^2}.$$

Choosing $\varepsilon_n\downarrow 0$ so that $\Sigma\varepsilon_n^{-2}n^2<\infty$, we get

$$\mathbf{P}\{\lim_{n\to\infty}\sup_{t\le T}\left|at-\frac{1}{n^2}\nu(tn^2)\right|=0\}=1.$$

□

26°. *Let $\nu(t)$ be a Poisson process with the parameter 1, $w(t)$ be an independent of $\nu(t)$ Wiener process, i.e., a homogeneous continuous process with independent increments with the characteristic function*

$$\mathbf{M}\exp\{izw(t)\}=\exp\{-\frac{1}{2}tz^2\}.$$

Then $w(\frac{1}{n^2}\nu(n^2t))$ is a composed Poisson process and

$$\mathbf{P}\{\cap_{T>0}\{\lim_{n\to\infty}\sup_{t\le T}\left|w(t)-w\left(\frac{1}{n^2}\nu(n^2t)\right)\right|=0\}\}=1. \tag{4.84}$$

Proof. Consider the process $\eta_n(t)=w(n^{-2}\nu(n^2t))$. Monotonicity of ν implies that $\eta_n(t)$ has independent increments, $\eta_n(0)=0$ and

$$\mathbf{P}\{\eta_n(t)\in A\}=\sum_{k=0}^{\infty}\mathbf{P}\{w\left(\frac{k}{n^2}\right)\in A\}\mathbf{P}\{\nu(n^2t)=k\},$$

$$\begin{aligned}
&\mathbf{P}\{\eta_n(s+t)-\eta_n(s)\in A\}\\
&\quad=\sum_{l,k=0}^{\infty}\mathbf{P}\{w\left(\frac{k+l}{n^2}\right)-w\left(\frac{l}{n^2}\right)\in A\}\\
&\quad\quad\times\mathbf{P}\{\nu(n^2s)=l\}\mathbf{P}\{\nu(n^2(t+s)-n^2s)=k\}\\
&\quad=\sum_{l,k=0}^{\infty}\mathbf{P}\{w\left(\frac{k}{n^2}\right)\in A\}\mathbf{P}\{\nu(n^2s)=l\}\mathbf{P}\{\nu(n^2t)=k\}\\
&\quad=\sum_{k=0}^{\infty}\mathbf{P}\{w\left(\frac{k}{n^2}\right)\in A\}\mathbf{P}\{\nu(n^2t)=l\}=\mathbf{P}\{\eta_n(t)\in A\}.
\end{aligned}$$

Hence, $\eta_n(t)$ is a homogenous process with independent increments. Moreover, it is a step one. Therefore, it is a composed Poisson process. (4.84) follows from the continuity of $w(t)$ and from (4.83).

□

27°. Theorem. *Let $\xi(t)$ be a homogeneous process with independent increments and the characteristic function* (4.82), *and let $\nu(t)$ be a Poisson process independent of $\xi(t)$ with the parameter 1. There exists the sequence of composed Poisson processes $\xi_n(t)$, which can be expressed in terms of $\xi(t)$ and $\nu(t)$, such that*

$$\mathbf{P}\{\bigcap_{T>0}\{\lim_{n\to\infty}\sup_{t\le T}|\xi(t)-\xi_n(t)|=0\}\}=1. \tag{4.85}$$

Proof. Let $\nu_s(A)$ be a number of jumps of the process $\xi(t)$ hitting the set A up to the time s and

$$\xi_0(t)=\xi(t)-\int_{|x|>1}x\nu_t(dx)-\lim_{c\to 0}\left[\int_{c<|x|\le 1}x\nu_t(dx)-t\int_{c<|x|\le 1}x\Pi_t(dx)\right],$$

$\xi_0(t)$ is a continuous component of $\xi(t)$ of the form

$$\xi_0(t)=at+\sqrt{b}w(t),$$

where $w(t)$ is a Wiener process. Further, let us choose a sequence $c_n\downarrow 0$ so that for all T

$$\lim_{n\to\infty}\sup_{t\le T}|\xi(t)-at-\sqrt{b}w(t)-\int_{|x|>1}x\nu_t(dx)$$
$$-\int_{c_n\le x\le 1}x[\nu_t(dx)-t\Pi(dx)]|=0$$

with probability 1. This is so if for some sequence $\varepsilon_n\downarrow 0$

$$\sum_n\frac{1}{\varepsilon_n^2}\mathbf{D}\Big(\xi(T)-aT-\sqrt{b}w(T)-\int_{|x|>1}x\nu_T(dx)$$
$$-\int_{c_n\le|x|\le 1}x\left[\nu_T(dx)-T\Pi(dx)\right]\Big)$$
$$=\sum_n\frac{T}{\varepsilon_n^2}\int_{|x|\le c_n}x^2\Pi(dx)<\infty$$

(this follows from Kolmogorov's inequality). Hence, to prove the theorem it suffices to show that there exists a sequence of composed Poisson processes $\xi_m(t)$ such that for all $T>0$,

$$\lim_{m\to\infty}\sup_{t\le T}\left|\left(at+\sqrt{b}w(t)+\int_{|x|>c_m}x\nu_t(dx)+t\int_{c_m\le|x|\le 1}x\Pi(dx)-\xi_m(t)\right)\right|=0.$$

with probability 1. Then 25° and 26° imply that $\xi_m(t)$ can be chosen as follows:

$$\begin{aligned}\xi_m(t) = \int_{|x|\geq c_m} x\nu_t(dx) + \sqrt{b}w\left(\frac{1}{n_m}\nu(n_m t)\right) \\ + \left(a + \int_{c_m\leq |x|\leq 1} x\Pi(dx)\right)\frac{1}{n_m}\nu(n_m t),\end{aligned} \tag{4.86}$$

where n_m is a sequence such that for all T,

$$\left(1 + \int_{c_m\leq|x|\leq 1} x\Pi(dx)\right)\sup_{t\leq T}\left|\frac{1}{n_m}\nu(n_m t) - t\right| \to 0.$$

with probability 1.

□

Let us use the theorem proved to find the distributions of $\sup_{s\leq t}\xi(s)$, $\inf_{s\leq t}\xi(s)$, τ^x and γ_x, $x > 0$.

28°. Theorem. *Let*

$$F_t(x) = \mathbf{P}\{\xi(t) < x\},$$

$$q_+(\lambda, x) = \int_0^\infty e^{-\lambda t}\mathbf{P}\{\sup_{s\leq t}\xi(s) < x\}dt,$$

$$q_-(\lambda, x) = \int_0^\infty e^{-\lambda t}\mathbf{P}\{\inf_{s\leq t}\xi(s) < x\}dt,$$

$$\tilde{q}_\pm(\lambda, z) = \int e^{izx}dq_\pm(\lambda, x),$$

$$n(\lambda, y, x) = \int_0^\infty e^{-\lambda t}d_t\Pi\{\tau^x < t, \gamma_x \geq y\}.$$

The following equalities hold:

$$\tilde{q}_+(\lambda, z) = \frac{1}{\lambda}\exp\{\int_0^\infty \frac{e^{-\lambda t}}{t}\int_0^\infty (e^{izx} - 1)d_xF_t(x)dt\}, \tag{4.87}$$

$$\tilde{q}_-(\lambda, z) = \frac{1}{\lambda}\exp\{\int_0^\infty \frac{e^{-\lambda t}}{t}\int_{-\infty}^0 (e^{izx} - 1)d_xF_t(x)dt\}, \tag{4.88}$$

$$n(\lambda, y, x) = \lambda\int\left(\int_{-\infty}^0 \Pi([x + y - u - z, \infty[)dq_-(\lambda, z)\right)dq_-(\lambda, u). \tag{4.89}$$

Proof. Suppose that $\xi_m(t)$ is given by (4.86). Then for all t,

$$\lim_{m\to\infty}\sup_{s\leq t}\xi_m(s) = \sup_{s\leq t}\xi(s)$$

with probability 1, and thus

$$\begin{aligned}\tilde{q}_+(\lambda, z) &= \int_0^\infty e^{-\lambda t}\mathbf{M}\exp\{iz\sup_{s\le t}\xi(s)\}dt\\ &= \int_0^\infty e^{-\lambda t}\lim_{m\to\infty}\mathbf{M}\exp\{iz\sup_{s\le t}\xi_m(s)\}dt\\ &= \lim_{m\to\infty}\int_0^\infty e^{-\lambda t}\mathbf{M}\exp\{iz\sup_{s\le t}\xi_m(s)\}dt.\end{aligned}$$

By virtue of Theorem 21°, the expression to the right equals

$$\frac{1}{\lambda}\exp\{\int_0^\infty \frac{1}{t}e^{-\lambda t}\int_0^\infty (e^{izx}-1)d_x F_t^{(m)}(x)\},$$

where $F_t^{(m)}(x) = \mathbf{P}\{\xi_m(t) < x\}$. Hence, to prove (4.87) is suffices to show that

$$\lim_{m\to\infty}\int_0^\infty \frac{1}{t}e^{-\lambda t}\int_0^\infty (e^{izx}-1)d_x F_t^{(m)}(x) = \int_0^\infty \frac{1}{t}e^{-\lambda t}\int_0^\infty (e^{izx}-1)d_x F_t(x)$$

or

$$\lim_{m\to\infty}\int_0^\infty \frac{1}{t}e^{-\lambda t}\mathbf{M}[(e^{iz\xi_m(t)}-1)I_{\{\xi_m(t)>0\}} - (e^{iz\xi(t)}-1)I_{\{\xi(t)>0\}}]dt = 0.$$

The function $(e^{izx}-1)I_{\{x>0\}}$ satisfies the Lipschitz condition

$$|(e^{izx}-1)I_{\{x>0\}} - (e^{izy}-1)I_{\{y>0\}}| \le |z||x-y|.$$

(4.87) follows from the equality

$$\lim_{m\to\infty}\int_0^\infty \frac{1}{t}e^{-\lambda t}\mathbf{M}|\xi(t)-\xi_m(t)|dt = 0. \tag{4.90}$$

Then (4.86) yields

$$\begin{aligned}\mathbf{M}|\xi(t)-\xi_m(t)| &\le \sqrt{b}\mathbf{M}\left|w\left(\frac{1}{n_m}\nu(n_m t)\right) - w(t)\right|\\ &+ \left(|a| + \left|\int_{c_m\le|x|\le 1} x\Pi(dx)\right|\right)\mathbf{M}\left|\frac{1}{n_m}\nu(n_m t) - t\right|.\end{aligned}$$

We have

$$\mathbf{M}\left|\frac{1}{n_m}\nu(n_m t)-t\right| \le \sqrt{\mathbf{D}\left(\frac{1}{n_m}\nu(n_m t)\right)}=\sqrt{\frac{t}{n_m}},$$

$$\begin{aligned}\mathbf{M}&\left|w\left(\frac{1}{n_m}\nu(n_m t)\right)-w(t)\right|\\ &=\sum_{k=0}^{\infty}\mathbf{M}\left|w\left(\frac{k}{n_m}\right)-w(t)\right|\mathbf{P}\{\nu(n_m t)=k\}\\ &\le\sum_{k=0}^{\infty}\left(\mathbf{M}\left|w\left(\frac{k}{n_m}\right)-w(t)\right|^2\right)^{1/2}\mathbf{P}\{\nu(n_m t)=k\}\\ &=\sum_{k=0}^{\infty}\sqrt{\left|\frac{k}{n_m}-t\right|}\mathbf{P}\{\nu(n_m t)=k\}\\ &=\mathbf{M}\sqrt{\left|\frac{1}{n_m}\nu(n_m t)-t\right|}\le\sqrt[4]{\mathbf{D}\left(\frac{1}{n_m}\nu(n_m t)\right)}\le\left(\frac{t}{n_m}\right)^{1/4}.\end{aligned}$$

Thus,

$$\mathbf{M}|\xi(t)-\xi_m(t)|\le\sqrt{b}\,n_m^{-1/4}t^{1/4}+\left(|a|+\left|\int_{c_m\le|x|\le 1}x\Pi(dx)\right|\right)n_m^{-1/2}t^{1/2}.$$

If $n_m\to\infty$ is such that $\left|\int_{c_m\le x\le 1}x\Pi(dx)\right|^2 n_m^{-1}\to 0$, then (4.90) is satisfied. (4.88) is obtained similarly.

Assume that the characteristic function of $\xi_m(t)$ has the form

$$\mathbf{M}\exp\{iz\xi_m(t)\}=\exp\{t\int(e^{izx}-1)\Pi^{(m)}(dx)\}.$$

Denote by τ_m^x the time of the first jump of the process $\xi_m(t)$ over the level x. We have $\tau_m^x\to\tau^x$, $\xi_m(\tau_m^x)\to\xi(\tau^x)$ for $\xi(\tau^x)>x$. Thus, for almost all x and y,

$$n^{(m)}(\lambda,y,x)=\int_0^\infty e^{-\lambda t}d_t\mathbf{P}\{\tau_m^x<t,\xi(\tau_m^x)-x\ge y\}\to n(\lambda,y,x).$$

Note that for $z>0$,

$$\begin{aligned}\Pi^{(m)}([z,\infty[)&=\Pi([z\vee c_m,\infty[)\\ &\quad+n_m I_{\left\{\frac{1}{n_m}\left(a+\int_{c_m\le x\le 1}x\Pi(dx)\right)>z\right\}}+\mathbf{P}\{w(1/n_m)>z/\sqrt{b}\}\end{aligned}$$

and $\Pi^{(m)}([z,\infty[)$ uniformly converge to $\Pi([z,\infty[)$ as $m \to \infty$. By virtue of Theorem 23°, we can write the equality

$$n^{(m)}(\lambda,y,x) = \lambda \int_0^x \left(\int_{-\infty}^0 \Pi^{(m)}(x+y-u-z)dq_-^{(m)}(\lambda,z) \right) dq_+^{(m)}(\lambda,z), \tag{4.91}$$

where

$$q_+^{(m)}(\lambda,x) = \int_0^\infty e^{-\lambda t}\mathbf{P}\{\sup_{s\le t}\xi^{(m)}(s) < x\}dt,$$

$$q_-^{(m)}(\lambda,x) = \int_0^\infty e^{-\lambda t}\mathbf{P}\{\inf_{s\le t}\xi^{(m)}(s) < x\}dt,$$

and use the already proved convergence of $q_\pm^{(m)}(\lambda,x)$ to $q_\pm(\lambda,x)$. The uniform convergence of $\Pi^{(m)}$ to Π yields

$$\int_{-\infty}^0 \Pi^{(m)}(x+y-u-z)dq_-^{(m)}(\lambda,z) \to \int_{-\infty}^0 \Pi(x+y-u-z)dq_-(\lambda,z). \tag{4.92}$$

Note that in the case when $\xi(t)$ is not a composed Poisson process, the function $q_-(\lambda,z)$ may have the only discontinuity point at the point 0. This follows from (4.88) and from the fact that the measure

$$\int_0^\infty e^{-\lambda t}\frac{1}{t}\mathbf{P}\{\xi(t)\in A\}dt,$$

where A stays at a positive distance from the point 0, has no atoms (if $\mathbf{P}\{\xi(s)=c\}>0$ for $s<t$ and $\mathbf{P}\{\xi(t)=c\}>0$, then $\mathbf{P}\{\xi(t)-\xi(s)=0\}>0$). If $q_-(\lambda,0+)-q_-(\lambda,0)>0$, then $q_-^{(m)}(\lambda,0+)-q_-^{(m)}(\lambda,0) \to q_-(\lambda,0+)-q_-(\lambda,0)$ and $q_-^{(m)}(\lambda,x) \to q_-(\lambda,x)$ uniformly for $x \le 0$. This implies that the convergence in (4.92) is also uniform, and we can proceed in (4.91) to the limit as $m \to \infty$.

□

Remark. Suppose that $x<0$. Let us denote $\tau^x = \inf\{t : \xi(t) < x\}$, $\gamma_x = \xi(\tau_x+0)-x$. Changing the sign of the process $\xi(t)$ we can get, in exactly the same way as (4.89), the following relation: for $x<0$ and $y<0$,

$$\int_0^\infty e^{-\lambda t}d_t\mathbf{P}\{\tau^x<t,\gamma_x\le y\}$$
$$= \lambda \int_x^0 \left(\int_0^\infty \Pi(]-\infty,x+y-z-u])dq_+(\lambda,z) \right) dq_-(\lambda,u). \tag{4.93}$$

Joint Distribution of the Process Suppremum and Infimum

Let $a < 0$, $b > 0$, $a \le \alpha < \beta \le b$. Write

$$Q(t;a,b;\alpha,\beta) = \mathbf{P}\{\inf_{s\le t} \xi(s) \ge a, \sup_{s\le t} \xi(s) \le b, \alpha < \xi(t) < \beta\}. \qquad (4.94)$$

Our aim is to express $Q(t;a,b;\alpha_1,\beta)$ in terms of distributions of the ladder functions τ^x and $\xi(\tau^x)$. We put

$$\Gamma(x,dt,dy) = \mathbf{P}\{\tau^x \in dt, \xi(\tau^x) \in dy\}$$

(if $x > 0$, the measure Γ is concentrated on $R_+ \times [x,\infty[$, if $x < 0$, it is concentrated on $R_+ \times]-\infty, x]$).

29°. Theorem. *Let $F_t(A)$ be a distribution of the variable $\xi(t)$; $c_k = a$ if k is even; and $c_k = b$ if k is odd. Then*

$$\begin{aligned}
&Q(t;a,b;\alpha,\beta)\\
&= F_t((\alpha,\beta)) - \sum_{k=0}^{\infty}(-1)^k \Big[\int_{0<t_1<\cdots<t_{k+1}}\!\!\cdots\int \Gamma(b,dt_1,dy_1)\Gamma(a-y_1,dt_2,dy_2)\ldots\\
&\quad \times \Gamma(c_{k+1}-y_k,dt_{k+1},dy_{k+1})F_{t-t_k}(\alpha-y_{k+1},\beta-y_{k+1})\\
&\quad + \int_{0<t_1<\cdots<t_{k+1}}\!\!\cdots\int \Gamma(a,dt_1,dy_1)\ldots\\
&\quad \times \Gamma(c_k-y_k,dt_{k+1},dy_{k+1})F_{t-t_{k+1}}((\alpha-y_{k+1},\beta-y_{k+1}))\Big].
\end{aligned} \qquad (4.95)$$

Proof. We introduce the events $\mathfrak{A}_k^+$ and $\mathfrak{A}_k^-$ as follows: The event $\mathfrak{A}_k^+$ means that before the time t, the process $\xi(s)$ visits the interval $]b,\infty[$ earlier than $]-\infty,a[$, and then crosses the segment $[a,b]$ at least k times also before the time t, and $\xi(t) \in (\alpha,\beta)$. The event $\mathfrak{A}_k^-$ means that $\xi(s)$ crosses the segment $[a,b]$ at least k times before the time t, visiting $]-\infty,a[$ first, and $\xi(t) \in (\alpha,\beta)$. Then

$$Q(t;a,b;\alpha,\beta) = F((\alpha,\beta)) - \mathbf{P}\{\mathfrak{A}_0^+\} - \mathbf{P}\{\mathfrak{A}_0^-\}. \qquad (4.96)$$

We put $\mathcal{B}_k^+ = \mathfrak{A}_k^+ \cup \mathfrak{A}_{k+1}^-$. The event $\mathcal{B}_k^+$ means that before the time t, the process visits $]b,\infty[$, crosses $[a,b]$ at least k times, and finds itself in (α,β) at the time t. The event $\mathcal{B}_k^- = \mathfrak{A}_k^- \cup \mathfrak{A}_{k+1}^+$ means that before the time t, the process gets into $]-\infty,a[$, crosses $[a,b]$ at least k times, and gets into (α,β) at the time t. Since

$$\mathbf{P}\{\mathfrak{B}_k^+\} + \mathbf{P}\{\mathfrak{B}_k^-\} = \mathbf{P}\{\mathfrak{A}_k^+\} + \mathbf{P}\{\mathfrak{A}_k^-\} + \mathbf{P}\{\mathfrak{A}_k^-\} + \mathbf{P}\{\mathfrak{A}_{k+1}^-\}$$

and $\mathbf{P}\{\mathfrak{A}_k^+\} + \mathbf{P}\{\mathfrak{A}_k^-\} \to 0$ (the number of crossings of the segment $[a,b]$ during $[0,t]$ is finite), we have

$$\mathbf{P}\{\mathfrak{A}_0^+\} + \mathbf{P}\{\mathfrak{A}_0^-\} = \sum_{k=0}^{\infty}(-1)^k[\mathbf{P}\{\mathfrak{B}_k^+\} + \mathbf{P}\{\mathfrak{B}_k^-\}]. \qquad (4.97)$$

Let us calculate $\mathbf{P}\{\mathcal{B}_k^+\}$. Let τ_1 denote the time when the process hits $]b, \infty[$ for the first time, let τ_2 denote the first time after τ_1 that the process hits $]-\infty, a[$, etc. For odd i, $\tau_{i+1}-\tau_i$ is the first time that the process $\xi_i(s) = \xi(s+\tau_i)-\xi(\tau_i)$ hits $]-\infty, a-\xi(\tau_i)[$, and the process $\xi_i(s)$ is independent of τ_i and $\xi(\tau_i)$. For even i, $\tau_{i+1}-\tau_i$ is the first time that the process $\xi_i(s)$ hits $]b-\xi(\tau_i), \infty[$. Thus,

$$\mathbf{P}\{\tau_{i+1} - \tau_i \in ds, \xi(\tau_{i+1}) \in dy/\mathcal{F}_{\tau_i}\} = \Gamma(c_i - \xi(\tau_i), ds, dy).$$

This equality yields

$$\mathbf{P}\{\mathfrak{B}_k^+\} = \int_{0<t_1<\cdots<t_k<t_{k+1}<t} \int \Gamma(b, dt_1, dy_1)\Gamma(a - y_1, dt_2, dy_2) \\ \ldots \Gamma(c_{k+1} - y_k, dt_{k+1}, dy_{k+1}) F_{t-t_{k+1}}((\alpha - y_{k+1}, \beta - y_{k+1})). \tag{4.98}$$

Similarly,

$$\mathbf{P}\{\mathfrak{B}_k^-\} = \int_{0<t_1<\cdots<t_k<t_{k+1}<t} \int \Gamma(a, dt_1 dy_1)\Gamma(b - y_1, dt_2, dy_2) \ldots \\ \times \Gamma(c_{k+1} - y_k, dt_{k+1}, dy_{k+1}) F_{t-t_{k+1}}((\alpha - y_{k+1}, \beta - y_{k+1})). \tag{4.99}$$

Substituting (4.98) and (4.99) into (4.97) and the expression obtained into (4.96), we get (4.95).

□

Remark. A simpler expression can be obtained for the Laplace transform of $Q(t; a, b; \alpha, \beta)$ with respect to t. Let

$$q(\lambda; a, b; \alpha, \beta) = \int_0^\infty e^{-\lambda t} Q(t; a, b; \alpha, \beta) dt,$$

$$r(\lambda, \alpha, \beta) = \int_0^\infty e^{-\lambda t} F_t((\alpha, \beta)) dt,$$

$$\gamma(\lambda, x, dy) = \int_0^\infty e^{-\lambda t} \Gamma(x, dt, dy).$$

Finally, by introducing the functions

$$\gamma_+^{a,b}(\lambda, y, dz) = \int \gamma(\lambda, a - y, dy_1)\gamma(\lambda, b - y_1, dz),$$

$$\gamma_-^{a,b}(\lambda, y, dz) = \int \gamma(\lambda, b - y, dy_1)\gamma(\lambda, a - y_1, dz),$$

we get for even k,

$$\int_0^\infty \mathbf{P}\{\mathfrak{B}_k^+\}e^{-\lambda t}dt = \int \gamma(\lambda, b, dy_1)\gamma(\lambda, a-y_1, dy_2)$$
$$\dots\gamma(\lambda, c_{k+1}-y_k, dy_{k+1})r(\lambda, \alpha-y_{k+1}, \beta-y_{k+1})$$
$$= \int \gamma(\lambda, b, dy_1)\gamma_+^{a,b}(\lambda, y_1, dy_2)\gamma_+^{a,b}(\lambda, y_2, dy_3)\dots$$
$$\times\, \gamma_+^{a,b}(\lambda, y_l, dy_{l+1})r(\lambda, \alpha-y_{l+1}, \beta-y_{l+1}),$$

where $l = k/2$, and for odd k,

$$\int_0^\infty \mathbf{P}\{\mathfrak{B}_k^+\}e^{-\lambda t}dt$$
$$= \int \gamma_+^{a,b}(\lambda, 0, dy_1)\dots\gamma_+^{a,b}(\lambda, y_1, y_{l+1})r(\lambda, \alpha-y_{l+1}, \beta-y_{l+1}),$$
$$l = (k-1)/2.$$

Similarly,

$$\int_0^\infty \mathbf{P}\{\mathfrak{B}_{2l}^-\}e^{-\lambda t}dt = \int \gamma(\lambda, a, dy_1)\gamma_-^{a,b}(\lambda, y_1, dy_2)$$
$$\dots\gamma_-^{a,b}(\lambda, y_l, dy_{l+1}r(\lambda, \alpha-y_{l+1}, \beta-y_{l+1}),$$
$$\int_0^\infty \mathbf{P}\{\mathfrak{B}_{2l+1}^-\}e^{-\lambda t}dt = \int \gamma_-^{a,b}(\lambda, 0, dy_1)\gamma_-^{a,b}(\lambda, y_1, dy_2)$$
$$\dots\gamma_-^{a,b}(\lambda, y_l, dy_{l+1}r(\lambda, \alpha-y_{l+1}, \beta-y_{l+1}).$$

We put

$$K_\lambda^\pm(a, b; y, A) = \sum_{n=1}^\infty \int\dots\int \gamma_\pm^{a,b}(\lambda, y, dy_1)\dots\gamma_\pm^{a,b}(\lambda, y_{n-1}, A).$$

Then

$$q(\lambda; a, b; \alpha, \beta) = r(\lambda, \alpha, \beta) - \int [\gamma(\lambda, a, dy) + \gamma(\lambda, b, dy]$$
$$\times\, r(\lambda, \alpha-y, \beta-y) + \int [K_\lambda^+(a, b; 0, dy) + K_\lambda^-(a, b; 0, dy)]$$
$$\times\, r(\lambda, \alpha-y, \beta-y) - \int (\gamma(\lambda, a, dy_1)K_\lambda^-(a, b; y_1, dy_2)$$
$$+\, \gamma(\lambda, b, dy_1)K_\lambda^+(a, b; y_1, dy_2))r(\lambda, \alpha-y_2, \beta-y_2). \tag{4.100}$$

Poisson Process with Drift

Let $\nu(t)$ be a homogeneous Poisson process with a parameter a and let $c \in R$. Consider the process $\xi(t) = \nu(t) + ct$. Let us find distributions of some functionals of this process using the results proved above. Suppose first that $c > 0$. Then $\sup_{0\le s\le t}\xi(s) = \xi(t)$, $\inf_{0\le s\le t}\xi(s) = 0$. The only nontrivial thing is to find the joint distribution of τ^x and γ_x. But γ_x is completely determined by the variable τ^x. If $x - c\tau^x$ is an integer, then $\xi(t)$ crosses the level x continuously and $\gamma_x = 0$, but if we have $k < x - c\tau^x < k+1$ for some k, then $\gamma_x = c\tau^x - x - k$. Thus, γ_x coincides with the fractional part of $c\tau^x - x$. The joint distribution of τ^x and γ_x is determined by the distribution of τ^x. However,

$$\mathbf{P}\{\tau^x < t\} = \mathbf{P}\{\xi(t) > x\} = \mathbf{P}\{\nu(t) > x - ct\} = \sum_{k > x-ct} \frac{(at)^k}{k!} e^{-at},$$

Hence, for $0 \le y < 1$,

$$\begin{aligned}
\mathbf{P}\{\tau^x < t, \gamma_x \ge y\} &= \sum_{l=0}^{\infty} \mathbf{P}\{\tau^x < t, l + y \le x - c\tau^x \le l+1\} \\
&= \sum_{l=0}^{\infty} \mathbf{P}\{\tau^x < t, \frac{x-l-1}{c} \le \tau^x \le \frac{x-l-y}{c}\} \\
&= \sum_{\substack{k > x-ct \\ x-ct-1 \le l < x-y}} \frac{1}{k!}\left[\left(a\frac{x-l-y}{c}\lambda t\right)^k \exp\{-a\left(\frac{x-l-y}{c}\lambda t\right)\}\right. \\
&\quad \left. - \left(a\frac{x-l-1}{c}\right)^k \exp\{-a\frac{x-l-1}{c}\}\right].
\end{aligned}$$

Now let $c < 0$. Then for $x > 0$,

$$\begin{aligned}
\int_0^\infty \frac{1}{t} e^{-\lambda t}\mathbf{P}\{\xi(t) > x\}dt &= \int_0^\infty \frac{1}{t} e^{-\lambda t} \sum_{k > x-ct} \frac{(at)^k}{k!} e^{-at} dt \\
&= \sum_{k > x} \int_0^{\frac{k-x}{c}} \frac{1}{k!} e^{-t(a+\lambda)} a^k t^{k-1} dt.
\end{aligned}$$

The distribution of $\sup_{s\le t}\xi(s)$ is defined by

$$\begin{aligned}
&\int_0^\infty e^{-\lambda t}\mathbf{M}\exp\{iz\sup_{s\le t}\xi(s)\}dt \\
&= \frac{1}{\lambda}\exp\left\{\sum_{k=1}^{\infty}\int_0^k (e^{izx}-1)(k-x)^{k-1}\exp\{-\frac{a+\lambda}{-c}(k-x)\}dx\right\}.
\end{aligned}$$

For $x < 0$,

$$\int_0^\infty \frac{e^{-\lambda t}}{t}\mathbf{P}\{\xi(t) < x\}dt = \int_0^\infty \frac{e^{-\lambda t}}{t} \sum_{k<x<ct} \frac{(at)^k}{k!} e^{-at}dt$$
$$= \sum_k \int_0^\infty I_{\{t>\frac{k-x}{-c}\}} e^{-t(a+\lambda)}\frac{(at)^k}{k!}dt.$$

Hence,

$$\int_0^\infty e^{-\lambda t}\mathbf{M}\exp\{iz\inf_{s\le t}\xi(s)\}dt$$
$$= \frac{1}{\lambda}\exp\{\sum_{k=1}^\infty \int_{-\infty}^0 (e^{izx}-1)\exp\{-\frac{a+\lambda}{-c}(k-x)\}$$
$$\times \frac{1}{k!}\left(\frac{a}{-c}\right)^k (k-x)^{k-1}dx\}.$$

Semicontinuous Processes

The processes are called semicontinuous if they either have no positive jumps (upper semicontinuous) or have no negative jumps (lower semicontinuous). To study them, the Laplace transform, i.e., the function $\mathbf{M}e^{z\xi(t)}$ for complex z, is used instead of the Fourier transform. We now give some results on the existence of these transforms.

30°. Lemma. (1) *If $\xi_0(t)$ is a continuous Gaussian process, then*

$$\mathbf{M}\exp\{z\xi_0(t)\} = \exp\{z\mathbf{M}\xi_0(t) + \frac{z^2}{2}\mathbf{D}\xi_0(t)\}$$

This is an entire analytic function; (2) *if the characteristic function of $\xi^1(t)$ is defined by*

$$\mathbf{M}e^{iz\xi^1(t)} = \exp\{t\int_{|x|\le c}(e^{izx}-1-izx)\Pi(dx)\},$$

then $\mathbf{M}e^{z\xi^1(t)}$ is an entire analytic function:

$$\mathbf{M}e^{z\xi^1(t)} = \exp\{t\int_{|x|\le c}(e^{zx}-1-zx)\Pi(dx)\};$$

(3) *if, for some $c > 0$, the characteristic function of the process $\xi_1(t)$ has the form*

$$\mathbf{M}e^{iz\xi_1(t)} = \exp\{t\int_{-\infty}^{-c}(e^{izx}-1)\Pi(dx)\},$$

then $\mathbf{M}e^{z\xi_1(t)} = \exp\{t\int_{-\infty}^{-c}(e^{zx}-1)\Pi(dx)\}$; *it is an analytic function for* $\operatorname{Re} z > 0$ *and a continuous function for* $\operatorname{Re} z \geq 0$.

Proof. Analyticity of the functions in corresponding domains follow in each case from the above formulas. The formula for the Gaussian case arises from the form of the Gaussian distribution, whereas in the cases of (2) and (3), the corresponding formulas follow from the Poisson distribution formulas, in exactly the same way as in Theorem 12°, Chapter 3.

□

Corollary. *If* $\xi(t)$ *is a homogeneous upper semicontinuous process with independent increments, then its Laplace transform is given by*

$$\mathbf{M}e^{z\xi(t)} = \exp\{t\left[az + \frac{b}{2}z^2 + \int_{-\infty}^{0}(e^{zx} - 1 - zxI_{\{x>-1\}})\Pi(dx)\right]\}. \quad (4.101)$$

If $K(z)$ *is a cumulant of the process* $\xi(t)$ *(see (4.2)), the Laplace transform (4.101) has the form* $\exp\{tK(-iz)\}$.

31°. Lemma. *We put* $a_1 = a + \int_{-\infty}^{-1} x\Pi(dx) = \lim_{z\to 0}\frac{1}{z}\cdot\frac{1}{t}\ln \mathbf{M}e^{z\xi(t)}$ *(this limit can be equal to* $-\infty$*). Then* $\mathbf{P}\{\sup_t \xi(t) = +\infty\} = 1$ *for* $a_1 \geq 0$ *and* $\mathbf{P}\{\sup_t \xi(t) = +\infty\} = 0$ *for* $a_1 < 0$.

Proof. If $a_1 \neq 0$, the proof follows from the strong law of large numbers. For $a_1 = 0$, we consider the sequence of sums of independent equally distributed variables

$$\xi(n) = \sum_{k=1}^{n}(\xi(k) - \xi(k-1)),$$

for which $\mathbf{M}(\xi(k) - \xi(k-1)) = a_1 = 0$. By virtue of 64° (Section 1.5), we have

$$\mathbf{P}\{\sup_n \xi(n) = +\infty\} = 1.$$

□

Let $a + \int_{-\infty}^{-1} x\Pi(dx) \geq 0$. Denote by τ_y, $y > 0$, the time when the process $\xi(t)$ with the Laplace transform (4.101) reaches the level y for the first time (since the process is upper semicontinuous, we have $\xi(\tau_y) = y$). Under accepted assumptions, τ_y is finite with probability 1.

32°. Theorem. *If for all* $y > 0$, τ_y *is finite with probability* 1, *then as a function of* y *it is a homogeneous process with independent increments, it is nondecreasing and*

$$\mathbf{M}e^{-z\tau_y} = e^{-yB(z)}; \quad (4.102)$$

moreover, $B(\lambda)$ is (for $\lambda > 0$) a unique positive solution of the equation $K(-iB(\lambda)) = 0$.

Proof. If $\xi_y(t) = \xi(t + \tau_y) - \xi(\tau_y)$, then the process $\xi_y(t)$ is independent of the σ-algebra $\mathcal{F}_{\tau_y}$ with respect to which τ_{y_1} are measurable for $y_1 < y$ (since $\tau_{y_1} < \tau_y$). Let τ_z^y be the time when the process $\xi_y(t)$ reaches the level $z > 0$ for the first time. Then $\tau_y + \tau_z^y = \tau_{y+z}$, $\tau_{y+z} - \tau_y = \tau_z^y$ has the same distribution as τ_z and does not depend on τ_{y_1} for $y_1 < y$. Thus, we have proved that τ_y is a homogeneous process with independent increments with respect to y. Since τ_y is a nondecreasing process (i.e., lower semicontinuous), we have

$$\mathbf{M}e^{iz\tau_y} = \exp\{y[icz + \int_0^\infty (e^{izx} - 1)\Pi_1(dx)]\},$$

where $c \geq 0$; and hence

$$\mathbf{M}e^{-\lambda\tau_y} = \exp\{-yB(\lambda)\},$$

where

$$B(\lambda) = \lambda c - \int_0^\infty (e^{-\lambda x} - 1)\Pi_1(dx).$$

Note that

$$\begin{aligned}\mathbf{M}e^{-\lambda\tau_y} &= -\int_0^\infty e^{-\lambda t} d_t \mathbf{P}\{\sup_{s\leq t}\xi(s) < y\} \\ &= 1 - \lambda\int_0^\infty e^{-\lambda t}\mathbf{P}\{\sup_{s\leq t}\xi(s) < y\}dt = 1 - \lambda q_+(\lambda, y),\end{aligned}$$

$$\begin{aligned}\lambda\int_0^\infty e^{izy} d_y q_+(\lambda, y) &= \int_0^\infty e^{izy} d_y(1 - \mathbf{M}e^{-\lambda\tau_y}) \\ &= \int_0^\infty e^{izy} d_y(e^{-yB(\lambda)}) = \frac{B(\lambda)}{B(\lambda) - iz},\end{aligned}$$

i.e.,

$$\lambda\tilde{q}_+(\lambda, z) = \frac{B(\lambda)}{B(\lambda) - iz}.$$

Here $q_\pm(\lambda, y)$ and $\tilde{q}_\pm(\lambda, z)$ are the same as in Theorem 28°. Formulas (4.87) and (4.88) and Theorem 4° imply the equality $\lambda\tilde{q}_+(\lambda, z)\lambda\tilde{q}_-(\lambda, z) = \frac{\lambda}{\lambda - K(z)}$, so

$$\lambda\tilde{q}_-(\lambda, z) = \frac{\lambda(B(\lambda) - iz)}{B(\lambda)(\lambda - K(z))},$$

$$\frac{\lambda(B(\lambda) - iz)}{B(\lambda)(\lambda - K(z))} = \exp\{\int_0^\infty \frac{e^{-\lambda t}}{t}\int_{-\infty}^0 (e^{izx} - 1)d_x F_t(x)dt\}.$$

Since both of these functions allow analytic extension onto the region $\operatorname{Im} z < 0$, they coincide in it. The r.h.s. is analytic with respect to z in this region. The function

$$K(-iz) = az + \frac{1}{2}bz^2 + \int_{-\infty}^{0} (e^{zx} - 1 - zxI_{\{x>-1\}})\Pi(dx)$$

satisfies the condition

$$\frac{d^2}{dz^2}K(-iz) = b + \int_{-\infty}^{0} x^2 e^{zx}\Pi(dx) \geq 0,$$

and hence, it is convex towards below. If $K(-iz)$ is not identically zero, then by virtue of the condition

$$\frac{d}{dz}K(-iz)|_{z=0} = a + \int_{-\infty}^{-1} x\Pi(dx) \geq 0,$$

it increases up to $+\infty$, and there exists z_0 such that $K(-iz_0) = \lambda$. Then we also have $B(\lambda) - i(-iz_0) = 0$, because otherwise the l.h.s. is unbounded; i.e., $B(\lambda) = z_0$, $K(-iB(\lambda)) = \lambda$.

□

We now find the relation between the distributions of τ_x and $\xi(t)$.

33°. Theorem. *Let* $\Phi_y(s) = \mathbf{P}\{\tau_y < s\}$. *Then for all points at which the function* $\int_0^x y d_y F_s(y)$ *is continuous with respect to* s, *the following equality holds:*

$$\frac{d}{ds}\int_0^x \Phi_y(s)dy = \frac{1}{s}\int_0^x y\, d_y F_s(y). \tag{4.103}$$

Proof. Let

$$v_\lambda(x) = \int_0^\infty \frac{e^{-\lambda t}}{t}\int_0^x y\, d_y F_t(y)dt.$$

Then

$$\begin{aligned}\int_0^\infty \frac{e^{-\mu x} - 1}{x} dv_\lambda(x) &= \int_0^\infty \int_0^\infty \frac{e^{-\lambda t}}{t} \cdot (e^{-\mu x} - 1)dF_t(x) \\ &= \ln\left(\lambda \int_0^\infty e^{-\mu x} d_x q_+(\lambda, x)\right) \\ &= \ln(\lambda \tilde{q}_+(\lambda, -i\mu)) = \ln\frac{B(\lambda)}{B(\lambda) + \mu}.\end{aligned}$$

Differentiating this relation with respect to μ, we get

$$\int_0^\infty e^{-\mu x} dv_\lambda(x) = \frac{1}{B(\lambda)+\mu}.$$

Since $v_\lambda(0) = 0$, we have

$$v_\lambda(x) = \int_0^x e^{-B(\lambda)y} dy = \int_0^\infty \mathbf{M} e^{-\lambda \tau_y} dy = \int_0^\infty e^{-\lambda s} \int_0^x d_s \Phi_y(s) dy.$$

Thus,

$$\int_0^\infty e^{-\lambda t} \frac{1}{t} \int_0^x y d_y F_t(y) dt = \int_0^\infty e^{-\lambda s} d_s \left(\int_0^x \Phi_y(s) dy \right).$$

Coincidence of the Laplace transforms of two nondecreasing functions implies that these functions coincide at all the continuity points, i.e.,

$$\int_0^s \frac{1}{t} \int_0^x y d_y F_t(y) dt = \int_0^x \Phi_y(s) dy.$$

And this equality implies the statement of the theorem.

□

Remark 1. Let $\xi^+(s) = \xi(s) I_{\{\xi(s)>0\}}$. Since $\xi^+(s) \le e^{\xi(s)}$, $\mathbf{M}(\xi^+(s))^k$ exists for any k, and $\mathbf{M}\xi^+(s)$ is continuous by virtue of the stochastic continuity of the process. The equality

$$\int_0^\infty e^{-\lambda t} \frac{1}{t} \mathbf{M}\xi^+(t) dt = \int_0^\infty e^{-\lambda s} d_s \int_0^\infty \Phi_y(s) dy$$

yields

$$\frac{d}{dt} \int_0^\infty \Phi_y(t) dy = \frac{1}{t} \mathbf{M}\xi^+(t).$$

Boundedness of the r.h.s. as $t \to 0$ follows from the relation

$$\overline{\lim}_{t\to 0} \frac{1}{t} \mathbf{M}\xi^+(t) \le \lim_{t\to 0} \frac{1}{t} \mathbf{M}(e^{\xi(t)} - 1) = K(-i) < \infty.$$

Remark 2.

$$\begin{aligned}\int_0^\infty e^{-\lambda t} \frac{1}{t} \mathbf{M}\xi^+(t) dt &= \int_0^\infty \int_0^\infty e^{-\lambda t} d_t \Phi_y(t) dy \\ &= \int_0^\infty dy \int_0^\infty e^{-\lambda t} d_t \Phi_y(t) = \frac{1}{B(\lambda)}.\end{aligned}$$

Remark 3. The equality

$$\tilde{q}_-(\lambda, z) = \frac{1}{\lambda - K(z)} - \frac{1}{B(\lambda)} \cdot \frac{iz}{\lambda - K(z)}$$

implies

$$\int e^{izx} d\mathbf{P}\{\inf_{s\le T} \xi(s) < x\}$$
$$= \int e^{izx} d_x F_t(x) - \int ize^{izx} d_x \int_0^t F_{t-s}(x) \frac{1}{s} \mathbf{M}\xi^+(s) ds,$$

from whence

$$\mathbf{P}\{\inf_{s\le t} \xi(s) < x\} = F_t(x) + \frac{d}{dx}\int_0^t \frac{1}{s}\mathbf{M}\xi^+(s) F_{t-s}(x) ds.$$

Stable Processes

We say that a process is stable if it is a homogeneous process with independent increments $\xi(t)$ with the characteristic function

$$\mathbf{M}e^{iz\xi(t)}$$
$$= \exp\{t\left[iaz + \int (e^{izx} - 1 - izxI_{\{|x|\le 1\}})(I_{\{x<0\}}c_1 + I_{\{x>0\}}c_2)\frac{dx}{|x|^{1+\alpha}}\right]\}, \tag{4.104}$$

where $a \in R$, $c_1, c_2 \in R_+$, $0 < \alpha < 2$. The last condition follows from

$$\int \frac{x^2}{1+x^2} \cdot \frac{dx}{|x|^{1+\alpha}} < \infty,$$

which is necessary and sufficient for the existence of the integral in (4.104).

34°. Theorem. *The characteristic function of a stable process can be represented as follows:*

$$\mathbf{M}e^{iz\xi(t)} = \exp\{t[iaz - c|z|^\alpha \omega(\alpha,\beta,z)]\}, \tag{4.105}$$

where $a \in R$, $c \in R_+$, $\alpha \in (0,2)$, $\beta \in [-1,1,], \omega(\alpha,\beta,z) = \left(1 - i\beta\frac{z}{|z|} tg\frac{\pi}{2}\alpha\right)$ *for* $\alpha \neq 1$ *and* $\omega(1,\beta,z) = 1 - \frac{2i\beta z}{\pi|z|}\ln|z|$.

Proof. Let $0 < \alpha < 1$. Then

$$\int_{-1}^1 |x|\frac{dx}{|x|^{1+\alpha}} = \int_{-1}^1 |x|^{-\alpha} dx < \infty,$$

and

$$\mathbf{M}e^{iz\xi(t)} = \exp\{t\left[i\left(a + \int_{-1}^{1}\frac{x}{|x|^{1+\alpha}}dx\right)z + \int_{-\infty}^{0}(e^{izx}-1)\frac{c_1}{|x|^{1+\alpha}}dx + \int_{0}^{\infty}(e^{izx}-1)\frac{c_2}{x^{1+\alpha}}dx\right]\},$$

$$\begin{aligned}\int_{-\infty}^{0}(e^{izx}-1)\frac{dx}{|x|^{1+\alpha}} &= \int_{-\infty}^{0}(\exp\{i(\mathrm{sign} z)u\}-1)\frac{du}{|u|^{1+\alpha}}|z|^\alpha \\ &= |z|^\alpha\int_0^\infty(e^{-i(\mathrm{sign} z)u}-1)\frac{du}{u^{1+\alpha}} \\ &= -|z|^\alpha\int_0^\infty\frac{1-\cos u}{u^{1+\alpha}}du - i|z|^\alpha \mathrm{sign} z\int_0^\infty\frac{\sin u}{u^{1+\alpha}}du,\end{aligned}$$

$$\int_0^\infty(e^{izx}-1)\frac{dx}{x^{1+\alpha}} = |z|^\alpha\int_0^\infty\frac{1-\cos u}{u^{1+\alpha}}du + i|z|^\alpha \mathrm{sign} z\int_0^\infty\frac{\sin u}{u^{1+\alpha}}du.$$

Hence,

$$\mathbf{M}e^{iz\xi(t)} = \exp\left\{t\left[ia_1 z - c|z|^\alpha\left(1 - ib\frac{z}{|z|}\right)\right]\right\},$$

where

$$c = (c_1+c_2)\int_0^\infty\frac{1-\cos u}{u^{1+\alpha}}du,$$

$$b = \frac{c_2-c_1}{c_1+c_2}\left(\int_0^\infty\frac{1-\cos u}{u^{1+\alpha}}du\right)^{-1}\int_0^\infty\frac{\sin u}{u^{1+\alpha}}du.$$

We have

$$\int_0^\infty\frac{1-\cos u}{u^{1+\alpha}}du = \frac{\Gamma(1-\alpha)}{\alpha}\cos\frac{\pi}{2}\alpha,$$

$$\int_0^\infty\frac{\sin u}{u^{1+\alpha}}du = \frac{\Gamma(1-\alpha)}{\alpha}\sin\frac{\pi}{2}\alpha,$$

and thus $b = \beta tg\frac{\pi}{2}\alpha$, where $\beta = \frac{c_1-c_2}{c_1+c_2}$; i.e., we have obtained (4.105) for $\alpha\in(0,1)$. For $1<\alpha<2$, there exists

$$\begin{aligned}\int_{-\infty}^{0}(e^{izx}-1-izx)\frac{\alpha x}{|x|^{1+\alpha}} &= \int_0^\infty(e^{-izx}-1+izx)\frac{dx}{x^{1+\alpha}} \\ &= -\frac{iz}{\alpha}\int_0^\infty(e^{-izx}-1)\frac{dx}{x^\alpha},\end{aligned}$$

$$\int_0^\infty(e^{izx}-1-izx)\frac{dx}{x^{1+\alpha}} = \frac{iz}{\alpha}\int_0^\infty(e^{izx}-1)\frac{dx}{x^\alpha}.$$

Using the previous calculation we can establish the validity of (4.105) in this case, too. We have

$$c = \frac{\Gamma(2-\alpha)}{\alpha(\alpha-1)}(c_1+c_2)\sin\frac{\pi}{2}(\alpha-1), \qquad \beta = \frac{c_2-c_1}{c_1+c_2}.$$

For $\alpha = 1$, we find

$$\begin{aligned}
&\int_0^1 (e^{izx} - 1 - izx)\frac{dx}{x^2} + \int_1^\infty (e^{izx} - 1)\frac{dx}{x^2} \\
&= iz\left[\int_0^1 \frac{e^{izx} - 1}{x}dx + \int_1^\infty \frac{e^{izx}}{x}dx + 1\right] \\
&= iz + iz\int_0^\infty \frac{i\sin zx}{x}dx + iz\int_0^1 \frac{\cos zx - 1}{x}dx + iz\int_1^\infty \frac{\cos xz}{x}dx \\
&= iz\left(1 + \int_0^1 \frac{\cos x - 1}{x}dx + \int_1^\infty \frac{\cos x}{x}dx\right) \\
&\quad + |z|\frac{\pi}{2} + iz\int_0^\infty \frac{\cos zx - \cos x}{x}dx \\
&= iza_1 - |z|\frac{\pi}{2} - iz\ln|z|.
\end{aligned}$$

This implies the validity of (4.105) for $\alpha = 1$ if

$$c = \frac{\pi}{2}(c_1 + c_2), \qquad \beta = \frac{c_1 - c_2}{c_1 + c_2}.$$

□

CHAPTER 5

Multiplicative Processes

§5.1 Definition and General Properties

Let G be some group (with an operation of product ab, $a \in G$, $b \in G$). Let us examine random processes with values in G. We choose in G a σ-algebra of subsets $\mathcal{B}$ satisfying the following condition: For all $A \in \mathcal{B}$ and $a \in G$, the sets $A_a = \{x : a^{-1}x \in A\}$ and $A'_a = \{x : xa^{-1} \in A\}$ belong to $\mathcal{B}$ (i.e., operations of multiplication by an element of G from the right or from the left are measurable). If G is a topological group, then we take a Borel σ-algebra to be $\mathcal{B}$, since it satisfies the aforesaid condition. If this condition is satisfied and the σ-algebra $\mathcal{B}$ is countably generated, then the product of two random elements in G is also a random element in G.

The process $\xi(t)$, $t \in R_+$, with values in G, such that for all $t_1 < t_2 < \cdots < t_n$ the random elements (in G) $\xi(t_1), \xi^{-1}(t_1)\xi(t_2), \ldots, \xi^{-1}(t_{n-1})\xi(t_n)$ are independent, is called a *(right) multiplicative process in* G. The process $\xi'(t)$, $t \in R_+$, such that for all $t_1 < t_2 < \cdots < t_n$ the random elements (in G) $\xi'(t_1), \xi'(t_2)(\xi'(t_1))^{-1}, \ldots, \xi'(t_n)(\xi'(t_{n-1}))^{-1}$ are independent, is called a *(left) multiplicative process in* G.

If X is a linear space, it can be considered as an additive group and then the above definitions reduce to the definition of a process with independent increments (in this case, the right and left multiplicative processes coincide due to commutativity). Thus, a multiplicative process is a more complicated construction than a process with independent increments. The essential role is played here by noncommutativity of the product (for locally compact Abelian groups, the theory of multiplicative processes is completely analogous to the theory of processes with independent increments—see Section 5.2). The most interesting examples of multiplicative processes are processes whose values are linear operators acting in some linear space. Processes whose values are operators in a finite-dimensional Euclidean space are considered in detail in Section 5.3.

Properties of Stochastically Continuous Multiplicative Processes

Let us consider G to be a topological group with a topology generated by a metric $r(x,y)$; the product operation is continuous in this metric. Assume also that there exists a "norm," i.e., a positive locally uniformly continuous function $N(x)$ given on G, with the following properties:

(1) $N(x) \geq 1$ and $N(e) = 1$ (e is the unit element of G);

(2) $N(xy) \leq N(x)N(y), \qquad N(x^{-1}) = N(x)$;

(3) $r(xy, xz) \leq N(x)r(x,z) \qquad r(yx, zx) \leq N(x)r(y,z)$,

(4) for some r_0

$$N(x) \leq 1 + r_0 r(e,x).$$

Stochastically continuous multiplicative processes will be considered on this group. To study them, we shall need some auxiliary assumptions for products of independent elements in G.

1°. Lemma. *Let $\xi_1, \xi_2, \ldots, \xi_n$ be random elements in G, $\eta_k = \xi_1 \ldots \xi_k$, $\eta_n^k = \xi_{k+1} \ldots \xi_n$, $k = 0, 1, \ldots, n$. If for some $\varepsilon > 0$, $\varepsilon < 1/r_0$,*

$$\mathbf{P}\{r(\eta_n^k, e) \geq \varepsilon\} \leq \alpha < 1,$$

then for all $c > \varepsilon/(1 - \varepsilon r_0)$, we have

$$\mathbf{P}\{\sup_k r(\eta_k, e) \geq c\} \leq \frac{1}{1-\alpha}\mathbf{P}\{r(\eta_n, e) \geq c - \varepsilon(1 + r_0 c)\}.$$

Proof. For $r(e, \eta_n^l) \leq \varepsilon$ and $l < n$, we have

$$\begin{aligned} r(\eta_l, e) \leq r(\eta_n, e) + r(\eta_l \eta_n^l, \eta_l) &\leq r(\eta_n, e) + N(\eta_l)\varepsilon \\ &\leq r(\eta_n, e) + \varepsilon(1 + r_0 r(\eta_l, e)), \\ r(\eta_l, e)(1 - \varepsilon r_0) - \varepsilon &\leq r(\eta_n, e). \end{aligned}$$

Moreover, if $r(\eta_l, e) \geq c$, then $r(\eta_n, e) \geq c - \varepsilon(1 + r_0 c)$. Consequently,,

$$\begin{aligned} &\mathbf{P}\{\sup_{k \leq n} r(\eta_k, e) \geq c\} \\ &\quad = \sum_{l=1}^{n} \mathbf{P}\{\sup_{k<l} r(\eta_k, e) < c, r(\eta_l, e) \geq c\} \\ &\quad \leq \frac{1}{1-\alpha} \sum_{l=1}^{n} \mathbf{P}\{\sup_{k<l} r(\eta_k, e) < c, r(\eta_l, e) \geq c, r(\eta_n^l, e) \leq \varepsilon\} \\ &\quad = \frac{1}{1-\alpha}\mathbf{P}\{\cup_{l=1}^{n}\{\sup_{k<l} r(\eta_k, e) < c, r(\eta_l, e) \geq c, r(\eta_n^l, e) \leq \varepsilon\}\} \\ &\quad \leq \frac{1}{1-\alpha}\mathbf{P}\{r(\eta_n, e) \geq c - \varepsilon(1 + r_0 c)\}. \quad \square \end{aligned}$$

Let $\{x_s, s \in V\}$, $V \in R_+$, be some collection of elements of G. We write

$$\nu_\delta(x_s, s \in V) = \sup\{k : \exists s_1, \ldots, s_{k+1} \in V, s_1 < s_2 < \cdots < s_{k+1};\\ r(e, x_{s_i}^{-1} x_{s_{i+1}}) \geq \delta, i = 1, \ldots, k\}$$

(the multiplicative analogue for a number of δ-oscillations).

2°. Lemma. *Under the conditions of Lemma 1° for* $\frac{\delta}{2+r_0\delta} \geq \frac{2}{1-r_0\varepsilon}$, $\alpha < 1/2$, *the inequality*

$$\mathbf{M}\nu_\delta \leq \frac{\alpha}{1-2\alpha}$$

holds for the variable $\nu_\delta = \nu_\delta(\eta_1, \ldots, \eta_n)$.

Proof. Let $\rho = \frac{\delta}{2+r_0\delta}$. Using the inequality $r(ab^{-1}, e) \leq N(b) r(a, b)$, we obtain that the event $\{\nu_\delta \geq k\}$ involves one of the events

$$\{r(e, \eta_1) < \rho, \ldots, r(e, \eta_{l-1}) < \rho, r(e, \eta_l) \geq \rho; \nu_\delta(\eta_{l+1}, \ldots, \eta_n) \geq k-1\}.$$

Since the events $\{r(e, \eta_1) < \rho, \ldots, r(e, \eta_{l-1}) < \rho, r(e, \eta_l) \geq \rho\}$ and $\{\nu_\delta(\eta_{l+1}, \ldots, \eta_n) \geq k-1\}$ are independent and since the events $\{r(e, \eta_1) < \rho, \ldots, r(e, \eta_{l-1}) < \rho, r(e, \eta_l) \geq \rho\}$ are disjoint for different l, we have

$$\begin{aligned}\mathbf{P}\{\nu_\delta \geq k\} &\leq \sum_{l=1}^{n} \mathbf{P}\{r(e, \eta_1) < \rho, \ldots, r(e, \eta_{l-1}) < \rho, r(e, \eta_l) \geq \rho\} \\ &\quad \times \mathbf{P}\{\nu_\delta(e, \eta_1, \ldots, \eta_n) \geq k-1\} \\ &= \mathbf{P}\{\sup_l r(e, \eta_l) \geq \rho\} \mathbf{P}\{\nu_\delta \geq k-1\} \\ &\leq \frac{1}{1-\alpha} \mathbf{P}\{r(e, \eta_n) \geq \rho - \varepsilon(1 - r_0\rho)\} \mathbf{P}\{\nu_\delta \geq k-1\}.\end{aligned}$$

Hence, $\mathbf{P}\{\nu_\delta \geq k\} < (\frac{\alpha}{1-\alpha})^k$, since

$$\mathbf{P}\{r(e, \eta_n) \geq \rho(1 - \varepsilon r_0) - \varepsilon\} \leq \mathbf{P}\{r(e, \eta_n) \geq \varepsilon\} \leq \alpha.$$

Thus,

$$\mathbf{M}\nu_\delta \leq \sum_{k=1}^{\infty} \mathbf{P}\{\nu_\delta = k\} \leq \sum_{k=1}^{\infty} \left(\frac{\alpha}{1-\alpha}\right) = \frac{\alpha}{1-2\alpha}.$$

□

3°. Theorem. *Let* $\xi(t)$, $t \in R_+$, *be a stochastically continuous multiplicative process in* G. *It has a modification continuous from the right with limits from the left.*

Proof. Consider a right multiplicative process and show that the process $\xi(t)$ has a finite number of ε-oscillations on $\Lambda \cap [0, T]$, for any T and any countable set Λ dense everywhere in R_+. If this is proved, then for every bounded increasing sequence $t_n \in \Lambda$, there exists $\lim \xi(t_n)$, and for every decreasing sequence $t_n \in \Lambda$, there exists $\lim \xi(t_n)$. We put $\xi'(t) = \lim_{t_n \downarrow t, t_n \in \Lambda} \xi(t_n)$. $\xi'(t)$ is a process continuous from the right with limits from the left. Its stochastic equivalence with the process $\xi(t)$ follows from the stochastic continuity of $\xi(t)$.

The stochastic continuity of $\xi(t)$ also implies that for every $\alpha > 0$, $\varepsilon > 0$, and $T > 0$, there exists $h > 0$ such that $\mathbf{P}\{r(\xi^{-1}(s)\xi(u), e) > \varepsilon\} \le \alpha$ for $|s - u| \le h$, $s, u \le T$. In fact, if it is not so, one can find sequences $s_n \to s_0$ and $u_n \to s_0$, $s_n < u_n$, such that for some $\varepsilon > 0$ and $\alpha > 0$,

$$\mathbf{P}\{r(\xi^{-1}(s_n)\xi(u_n), e) > \varepsilon\} \ge \alpha.$$

But by virtue of the stochastic continuity, $\xi^{-1}(s_n) \to \xi^{-1}(s_0)$, $\xi(u_n) \to \xi(s_0)$, $\xi^{-1}(s_n)\xi(u_n) \to e$, $r(\xi^{-1}(s_n)\xi(u_n), e) \to 0$ in probability, and we arrive at a contradiction. Assume that h is chosen so that

$$\begin{gathered}\mathbf{P}\{r(\xi^{-1}(s)\xi(u), e) > \varepsilon\} \le \alpha \\ \text{for } \quad |s - h| \le h, \quad s, u \le T.\end{gathered} \tag{5.1}$$

Whatever $t_1, t_2, \ldots, t_n \in [kh, kh + h] \subset [0, T]$ are taken, we have

$$\begin{aligned}&\mathbf{P}\{\sup_{i \le n} r(\xi^{-1}(kh)\xi(t_i), e) \bigvee r(\xi^{-1}(kh)\xi(kh + h), e) \ge c\} \\ &\qquad \le \frac{1}{1-\alpha}\mathbf{P}\{r(\xi^{-1}(kh)\xi(kh + h), e) \ge c - \varepsilon(1 + r_0 c)\}.\end{aligned} \tag{5.2}$$

by virtue of Lemma 1°. We have applied Lemma 1° to the independent variables $\xi^{-1}(kh)\xi(t_1), \xi^{-1}(t_1)\xi(t_2), \ldots, \xi^{-1}(t_n)\xi(kh+h)$. In (5.2), by passing to the limit with respect to n, we get

$$\begin{aligned}&\mathbf{P}\{\sup_{s \in \Lambda \cap [kh, kh+h]} r(\xi^{-1}(kh)\xi(s), e) \bigvee r(\xi^{-1}(kh)\xi(kh + h), e) \ge c\} \\ &\qquad \le \frac{1}{1-\alpha}\mathbf{P}\{r(\xi^{-1}(kh)\xi(kh + h), e) \ge c - \varepsilon(1 + r_0 c)\},\end{aligned}$$

and consequently,

$$\mathbf{P}\{\sup_{s \in \Lambda \cap [kh, kh+h]} r(\xi^{-1}(kh)\xi(s), e) \bigvee r(\xi^{-1}(kh)\xi(kh+h), e) < \infty\} = 1. \tag{5.3}$$

Then (5.3) yields

$$\mathbf{P}\{\sup_{s \in \Lambda \cap [0,T]} r(e, \xi(s)) < \infty\} = 1, \tag{5.4}$$

for all $T > 0$; i.e., the process $\xi(s)$, $s \in \Lambda$, is locally bounded with probability 1. Since

$$\begin{aligned}r(e, \xi^{-1}(s)) &\le N(\xi^{-1}(s))r(e, \xi(s)) \\ &\le N(\xi(s))r(e, \xi(s)) \le r(e, \xi(s)) + r_0 r^2(e, \xi(s)),\end{aligned}$$

we have that $\xi^{-1}(s)$, $s \in \Lambda$, is also locally bounded with probability 1.

Suppose that (5.1) is satisfied and $\alpha < \frac{1}{2}$. Now if $\Lambda = \bigcup_n \Lambda_n$, where $\Lambda_n \subset \Lambda_{n+1}$ and Λ_n are finite sets, then we have for $\delta/(2+r_0\delta) \geq 2\varepsilon/(1-r_0\varepsilon)$,

$$\begin{aligned}\mathbf{M}\nu_\delta(\xi(s), s \in \Lambda \cap [kh, kh+h]) &\leq \lim_{n\to\infty} \mathbf{M}\nu_\delta(\xi(s), s \in \Lambda_n \cap [kh, kh+h]) \\ &\leq \frac{\alpha}{1-2\alpha}\end{aligned}$$

by virtue of Lemma 2°. Thus, for all $\delta > 0$ and $T > 0$,

$$\mathbf{P}\{\nu_\delta(\xi(s), s \in \Lambda \cap [0,T]) < \infty\} = 1. \tag{5.5}$$

Now note that if $r(\xi(s), \xi(u)) > \varepsilon$ for some $s < u$, then

$$r(\xi^{-1}(s)\xi(u), e) \geq \frac{\varepsilon}{N(\xi(s))} \geq \frac{\varepsilon}{1 + r_0 r(\xi(s), e)}.$$

Hence, the number of ε-oscillations of the process $\xi(s)$ on $\Lambda \cap [0,T]$, defined as the maximal number of those k for which there exist $s_1 < s_2 < \cdots < s_{k+1} \in \Lambda \cap [0,T]$ such that $r(\xi(s_i), \xi(s_{i+1})) \geq \varepsilon$ $(i = 1, 2, \ldots, k)$, does not exceed $\nu_{\delta'}(\xi(s), s \in \Lambda \cap [0,T])$, where $\delta' = \varepsilon(1 + r_0 \sup_{s \in [0,T] \cap \Lambda} r(\xi(s), e))^{-1}$. Finiteness of the number of ε-oscillations of $\xi(s)$ on $\Lambda \cap [0,T]$ thus follows from (5.4) and (5.5).

If $\xi(s)$ is a left multiplicative process, then $\xi^{-1}(s)$ is a right one, and the foregoing proof is valid for it.

□

Denote by $\mathcal{B}_e$ a ring of Borel sets from G, lying at a positive distance from e. Further, we shall examine right processes, taking into account that in this case $\xi^{-1}(s)$ is a left one, and we can easily reformulate all the results for these processes.

Whatever $A \subset \mathcal{B}_e$ and $t \in R_+$ are taken, the variable

$$\nu_t(A) = \sum_{s \leq t} I_{\{\xi^{-1}(s-)\xi(s) \in A\}} \tag{5.6}$$

is defined for the multiplicative process $\xi(s)$ continuous from the right (here summation is taken over all discontinuity points of $\xi(s)$; their number is at most countable).

If the distance between A and e is greater than δ, then $\nu_t(A) \leq \nu_\rho(\xi(s), s \in [0,t])$; ($\rho$ is defined in Lemma 2°).

4°. Theorem. *$\nu_t(A)$ is a Poisson process for every $A \in \mathcal{B}_e$; as a function of A, it is a Poisson r.m.i.v. on $\mathcal{B}_e$ for all $t \in R_+$.*

Proof. Let us introduce σ-algebras $\mathcal{F}_t^s$ $(0 \leq s < t)$ generated by the variables $\xi^{-1}(s)\xi(u)$, $s \leq u \leq t$, and $\mathcal{F}_t$, $t \in R_+$, generated by the variables $\xi(s)$,

$s \le t$. If $0 < t_1 < \cdots < t_n$, then $\mathcal{F}_{t_1}, \mathcal{F}_{t_2}^{t_1}, \ldots, \mathcal{F}_{t_n}^{t_{n-1}}$ are independent. Let $A_1, A_2, \ldots, A_m$ be mutually disjoint sets from $\mathcal{B}_e$. We define on R_+ the r.m.i.v. $\mu_i([s,t[) = \nu_t(A_i) - \nu_s(A_i)$, $i = 1, 2, \ldots, m$. They can be extended onto the algebra of sets, generated by intervals. $\mu_i([s,t[)$ are $\mathcal{F}_t^s$-measurable. Clearly, they are counting measures; they are stochastically continuous with no common atoms. Thus, by virtue of Theorem 10°, Chapter 2, they are independent Poisson measures. This implies the proof of the theorem.

□

We shall call $\nu_t(A)$ a *Poisson measure corresponding to the multiplicative process* $\xi(t)$.

Corollary. *Let G be a countable (finite) group with a discrete topology. Then for every stochastically continuous multiplicative process $\xi(t)$, there exist mutually independent Poisson processes $\{\nu_x(t), x \in G\backslash\{e\}\}$ such that*

$$\xi(t) = \xi_0 \lim_{\max \Delta t_k \to 0} \prod_{0=t_0<t_1<\cdots<t_n=t} \left(\prod_{x \ne e} x^{\nu_x(t_{k+1}) - \nu_x(t_k)} \right); \qquad (5.7)$$

the first product on the r.h.s. of (5.7) is taken in increasing order of the index t_k, whereas the second one is independent of the factor order. We consider that $x^0 = e$. For sufficiently small $\Delta t_k = t_{k+1} - t_k$, there is at most one nonzero number among $\nu_x(t_{k+1}) - \nu_x(t_k)$.

Define a process $\xi_t(A)$, which is the product of jumps hitting the set A, as follows: If $\nu_t(A)$ has jumps at the points $\tau_1 < \tau_2 < \cdots < \tau_l \le t$ belonging to $[0,t]$, then

$$\xi_t(A) = \xi^{-1}(\tau_1-)\xi(\tau_1)\xi^{-1}(\tau_2-)\xi(\tau_2)\ldots\xi^{-1}(\tau_l-)\xi(\tau_l). \qquad (5.8)$$

5°. Lemma. *$\xi_t(A)$ is a stochastically continuous multiplicative process.*

Proof. Let us obtain another formula for $\xi_t(A)$:

$$\xi_t(A) = \lim_{n\to\infty} \prod_{k=0}^{n-1} \left(\xi^{-1}\left(\tfrac{kt}{n}\right)\xi\left(\tfrac{k+1}{n}t\right)\right)^{\nu_{(k+1)t/n}(A) - \nu_{kt/n}(A)}. \qquad (5.9)$$

If n is so large that at most one point τ_i, $i = 1, \ldots, l$, gets into each interval $]\frac{k-1}{n}t, \frac{k}{n}t]$, then in the product on the r.h.s. of (5.9), we have only l factors differing from e of the form

$$\xi^{-1}\left(\tfrac{k_i}{n}t\right)\xi\left(\tfrac{k_i+1}{n}t\right), \quad \tfrac{k_i}{n}t < \tau_i < \tfrac{k_i+1}{n}t.$$

This expression tends to $\xi^{-1}(\tau_i-)\xi(\tau_i)$ as $n \to \infty$. Formula (5.9) is proved. By analogy we find that

$$\begin{aligned} \xi_s^{-1}(A)\xi_t(A) = \lim_{n\to\infty} &\left(\xi^{-1}(s)\xi\left(s + \tfrac{t-s}{n}\right)\right)^{\nu_{s+(t-s)/n}(A) - \nu_s(A)} \\ &\times \cdots \times \left(\xi^{-1}\left(t - \tfrac{t-s}{n}\right)\xi(t)\right)^{\nu_t(A) - \nu_{t-(t-s)/n}(A)}. \end{aligned} \qquad (5.10)$$

for $s < t$. Since (5.9) and (5.10) yield that $\xi_s^{-1}(A)\xi_t(A)$ is $\mathcal{F}_t^s$-measurable and that $\xi_t(A)$ is $\mathcal{F}_t$-measurable, $\xi_t(A)$ is a multiplicative process. Its stochastic continuity follows from the relation

$$\mathbf{P}\{r(\xi_s^{-1}(A)\xi_t(A,e)) > 0\} \leq \mathbf{P}\{\nu_t(A) - \nu_s(A) \geq 1\}.$$

□

A *jump* multiplicative process is one for which the limit $\lim_{\delta \to 0} \nu_t(U_\delta) = \nu_t(G) < \infty$ exists for all $t \in R_+$, where $U_\delta = \{x \in G, r(x,e) \geq \delta\}$ and $\xi(t) = \xi(0) \lim_{\delta \to 0} \xi_t(U_\delta)$.

Remark. If $\xi(t)$ is a jump multiplicative process, then the following formula, similar to (5.9), is valid for it:

$$\xi(t) = \xi(0) \lim_{\max \Delta t_k \to 0} \prod_{0=t_0<\cdots<t_n=t} (\xi^{-1}(t_k)\xi(t_{k+1}))^{\nu_{t_{k+1}}(G)-\nu_{t_k}(G)} \tag{5.11}$$

(the product is taken in increasing order of the indices). To be convinced of the validity of this formula, it suffices to note that $\xi_t(U_\delta) = \xi(t)$ for sufficiently small δ.

In order to define representations of jump multiplicative processes, we introduce a multiplicative integral with respect to the Poisson measure. Let $\nu_t(A)$ be the Poisson measure, corresponding to the jump multiplicative process $\xi(t)$. We put $\nu([s,t],A) = \nu_t(A) - \nu_s(A)$ for $t \geq s$.

Let us fix a sequence of decompositions of $G \backslash \{e\}$ into disjoint sets $A_{n1}, A_{n2}, \ldots$ from $\mathcal{B}_l$ and choose $x_{ni} \in A_{ni}$. We consider that $\lim_{n\to\infty} \max_i \operatorname{diam} A_{ni} = 0$ ($\operatorname{diam} A = \sup\{r(x,y), x,y \in A\}$ is a diameter of A). Now if $\nu([s,t], G\backslash\{e\}) = 1$, then the following expression is defined:

$$\prod x^{\nu([s,t],dx)} = \lim_{n\to\infty} \Pi_i x_{ni}^{\nu([s,t],A_{ni})}. \tag{5.12}$$

The product on the r.h.s. is taken in increasing order of the indices i, $x^0 = e$, and only one number among $\nu([s,t],A_{ni})$ differs from zero. The limit on the r.h.s. of (5.12) exists, and now we shall find it. Assume that $\xi(u)$ has a jump at the point $\tau \in [s,t]$. Denote by i_n the index for which $\nu([s,t],A_{ni_n}) = 1$. This means that the variable $\eta = \xi^{-1}(\tau-)\xi(\tau) \in A_{ni_n}$. Then $r(\eta, x_{ni_n}) \leq \max \operatorname{diam} A_{ni} \to 0$, and therefore the limit on the r.h.s. equals η. If $\nu([s,t]) = 0$, then the product on the r.h.s. of (5.12) equals e, and the limit exists. Finally, we put $\Pi x^{\nu([s,t],dx)} = e$ if $\nu([s,t], G\backslash\{e\}) > 1$.

6°. Theorem. *If $\xi(t)$ is a jump multiplicative process and $\nu_t(A)$ is a Poisson measure corresponding to it, then*

$$\xi(t) = \xi(0) \lim_{\max \Delta t_k \to 0} \prod_{0=t_0<t_1<\cdots<t_n=t} \prod x^{\nu([t_k,t_{k+1}],dx)} \tag{5.13}$$

is valid.

The proof follows from the fact that $\nu([t_k, t_{k+1}], G\backslash\{e\}) \leq 1$, for sufficiently small $\max \Delta t_k$, and from (5.11).

□

Multiplicative Processes with Bounded Jumps

Consider a stochastically continuous process $\xi(t)$ continuous from the right. Let $\nu_t(A)$ be the Poisson measure corresponding to it and $\xi_A(t)$ be the process constructed according to (5.8). For $\delta > 0$, we define the process

$$\xi_\delta(t) = \xi(\tau_1^\delta -)\xi^{-1}(\tau_1^\delta)\xi(\tau_2^\delta -)\xi^{-1}(\tau_2^\delta)\ldots\xi^{-1}(\tau_l^\delta)\xi(t), \tag{5.14}$$

where $\tau_1^\delta < \tau_2^\delta < \cdots < \tau_l^\delta \leq t < t_{l+1}^\delta < \cdots$ are all those points where $\nu_s(U_\delta)$ has jumps.

7°. Lemma. *$\xi_\delta(t)$ is a stochastically continuous multiplicative process; it is continuous at discontinuity points of $\nu_t(U_\delta)$; if $\nu_t(U_\delta)$ is constant on the interval $[s, u]$, then $\xi^{-1}(s)\xi(t) = \xi_\delta^{-1}(s)\xi_\delta(t)$ for $t \in [s, u]$ and $\sup_t r(\xi_\delta^{-1}(t-)\xi_\delta(t), e) \leq \delta$.*

Proof. By analogy with (5.9), we can easily obtain

$$\begin{aligned}\xi_\delta(t) = \xi(0) \lim_{n\to\infty} &\left(\xi^{-1}(0)\xi\left(\frac{t}{n}\right)\right)^{1-\nu([0,t/n],U_\delta)} \\ &\times \left(\xi^{-1}\left(\frac{t}{n}\right)\xi\left(\frac{2t}{n}\right)\right)^{1-\nu([t/n,2t/n],U_\delta)} \\ &\cdots\left(\xi^{-1}\left(\frac{n-1}{n}t\right)\xi(t)\right)^{1-\nu([(n-1)t/n,t],U_\delta)},\end{aligned} \tag{5.15}$$

which implies that $\xi_\delta(t)$ is $\mathcal{F}_t$-measurable. Similarly we obtain that $\xi_\delta^{-1}(s)\xi_\delta(t)$ is $\mathcal{F}_t^s$-measurable for $s < t$. Thus, $\xi_\delta(t)$ is a multiplicative process. Continuity of $\xi_\delta(t)$ at the points τ_l^δ follows from (5.14):

$$\begin{aligned}\xi_\delta(\tau_l^\delta) = \xi_\delta(\tau_l^\delta +) &= \xi(\tau_1^\delta -)\xi^{-1}(\tau_1^\delta)\ldots\xi(\tau_l^\delta -) \\ &= \lim_{t\uparrow\tau_l^\delta} \xi(\tau_1^\delta -)\xi(\tau_1^\delta)\ldots\xi^{-1}(\tau_{l-1}^\delta)\xi(t) = \xi_\delta(\tau_l -).\end{aligned}$$

Further, if $\tau_l^\delta < s < t < \tau_{l+1}^\delta$, then

$$\begin{aligned}\xi_\delta(t) &= \xi(\tau_1^\delta-)\xi^{-1}(\tau_1^\delta)\dots\xi^{-1}(\tau_l^\delta)\xi(t)\\ &= \xi(\tau_1^\delta-)\xi^{-1}(\tau_1^\delta)\dots\xi^{-1}(\tau_l^\delta)\xi(s)\xi^{-1}(s)\xi(t) = \xi_\delta(s)\xi^{-1}(s)\xi(t).\end{aligned}$$

Hence, $\xi_\delta^{-1}(s)\xi_\delta(t) = \xi^{-1}(s)\xi(t)$ and $\xi_\delta^{-1}(t-)\xi_\delta(t) = \xi^{-1}(t-)\xi(t)$ for $t \neq \tau_k^\delta$. Thus,

$$\sup_t r(\xi_\delta^{-1}(t-)\xi_\delta(t), e) = \sup_{t, t\neq\tau_l^\delta, l=1,2,\dots} r(\xi^{-1}(t-)\xi(t), e) \leq \delta$$

(since $r(\xi^{-1}(t-)\xi(t), e) > \delta$ only for $t = \tau_l^\delta$).

□

8°. Lemma. *Let $\xi(t)$ be a stochastically continuous multiplicative process continuous from the right for which $\sup_t r(\xi^{-1}(t-)\xi(t), e) \leq \delta$. Then for all T and $\varepsilon > 0$,*

$$\lim_{\max \Delta t_k \to 0} \sum_{0=t_0<t_1<\dots<t_n=T} \mathbf{P}\{r(\xi^{-1}(t_k)\xi(t_{k+1}), e) > \delta + \varepsilon\} = 0. \tag{5.16}$$

Proof. It is evident that

$$\lim_{\max \Delta t_k \to 0} \max_{\substack{k\\ 0=t_0<t_1<\dots<t_n=T}} r(\xi^{-1}(t_k)\xi(t_{k+1}), e) = \sup_{t\leq T} r(\xi^{-1}(t-)\xi(t), e).$$

Hence,

$$\begin{aligned}1 &= \lim_{\max \Delta t_k \to 0} \mathbf{P}\{\max_{k\leq n-1} r(\xi^{-1}(t_k)\xi(t_{k+1}), e) \leq \delta + \varepsilon\}\\ &= \lim_{\max \Delta t_k \to 0} \Pi_{k=0}^{n-1}\mathbf{P}\{r(\xi^{-1}(t_k)\xi(t_{k+1}), e) \leq \delta + \varepsilon\}\\ &= \lim_{\max \Delta t_k \to 0} \Pi_{k=0}^{n-1}(1 - \mathbf{P}\{r(\xi^{-1}(t_k)\xi(t_{k+1}), e) > \delta + \varepsilon\})\\ &\leq \exp\{-\lim_{\max \Delta t_k \to 0} \sum \mathbf{P}\{r(\xi^{-1}(t_k)\xi(t_{k+1}), e) > \delta + \varepsilon\}\}.\end{aligned}$$

This implies (5.16).

□

9°. Lemma. *Let $\xi_1, \xi_2, \dots, \xi_n$ be independent elements in G such that $r(\xi_k, e) \leq \delta$ for all $k = 1, 2, \dots, n$ and $\mathbf{P}\{r(\eta_l^k, e) \geq \varepsilon\} \leq \alpha$ for $k < l$*

($\eta_l^k = \xi_{k+1}, \dots, \xi_l$). Write $\beta = 4\varepsilon/(1 - r_0\varepsilon)$, $\beta_1 = (1 + r_0\delta)^2(\beta + \delta)$, $\lambda = \frac{1}{(1+r_0\beta_1)^2}$. Then for $\lambda^m > \frac{\alpha}{1-\alpha}$, the following inequality

$$\mathbf{M} \sup_l r^m(\eta_l^0, e) \leq \frac{\beta_1^m}{(1-\lambda)^m} \cdot \frac{1-\alpha}{\lambda^m(1-\alpha) - \alpha} \tag{5.17}$$

holds.

Proof. For $c > \beta > \frac{4\varepsilon}{1-r_0\varepsilon}$, we have

$$\begin{aligned}
&\mathbf{P}\{\sup_{l \leq n} r(\eta_l^0, e) > c\} \\
&\quad \leq \sum_l \mathbf{P}\{\sup_{i<l} r(\eta_i^0, e) < \beta, r(\eta_l^0, e) \geq \beta, \sup_{k>l} r(\eta_l^0 \eta_k^l, e) > c\}.
\end{aligned}$$

Since $r(\eta_{l-1}^0, e) < \beta$, $r(\xi_l, e) \leq \delta$ then

$$\begin{aligned}
r(\eta_l^0, e) &\leq N(\xi_l) r(\eta_{l-1}^0, \xi_l^{-1}) \leq (1 + r_0\delta)[\beta + r(e, \xi_l^{-1})] \\
&\leq (1 + r_0\delta)[\beta + N(\xi_l^{-1})\delta] \\
&= (1 + r_0\delta)\beta + (1 + r_0\delta)^2\delta < (1 + r_0\delta)^2(\beta + \delta) = \beta_1.
\end{aligned}$$

Further,

$$\begin{aligned}
r(\eta_l^0 \eta_k^l, e) &\leq N(\eta_l^0)[r(\eta_k^l, e) + r((\eta_l^0)^{-1}, e)] \\
&\leq N(\eta_l^0)[r(\eta_k^l, e) + N(\eta_l^0) r(\eta_l^0, e)] \\
&\leq (1 + r_0 r(\eta_l^0, e)) r(\eta_k^l, e) + (1 + r_0 r(\eta_l^0, e))^2 r(\eta_l^0, e) \\
&\leq (1 + r_0\beta_1)^2[r(\eta_k^l, e) + \beta_1],
\end{aligned}$$

so the event $\{\sup_{k>l} r(\eta_l^0, \eta_k^l, e) > c\}$ involves the following one:

$$\left\{\sup_{k>l} r(\eta_k^l, e) > \frac{c}{(1 + r_0\beta_1)} - \beta_1\right\}.$$

Consequently,

$$\begin{aligned}
&\mathbf{P}\{\sup_{l \leq n} r(\eta_l^0, e) > c\} \leq \frac{\alpha}{1-\alpha} \sup_l \mathbf{P}\left\{\sup_{k>l} r(\eta_k^l, e) > \frac{c}{(1 + r_0\beta_1)^2} - \beta_1\right\}, \\
&\mathbf{P}\{\sup_k r(\eta_k^0, e) > \beta\} \leq \frac{\alpha}{1-\alpha}
\end{aligned}$$

(we have used the fact that $\mathbf{P}\{\sup_k r(\eta_k^0, e) > \beta\} \leq \frac{\alpha}{1-\alpha}$ for $\beta \geq \frac{4\varepsilon}{1+r_0\varepsilon}$). Similarly, we get

$$\mathbf{P}\{\sup_{k>l} r(\eta_k^l, e) > c_1\} \leq \frac{\alpha}{1-\alpha} \sup_j \mathbf{P}\left\{\sup_{k>j} r(\eta_k^j, e) > \frac{c_1}{(1 + r_0\beta_1)^2} - \beta_1\right\}$$

for all l and $c_1 > \beta$. Write $\lambda = \frac{1}{(1+r_0\beta_1)^2}$. Then for all l and i, we have

$$\begin{aligned}\sup_l \mathbf{P}\left\{\sup_{k>l} r(\eta_k^l, e) > \lambda^i c - \frac{1}{1-\lambda}\beta_1\right\} \\ \leq \frac{\alpha}{1-\alpha}\sup_l \mathbf{P}\left\{\sup_{k>l} r(\eta_k^l, l) > \lambda\left(\lambda^i c - \frac{1}{1-\lambda}\beta_1\right) - \beta_1\right\} \\ = \frac{\alpha}{1-\alpha}\sup_l \mathbf{P}\left\{\sup_{k>l} r(\eta_k^l, e) > \lambda^{i+1} c - \frac{1}{1-\lambda}\beta_1\right\}.\end{aligned}$$

Consequently,

$$\mathbf{P}\{\sup_{l\leq n} r(\eta_l^0, e) > c\} \leq \left(\frac{\alpha}{1-\alpha}\right)^i,$$

for $\lambda^i c - \frac{1}{1-\lambda}\beta_1 > 0$, i.e.,

$$\mathbf{P}\left\{\sup_{l\leq n} r(\eta_l^0, e) > \frac{\beta_1}{\lambda^i(1-\lambda)}\right\} \leq \left(\frac{\alpha}{1-\alpha}\right)^i.$$

But in this case

$$\begin{aligned}\mathbf{M}\sup_{l\leq n} r^m(\eta_l^0, e) &= \sum_{i=1}^{\infty}\left(\frac{\beta_1}{\lambda^i(1-\lambda)}\right)^m\left(\frac{\alpha}{1-\alpha}\right)^{i-1} \\ &\leq \frac{\beta_1^m}{\lambda^m(1-\lambda)^m}\,\frac{1}{1-\frac{\alpha}{\lambda^m(1-\alpha)}} \\ &= \frac{\beta_1^m}{(1-\lambda)^m}\cdot\frac{1-\alpha}{\lambda^m(1-\alpha)-\alpha}.\end{aligned}$$

□

10°. Lemma. *Given arbitrary $m > 0$ and $T > 0$, we have*

$$\sup_{t\leq T}\mathbf{M}r^m(\xi_\delta^{-1}(0)\xi_\delta(t), e) < \infty. \tag{5.18}$$

Proof. Note that property (3) of the norm implies the inequality

$$\begin{aligned}r(x_1\ldots x_k, y_1\ldots y_k) &\leq \sum_{i=1}^{k} N(y_1)\ldots N(y_{i-1})r(y_i, x_i)N(x_{i+1}\ldots x_k) \\ &\leq \sum_{i=1}^{k} N(y_1)\ldots N(y_{i-1})r(y_i, x_i)N(x_{i+1})\ldots N(x_k).\end{aligned} \tag{5.19}$$

for $x_i \in G$, $y_i \in G$, $i = 1, \ldots, k$.

Hence, to prove (5.18) it suffices to show that for every T, there exist $h > 0$ and $c > 0$ such that for $s < t < T$ and $t - s \leq h$, we have

$$\mathbf{M}r^m(\xi_\delta^{-1}(s)\xi_\delta(t), e) \leq c. \tag{5.20}$$

In fact, if $\eta_1, \ldots, \eta_k$ are independent and $\mathbf{M}r^m(\eta_i, e) \leq c$, then by virtue of (5.19), we have

$$\begin{aligned}\mathbf{M}r^m(\eta_1 \ldots \eta_k, e) &\leq \mathbf{M}\left(\sum_{i=l}^{k} N(\eta_1)\ldots N(\eta_{i-1})r(\eta_i, e))\right)^m \\ &\leq c_1 \mathbf{M}\Pi_{i=1}^k(1 + r(\eta_i, e))^m = c_1 \prod_{i=l}^{k} \mathbf{M}(1 + r(\eta_i, e))^m \\ &\leq c_2 + c_3 \prod_{i=1}^{k} \mathbf{M}r^m(\eta_i, e).\end{aligned}$$

Here c_1, c_2, c_3 are constants depending only on k and m.

Let us choose h so small that for $t - s \leq h$ and $t \leq T$, we find

$$\mathbf{P}\{r(\xi_\delta^{-1}(s)\xi_\delta(t), e) \geq \varepsilon/2\} < \alpha/2,$$

where α is such that $\frac{\alpha}{1-\alpha} < \lambda^m$ and $\lambda < 1$ is constructed for given ε and δ just as in Lemma 9°. Let $s < t < T$ and $t - s \leq h$. We choose a sequence of decompositions of the interval $[s, t] : s = t_{n0} < t_{n1} < \cdots < t_{nn} = t$ for which $\lim_{n\to\infty} \max_k(t_{nk+1} - t_{nk}) = 0$. We put

$$\xi_{nk} = \xi^{-1}(t_{nk-1})\xi(t_{nk}), \quad k = 1, 2, \ldots, \qquad \xi'_{nk} = \xi_{nk}^{I\{r(\xi_{nk}, e)\leq\delta\}}.$$

Assume first that $\mathbf{M}\nu_T(U'_\delta) = 0$, $U'_\delta = \{x \in G, r(e, x) = \delta\}$. The latter condition evidently holds for all $\delta > 0$, except for at most a countable set. Thus, it is clear that

$$\mathbf{P}\{\inf_{u\leq T} |r(\xi^{-1}(u-)\xi(u), e) - \delta| > 0\} = 1 \tag{5.21}$$

(i.e., jumps of $\xi(t)$ do not hit at the boundary of U_δ with probability 1). In this case,

$$r(\xi'_{ni}\xi'_{ni+1} \cdots \xi'_{nj}, \xi_\delta^{-1}(t_{ni-1})\xi_\delta(t_{nj})) \to 0 \tag{5.22}$$

as $n \to \infty$ uniformly with respect to $i < j \leq n$. Thus, for all sufficiently large n and for all $i < j \leq n$, we have

$$\mathbf{P}\{r(\xi'_{ni} \cdots \xi'_{nj}, e) \geq \varepsilon\} \leq \alpha.$$

Hence, the conditions of Lemma 9° hold for the variables $\xi'_{n1},\dots,\xi'_{nn}$; and thus,

$$\mathbf{M}r^m(\xi'_{n1},\dots,\xi'_{nn},e) \le c,$$

where c is independent of m. Proceeding to the limit as $n \to \infty$ and taking (5.22) into account, we get (5.20). For arbitrary $\delta > 0$, we can always find $\delta_n > \delta$, $\delta_n \to \delta$, for which (5.21) holds, and then use $\xi_\delta(t) = \lim \xi_{\delta_n}(t)$ (just for $\delta_n \downarrow \delta$).

□

11°. Lemma. *The processes $\xi_\delta(t)$ and $\xi_t(U_\delta)$ are independent.*

Proof. Since $\xi_\delta^{-1}(s)\xi_\delta(t)$ and $\xi_s^{-1}(U_\delta)\xi_t(U_\delta)$ are measurable for $s < t$ with respect to the σ-algebra $\mathcal{F}_t^s$, and since these σ-algebras are independent for disjoint intervals, it is sufficient to prove the independence of $\xi_\delta^{-1}(s)\xi_\delta(t)$ and $\xi_s^{-1}(U_\delta)\xi_t(U_\delta)$ for sufficiently small $s < t$ (cf. proof of 13°, Chapter 3).

Note that $\xi_s^{-1}(U_\delta)\xi_t(U_\delta)$ is defined completely by the measure $\nu_t(A)-\nu_s(A)$. Thus, to prove the lemma, it suffices to show that for uniformly continuous functions $g_1(x)$ and $g_2(x)$ from G into R ($g_2(x) = 0$ for $x \overline{\in} U_\delta$) and for an arbitrary integer $m > 0$, we have

$$\begin{aligned}\mathbf{M}g_1(\xi_\delta^{-1}(s)\xi_\delta(t)) \left(\int g_2(x)[\nu_t(dx) - \nu_s(dx)]\right)^m \\ = \mathbf{M}g_1(\xi_\delta^{-1}(s)\xi_\delta(t))\mathbf{M}\left(\int g_2(x)[\nu_t(dx) - \nu_s(dx)]\right)^m .\end{aligned} \tag{5.23}$$

As in the previous lemma, it is sufficient to examine $\delta > 0$ such that (5.21) is valid. Then we have

$$\xi_\delta^{-1}(s)\xi_\delta(t) = \lim_{n\to\infty} \xi'_{n1}\cdots\xi'_{nn},$$

$$\int g_2(x)[\nu_t(dx) - \nu_s(dx)] = \lim_{n\to\infty}\sum_{k=1}^n g_2(\xi_{nk}).$$

with probability 1. Consequently, to prove (5.23), it suffices to show that

$$\begin{aligned}\lim_{n\to\infty}\Bigg(\mathbf{M}g_1(\xi'_{n1}\cdots\xi'_{nn})\left(\sum_{k=1}^n g_2(\xi_{nk})\right)^m \\ - \mathbf{M}g_1(\xi'_{n1}\cdots\xi'_{nn})\mathbf{M}\left(\sum_{k=1}^n g_2(\xi_{nk})\right)^m\Bigg) = 0.\end{aligned} \tag{5.24}$$

Let us estimate the expression

$$\begin{aligned}\mathbf{M}g_1(\xi'_{n1}\cdots\xi'_{nn})g_2^{m_1}(\xi_{nk_1})\cdots g_2^{m_i}(\xi_{nk_i}) \\ - \mathbf{M}g_1(\xi'_{n1}\cdots\xi'_{nn})\mathbf{M}g_2^{m_1}(\xi_{nk_1})\cdots g_2^{m_i}(\xi_{nk_i}),\end{aligned} \tag{5.25}$$

(terms of this form are under the limit sign in (5.24)). If $g_2(\xi_{nk_j}) \neq 0$, then $\xi'_{nk_j} = e$ ($g_2(x) = 0$ for $x\overline{\in}U_\delta$). Hence, the l.h.s. of (5.25) coincides with

$$\mathbf{M}g_1([\xi'_{n1}\cdots\xi'_{nn}]_{k_1\ldots k_i})\mathbf{M}g_2^{m_1}(\xi_{nk_1})\cdots g_2^{m_i}(\xi_{nk_i});$$

where $[\ ]_{k_1\ldots k_i}$ is a product in which factors with the numbers $k_1,\ldots,k_i$ are omitted. (5.24) is valid if we prove that

$$\lim_{n\to\infty}\mathbf{M}(g_1([\xi'_{n1}\cdots\xi'_{nn}]_{k_1\ldots k_i}) - \mathbf{M}g_1(\xi'_{n1}\cdots\xi'_{nn})) = 0$$

uniformly in $k_1,\ldots,k_i$. This relation follows from, e.g., the equality

$$\lim_{n\to\infty}\mathbf{M}r([\xi'_{n1}\cdots\xi'_{nn}]_{k_1\ldots k_i}, \xi'_{n1}\cdots\xi'_{nn}) = 0 \tag{5.26}$$

uniformly in $k_1,\ldots,k_i$. Taking into account that

$$\sup_n \sup_{i<j\leq n}\mathbf{M}r(\xi'_{ni}\cdots\xi'_{nj}, e) < \infty, \qquad \lim_{n\to\infty}\sup_{i<n}\mathbf{M}r(\xi'_{ni}, e) = 0$$

by virtue of Lemma 10° and allowing for (5.19), we can easily prove that (5.26) is valid.

□

Remark. Let us write a formula allowing us to renew $\xi(t)$ when $\xi_\delta(t)$ and $\xi_t(U_\delta)$ are given. Since $\xi_t(U_\delta)$ is a jump process and $\xi_0(U_\delta) = e$, there exist sequences of times $\tau_k \uparrow \infty$ (the s.m. with respect to $\mathcal{F}_t$) and variables $\eta_k \in G$ such that

$$\xi_t(U_\delta) = \prod_k \eta_k^{I_{\{\tau_k\leq t\}}}$$

(the product is taken in increasing order of the indices). Then

$$\xi(t) = \xi_\delta(\tau_1)\eta_1\xi_\delta^{-1}(\tau_1)\xi_\delta(\tau_2)\eta_2\xi_\delta^{-1}(\tau_2)\ldots\eta_l\xi_\delta^{-1}(\tau_l)\xi_\delta(t) \tag{5.27}$$

for $\tau_l \leq t < \tau_{l+1}$. By virtue of Lemma 11°, the variables τ_k and η_k do not depend on the process $\xi_\delta(t)$.

In order to investigate the behaviour of $\xi_\delta(t)$ as $\delta \to 0$ and to construct a continuous component of a multiplicative process, in particular, more detailed information is needed about the group considered. These problems will be examined for some concrete classes of groups.

§5.2 Multiplicative Processes in Abelian Groups

Let us denote the operation of composition in Abelian groups by the sign $+$. We are only interested in topologic locally compact groups that can be represented as a direct sum of a discrete (countable) and continuous connected group, which is, in turn, a direct sum of some finite-dimensional torus and a finite-dimensional Euclidean space. In contrast to the general case, the processes on Abelian groups will be called additive instead of multiplicative. Additive processes in R^m have already been considered. We shall now examine discrete additive processes, processes on the m-dimensional torus T^m, and processes on direct sums of the above-mentioned groups.

Discrete Additive Processes

12°. Theorem. *Let G_d be a discrete Abelian group with elements $\{0, a_1, a_2, \dots\}$ (the neutral element of the Abelian group is denoted by 0), and let $\xi(t)$ be a stochastically continuous additive process continuous from the right with values in G_d. Then there exists a countable set of mutually independent Poisson processes $\nu_k(t)$ such that $\xi(t)$ can be represented as*

$$\xi(t) = \xi(0) + \sum_{k=1}^{\infty} \nu_k(t) a_k; \tag{5.28}$$

furthermore, for each t, at most a finite number of terms in this sum differ from zero.

Proof of this theorem follows directly from Corollary 4°.

Remark. If G_d is a finite group, then (5.28) contains only a finite number of Poisson processes (equal to a number of nonzero elements of the group).

We examine, for example, an additive process $\xi(t)$ on the group of residues modulo m. Then

$$\xi(t) = \xi(0) + \sum_{k=1}^{m-1} \nu_k(t) k \pmod m,$$

and for all $m > l \geq 0$, $s < t$, we have

$$\begin{aligned} \mathbf{M} \exp\left\{ i\frac{2\pi}{m} l[\xi(t) - \xi(s)] \right\} &= \mathbf{M} \exp\left\{ i\frac{2\pi}{m} l \sum_{k=1}^{m-1} [\nu_k(t) - \nu_k(s)] k \right\} \\ &= \exp\left\{ \sum_{k=1}^{m} (e^{i2\pi lk/m} - 1)[\lambda_k(t) - \lambda_k(s)] \right\}, \end{aligned} \tag{5.29}$$

where $\lambda_k(t) = \mathbf{M}\nu_k(t)$ is a nondecreasing continuous function. (5.28) is similar to the formula for the characteristic function of the increment of a process with independent increments.

Now consider additive processes on a direct sum $G_d \oplus G$ of the discrete group G_d and some group G. Elements of $G_d \oplus G$ are denoted by (a, x), where $a \in G_d$, $x \in G$; a distance in $G_d \oplus G$ is defined as follows:

$$r((a; x), (b; y)) = \delta_b^a + r(x, y),$$

where $\delta_b^a = 0$ for $a = b$ and $\delta_b^a = 1$ for $a \neq b$; $r(x, y)$ is a distance in G. Let us denote by $G(a)$ a subset of $G_d \oplus G$ consisting of the elements (a, x), $x \in G$. For $a \neq 0$, the set $G(a)$ lies at a distance from $(0; 0)$ not less than 1.

13°. Theorem. *Let $(\xi^1(t), \xi^2(t))$ be a stochastically continuous additive process on $G_d \oplus G$ ($\xi^1 \in G_d$, $\xi^2 \in G$) continuous from the right. Then we have*

$$\begin{aligned} \xi^1(t) &= \xi^1(0) + \sum_{k=1}^{\infty} \nu_k(t) a_k, \\ \xi^2(t) &= \xi^2(0) + \sum_{k=1}^{\infty} \xi_k(t) + \xi_0(t), \end{aligned} \tag{5.30}$$

where $\{a_k\} \in G_d$ are all nonzero elements of G_d; $\xi_k(t)$, $k = 0, 1, 2, \ldots$, are stochastically continuous additive processes on G; $\xi_k(t)$ for $k > 0$ is a jump process; the pairs of processes $(\nu_k(t); \xi_k(t))$, $k \geq 1$, on $R \oplus G$ and $\xi_0(t)$ are mutually independent; $\xi(t)$ has jumps only at the jump times of $\nu_k(t)$; $\sum_k \nu_k(t) < \infty$ for all $t \in R_+$.

Proof. Denote a pair $(\xi^1(t), \xi^2(t))$ by $\xi(t)$. Taking into account that $r((0,0), G(a_k)) \geq 1$, we construct processes $\bar{\xi}_t(G(a_k))$ (sums of jumps of $\bar{\xi}(t)$ hitting $G(a_k)$). Taking some $0 < \delta < 1$, we also construct a process $\bar{\xi}_\delta(t)$. Then by virtue of Lemma 11°, the processes $\bar{\xi}_t(G(a_k))$, $k = 1, \ldots$, and $\bar{\xi}_\delta(t)$ are independent, since $G(a_k) \subset U_\delta$. Independence of $\bar{\xi}_t(G(a_k))$ for different k follows from Theorem 4° and from the fact that $\bar{\xi}_t(G(a_k))$ can be represented in terms of the Poisson measure $\nu_t(A)$, which corresponds to $\bar{\xi}(t)$ on disjoint sets $G(a_k)$.

Lemmas 5° and 7° imply that all processes $\bar{\xi}_t(G(a_k))$ and $\bar{\xi}_\delta(t)$ are stochastically continuous additive processes.

Let $\bar{\xi}_\delta(t) = (\xi_0^1(t); \xi_0(t))$. The process $\xi_0^1(t)$ is an additive process in G_d continuous from the right; moreover, it has no jumps, since $1 > \delta \geq r(\bar{\xi}_\delta(t-), \bar{\xi}_\delta(t)) \geq \delta_{\xi_0^1(t)}^{\xi_0^1(t-)}$. Thus, $\xi_0^1(t) = \xi_0^1(0) = \xi^1(0)$ $(\bar{\xi}_0(G(a_k)) = (0;0))$. If $\bar{\xi}_t(G(a_k)) = (\xi_k^1(t); \xi_k(t))$, then $\xi_k^1(t)$ is an additive process on G_d, and all its jumps are equal to a_k. Taking Theorem 12° into account, we get $\xi_k^1(t) = \nu_k(t) a_k$. It remains to note that $\sum \nu_k(t) = \nu_t(U_\delta) < \infty$ and that by virtue of commutativity

$$\bar{\xi}(t) = \bar{\xi}_\delta(t) + \bar{\xi}_t(U_\delta) = \bar{\xi}_\delta(t) + \sum_k \bar{\xi}_t(G(a_k)). \quad \square$$

The theorem proved allows us to reduce our investigations of additive processes on $G_d \otimes G$ to the analysis of additive processes on $R \oplus G$.

Additive Processes on a Torus

A group of real numbers $[0, 1[$ with the addition modulo 1 is called a one-dimensional torus, denoted by T. The distance between x and $y \in T$ is defined as follows:

$$r(x, y) = \min(|x - y|; 1 - |x - y|).$$

Consider a stochastically continuous additive process $\xi(t)$ on T continuous from the right. We now construct a transformation, setting the correspondence between an additive process in R and an additive process on T continuous from the right.

Let $x(t)$, $t \in R_+$, be an arbitrary function in T continuous from the right without discontinuities of the second kind. Let us define a sequence of times τ_k as follows: $\tau_0 = 0$ and

$$\tau_{k+1} = \inf[t : t > \tau_k, r(x(\tau_k), x(t)) \geq 1/4] \tag{5.31}$$

if τ_k is given. Since the number of 1/2-oscillations $x(t)$ on each finite interval is finite, $\tau_n \to \infty$ as $n \to \infty$ (if the set in brackets to the right is empty, we take $\tau_{k+1} = +\infty$; therefore, $\tau_n = +\infty$ for all $n \geq k+1$). Now define $Sx(t)$ by

$$\begin{aligned} Sx(t) = \chi(x(0)) &+ \sum_{1\leq i\leq k} [\chi(x(\tau_i-) - x(\tau_{i-1})) + \chi(x(\tau_i) - x(\tau_i-))] \\ &+ \chi(x(t) - x(\tau_k)) \end{aligned} \tag{5.32}$$

for $\tau_k \leq t < \tau_{k+1}$. Here $Sx(t)$ is a function with values in R, and χ is a mapping of T into R defined by

$$\chi(x) = \begin{cases} x & \text{for } x < 1/2, \\ x - 1 & \text{for } x \geq 1/2. \end{cases}$$

Summation in (5.32) is taken in R.

14°. Theorem. *If $\xi(t)$ is an additive process in T continuous from the right, then $S\xi(t)$ is a process with independent increments in R continuous from the right; furthermore, $S\xi(t) = \xi(t)$ (mod 1).*

Proof. The continuity from the right of $S\xi(t)$ follows from (5.32). Using (5.32), we find $S\xi(t) - S\xi(s)$ for $s < t$. If $\tau_k \leq s < \tau_{k+1} \leq \tau_l \leq t < \tau_{l+1}$, then

$$\begin{aligned} &S\xi(t) - S\xi(s) \\ &\quad = \chi(\xi(t) - \xi(\tau_l)) + \sum_{k<i\leq l} [\chi(\xi(\tau_i) - \xi(\tau_i-)) + \chi(\xi(\tau_i) - \xi(\tau_{i-1}))] \\ &\qquad - \chi(\xi(s) - \xi(\tau_k)) \\ &\quad = \chi(\xi(\tau_{k+1}) - \xi(s)) + \sum_{k+1\leq i\leq l} \chi(\xi(\tau_i) - \xi(\tau_i-)) \\ &\qquad + \sum_{k+1<i\leq l} \chi(\xi(\tau_i-) - \xi(\tau_{i-1})) + \chi(\xi(t) - \xi(\tau_l)). \end{aligned} \tag{5.33}$$

We have used $\chi(x+y) = \chi(x) + \chi(y)$ for $r(x,0) < 1/4$, $r(y,0) \leq 1/4$. The same arguments yield

$$S\xi(t) - S\xi(s) = \chi(\xi(t) - \xi(s)) \tag{5.34}$$

for $\tau_k \le s < t < \tau_{k+1}$. Thus, if $\mathcal{F}_t^s$ is a σ-algebra generated by the increments $\xi(u) - \xi(s)$ for $s \le u \le t$, then $S\xi(t) - S\xi(s)$ is $\mathcal{F}_t^s$-measurable ((5.33) and (5.34) show that it can be represented in terms of $\xi(u)$ increments on $[s,t]$). This implies that $S\xi(t)$ is a process with independent increments. The equality $\chi(x) = x \pmod 1$ yields $S\xi(t) \equiv \xi(t) \pmod 1$.

□

Remark. If $\xi(t)$ is a stochastically continuous process, the same is true for $S\xi(t)$. Hence, for every stochastically continuous additive process $\xi(t)$ continuous from the right, there exists the unique stochastically continuous process $S\xi(t)$ continuous from the right such that $\xi(t) = \{S\xi(t)\}$ (here $\{x\} = x - \operatorname{Ent} x$ is a fractional part of x).

Corollary. *Let $\xi(t) = (\xi_1(t);\ldots;\xi_m(t))$ be an additive process on T^m. Provided that it is stochastically continuous and continuous from the right, we have that*

$$S\xi(t) = (S\xi_1(t);\ldots;S\xi_m(t))$$

is a stochastically continuous process with independent increments in R^m continuous from the right.

Additive Processes on Groups $G_d \otimes R^m \otimes T^l$

Consider the additive processes

$$\xi(t) = (\xi^d(t);\xi^1(t);\xi^2(t)), \tag{5.35}$$

where $\xi^d \in G_d = \{0;a_1,\ldots\}$, G_d is a discrete group, $\xi^1(t)$ is a process in R^m, and $\xi^2(t)$ is a process on T^l. Stochastically continuous processes continuous from the right are described by the following theorem.

15°. Theorem. *Assume that $\xi(t)$ is an additive process of the form (5.35), which is stochastically continuous and continuous from the right. Then there exist Poisson processes $\nu_i(t)$, $i = 1,2,\ldots$, and mutually independent processes $(\eta_1^{(i)}(t),\ldots,\eta_m^{(i)}(t),\zeta_1^{(i)}(t),\ldots,\zeta_l^{(i)}(t))$ in R^{m+l}, $i = 0,1,2,\ldots$, stochastically continuous and continuous from the right, such that $(\nu_i(t),\eta_1^{(i)}(t),\ldots,\zeta_l^{(i)}(t))$ and $(\eta_1^{(0)}(t),\ldots,\zeta_l^{(0)}(t))$ are mutually independent processes with independent increments; $(\nu_i(t),\eta_1^{(i)}(t),\ldots,\zeta_l^{(i)}(t))$ is a jump process for $i > 0$; the jumps of the processes $\eta_k^{(i)}(t)$, $\xi_j^{(i)}(t)$ happen only at the jump times of $\nu_i(t)$, moreover,*

$\sum_i \nu_i(t) < \infty$; and $\xi(t)$ can be represented in terms of these processes as follows:

$$\xi^d(t) = \xi(0) + \sum_{i=1}^{\infty} \nu_i(t) a_i, \qquad \xi_j^1(t) = \sum_{i\geq 0} \eta_j^{(i)}(t), \quad j = 1, \ldots, m;$$
$$\xi_k^2(t) = \{\sum_{i\geq 0} \zeta_k^{(i)}(t)\}, \quad k = 1, \ldots, l$$

(here $\{\cdot\}$ is a fractional part of a real number).

Proof. Using $\xi(t)$, let us construct the process $S\xi(t)$ on $G_d \oplus R^m \oplus R^l$ as follows:

$$S\xi(t) = (\xi^d(t); \xi^1(t); S\xi^2(t)).$$

Just as in Theorem 14°, we find $S\xi(t)$ to be an additive process stochastically continuous and continuous from the right. By using Theorem 13°, we complete the proof.

□

Multiplicative Processes in R

Consider a multiplicative process $\eta(t)$ on the group $R\backslash\{0\}$ of nonzero integers with the operation of product. Under transformation of $\eta(t)$ into the pair $\left(\frac{1-\operatorname{sign}\eta(t)}{2}; \ln|\eta(t)|\right)$, the process $\eta(t)$ turns into the additive process on the group $G_2 \oplus R$, where G_2 is a group of integers with the operation of addition modulo 2. It was proved that if $\eta(t)$ is a stochastically continuous process continuous from the right, then there exist independent processes $\xi_0(t)$ in R and $(\nu(t); \xi_1(t))$ in R^2 such that $(\nu(t); \xi_1(t))$ is a jump process in R^2; moreover, jumps of $\xi_1(t)$ occur only at the jump points of $\nu(t)$. Furthermore,

$$\frac{1 - \operatorname{sign}\eta(t)}{2} \equiv \nu(t) \pmod 2, \qquad \ln|\eta(t)| = \xi_0(t) + \xi_1(t).$$

If $\nu_1(t)$ is a number of jumps of $\xi_1(t)$, then $\nu_1(t) \leq \nu(t)$, and $\nu(t) - \nu_1(t) = \nu_0(t)$ is also a Poisson process independent of $\xi_1(t)$ (their jumps happen at different times). Finally, we obtain

$$\begin{aligned} \eta(t) = \operatorname{sign}\eta(t) e^{\ln|\eta(t)|} &= (-1)^{\nu(t)} e^{\ln|\eta(t)|} \\ &= (-1)^{\nu_0(t)}((-1)^{\nu_1(t)} e^{\xi_1(t)}) e^{\xi_0(t)}, \end{aligned} \tag{5.36}$$

Here $\nu_1(t)$ is a number of jumps of $\xi_1(t)$, and $\nu_0(t)$, $\xi(t)$, and $\xi_0(t)$ are mutually independent.

§5.3 Stochastic Semigroups of Linear Operators in R^d

Let us consider processes with values in $L(R^d)$, i.e., in the space of linear operators acting in the d-dimensional Euclidean space R^d. A product of elements is defined in $L(R^d)$, whereas the inverse element does not always exist. Hence, here it is more natural to examine stochastic semigroups in $L(R^d)$ instead of multiplicative random processes, which are not defined in $L(R^d)$. A definition of stochastic semigroups in $L(R^d)$ is given below.

16°. Definition. Let σ-algebras $\mathcal{F}_t^s$ and random elements $U_t^s(\omega)$ in $L(R^d)$ ($L(R^d)$ is a linear space with a natural norm: $\|A\| = \sup_{|x|\le 1} |Ax|$ for $A \in L(R^d)$; here $|\cdot|$ is the Euclidean norm in R^d) be given for all $0 \le s \le t < \infty$ so that the following conditions are satisfied:

(1) $\mathcal{F}_t^s \supset \mathcal{F}_{t_1}^{s_1}$ for $s \le s_1 < t_1 \le t$,

(2) The σ-algebras $\mathcal{F}_s^0$ and $\mathcal{F}_t^s$ are independent for $0 < s < t$,

(3) $U_t^s(\omega)$ is $\mathcal{F}_t^s$-measurable,

(4) $U_t^s(\omega)U_u^t(\omega) = U_u^s(\omega)$ for $s < t < u$ and $U_s^s(\omega) = I$ with probability 1, I being the unit operator in $L(R^d)$.

Then this family of random operators $\{U_t^s(\omega), 0 \le s \le t < \infty\}$ is called a *(right) stochastic semigroup.*

If, instead of condition (4), we have the following condition (4′):

(4′) $U_u^t(\omega)U_t^s(\omega) = U_u^s(\omega)$, $U_t^t(\omega) = I$ for $s < t < u$,

then this stochastic semigroup is called a *left semigroup.* If $U_t^s(\omega)$ is a left stochastic semigroup, we can obtain the right one by passing to the conjugated operators. Therefore, we shall consider only right stochastic semigroups from here on.

A semigroup is called *stochastically continuous* if for all $u, s < u < t$

$$\|U_t^s(\omega) - I\| \to 0 \text{ in probability for } s \uparrow u, t \downarrow u.$$

We shall consider stochastically continuous stochastic semigroups in $L(R^d)$. $L(R^d)$ is a semigroup with the operation of product (if $\|A\| = N(A)$ and $r(A,B) = \|A - B\|$, then conditions (1)–(4) of Section 5.1 are satisfied except for the relation $N(A) = N(A^{-1})$ in condition (2), since A^{-1} is not defined here). Thus, some results of Section 5.1 hold for stochastic semigroups.

Properties of Sample Functions

17°. Lemma. *For every t_0, $\varepsilon > 0$, $\delta > 0$, there exists h such that for $s < t \le t_0$, $t - s \le h$, we have*

$$\mathbf{P}\{\|U_t^s(\omega) - I\| > \varepsilon\} < \delta.$$

(It is natural to call this property the local uniform stochastic continuity of the semigroup $U_t^s(\omega)$).

Proof. Assume the contrary. Then we can find a point $u \le t_0$ and sequences $s_n < t_n$ such that $s_n \to u_0$, $t_n \to u_0$, but

$$\mathbf{P}\{\|U_{t_n}^{s_n}(\omega) - I\| > \varepsilon\} \ge \delta. \tag{5.37}$$

This is impossible if $s_n < u < t_n$. Suppose that $s_n < t_n < u$. Then choosing $u_n \downarrow u$, we get

$$\|U_{t_n}^{s_n}(\omega)U_{u_n}^{t_n}(\omega) - I\| \to 0$$

in probability and $\|U_{u_n}^{t_n}(\omega) - I\| \to 0$. This implies that $\|(U_{u_n}^{t_n}(\omega))^{-1} - I\| \to 0$ in probability, since A^{-1} exists for $\|A - I\| < 1$; and therefore $\|U_{t_n}^{s_n}(\omega) - I\| \to 0$ in probability. The last relation contradicts (5.37).

The case $u_n < s_n < t_n$ can be examined similarly.

□

18°. Lemma. *Let $\{U_t^s(\omega), 0 \le s \le t\}$ be a stochastically continuous stochastic semigroup separable as a function of parameters s and t. Then $U_t^s(\omega)$ have no discontinuities of the second kind as a function of s for $s \in [0, t[$ and as a function of t for $t \in]s, \infty[$.*

Proof. Lemma 2° implies that $U_t^s(\omega)$ has a finite number of ε-oscillations (multiplicative) on any countable set and thus on the set of separability. The proof follows directly from this fact.

19°. Theorem. *A stochastically continuous stochastic semigroup $\{U_t^s(\omega), 0 \le s \le t\}$ has a modification $\widetilde{U}_t^s(\omega)$ satisfying the following conditions:*

(1) *it is continuous from the right with respect to both arguments $\widetilde{U}_{t+}^{s+}(\omega) = \widetilde{U}_t^s(\omega)$,*

(2) *the following limits exist:*

$$\widetilde{U}_t^{s-}(\omega), \quad s < t, \qquad \widetilde{U}_{t-}^{s-}(\omega), \quad s < t,$$
$$\widetilde{U}_{t-}^{s}(\omega), \quad s < t, \qquad \widetilde{U}_{s-}^{s-}(\omega) = I,$$

(3) *the equalities*

$$\widetilde{U}_t^s(\omega)\widetilde{U}_u^t(\omega) = \widetilde{U}_u^s(\omega), \qquad \widetilde{U}_s^{s-}(\omega)\widetilde{U}_t^s(\omega) = \widetilde{U}_t^{s-}(\omega),$$
$$\widetilde{U}_{t-}^s(\omega)\widetilde{U}_t^{t-}(\omega) = \widetilde{U}_t^s(\omega), \qquad \widetilde{U}_s^{s-}(\omega)\widetilde{U}_{t-}^s(\omega) = \widetilde{U}_{t-}^{s-}(\omega),$$
$$\widetilde{U}_t^s(\omega)\widetilde{U}_{u-}^t(\omega) = \widetilde{U}_{u-}^s(\omega)$$

hold for $s < t < u$ and for all ω.

Proof. Consider a separable modification of $U_t^s(\omega)$ (denoted by the same symbol). Let us reconstruct $U_t^s(\omega)$ in the following way. Let Ω_1 be a set of those ω for which $U_u^s(\omega) = U_t^s(\omega)U_u^t(\omega)$ for all $s, t, u \in \Lambda$, where Λ is a set of dyadic rational numbers from R_+. Clearly, $\mathbf{P}(\Omega_1) = 1$. Let Ω_2 be a set of those ω for which the limits

$$\lim_{\substack{s_n \downarrow s,\ s_n \in \Lambda \\ t_n \downarrow t,\ t_n \in \Lambda}} U_{t_n}^{s_n}(\omega) = U_{t+}^{s+}(\omega)$$

exist. Then $\mathbf{P}(\Omega_2) = 1$. We put $\widetilde{U}_t^s(\omega) = U_{t+}^{s+}(\omega)$ for $\omega \in \Omega_1 \cap \Omega_2$ and $\widetilde{U}_t^s(\omega) = I$ for $\omega \overline{\in} \Omega_1 \cap \Omega_2$. This is just the modification desired. Taking into account that $\omega \in \Omega_1$, we check condition (3).

□

The variable $U_t^{t-}(\omega) - I$ is called a *value of the stochastic semigroup jump at the point t*.

20°. Lemma. *Provided that conditions (1)–(3) of Theorem 19° hold for the semigroup $U_t^s(\omega)$, the number of points $t \leq T$ for which $\|U_t^{t-}(\omega) - I\| \geq \varepsilon$ is finite for all T and $\varepsilon > 0$.*

The proof follows from the fact that $U_t^s(\omega)$ has no discontinuities of the second kind.

Poisson Measure Corresponding to a Stochastic Semigroup

Consider the stochastically continuous stochastic semigroups for which properties (1)–(3) of Theorem19° are satisfied. We shall call these semigroups *regular.*

For the Borel sets $C \subset L(R^d)$ lying at a positive distance from I, define the variable

$$\nu_t(C) = \sum_{s \leq t} I_{\{U_s^{s-}(\omega) \in C\}} \tag{5.38}$$

(the sum is taken over stochastic semigroup discontinuity points) and

$$Z_t(C) = \sum_{s \leq t} (U_s^{s-}(\omega) - I) I_{\{U_s^{s-}(\omega) \in C\}}. \tag{5.39}$$

Let $\mathcal{B}_0$ be a ring of Borel sets located at a positive distance from I. $\nu_t(C)$ and $Z_t(C)$ are countably additive functions of the set with values in Z_+ and $L(R^d)$, respectively. Denote $S_\varepsilon = \{A \in L(R^d) : \|A - I\| > \varepsilon\}$. There exists a sequence of times $\tau_k^\varepsilon \uparrow \infty$ such that $U_t^{t-}(\omega) \overline{\in} S_\varepsilon$ for $t \overline{\in} \{\tau_1^\varepsilon, \tau_2^\varepsilon, \dots\}$ and $U_{\tau_k^\varepsilon}^{\tau_k^\varepsilon -}(\omega) \in S_\varepsilon$. Using $U_t^s(\omega)$, we construct a "curtailed" semigroup $U_t^s(\varepsilon, \omega)$ as follows: If $\tau_k^\varepsilon \leq s \leq \tau_{k+1}^\varepsilon \leq \tau_l^\varepsilon \leq t < \tau_{l+1}^\varepsilon$, then

$$\begin{aligned} U_t^s(\varepsilon, \omega) &= U_{\tau_{k+1}^\varepsilon -}^s(\omega) U_{\tau_{k+2}^\varepsilon -}^{\tau_{k+1}^\varepsilon}(\omega) \dots U_{\tau_l^\varepsilon -}^{\tau_{l-1}^\varepsilon}(\omega) U_t^{\tau_l^\varepsilon}(\omega), \\ U_t^s(\varepsilon, \omega) &= U_t^s(\omega), \quad \tau_k \leq s < t < \tau_{k+1}. \end{aligned} \tag{5.40}$$

21°. Theorem. (a) *$\nu_t(C)$ is a stochastically continuous Poisson process as a function of t; if $C_1, C_2, \ldots, C_n$ are mutually disjoint, then $\nu_t(C_1), \ldots, \nu_t(C_n)$ are mutually independent;*

(b) *$Z_t(C)$, $C \in \mathcal{B}_0$, is a stochastically continuous process with independent increments in $L(R^d)$ ($L(R^d)$ is a linear space); moreover, as a function of C_1, $\nu_t(\{C_1 + I\} \cap C)$ is a jump Poisson measure for $Z_t(C)$;*

(c) *the process $Z_t(C)$ for $C \subset S_\varepsilon$ and the stochastic semigroup $U_t^s(\varepsilon, \omega)$ are independent;*

(d) $\sup_t \|U_t^{t-}(\varepsilon, \omega) - I\| \le \varepsilon$;

(e) *$(U_t^s(\varepsilon, \omega))^{-1}$ exists for $\varepsilon < 1$ and for all $0 \le s < t$; it is a regular (left) stochastic semigroup.*

Proof. Statement (a) follows from Theorem 4°. In order to prove statement (b), we use the formula

$$Z_t(C) = \lim_{n \to \infty} \sum_{k/2^n < t} I_{\{U^{k/2^n}_{(k+1)/2^n}(\omega) \in C\}} (U^{k/2^n}_{(k+1)/2^n}(\omega) - I),$$

which is valid when the set C with boundary C' is such that $\nu_t(C') = 0$ with probability 1 for all t. This formula implies that $Z_t(C)$ is a process with independent increments, whereas its stochastic continuity follows from (5.39). The relation between ν_t and Z_t is determined by (5.38) and (5.39).

Statement (c) is proved by analogy with the proof of Lemma 11°, and (d) follows from (5.40). To prove (e), note that by virtue of (d) we can always find $\delta > 0$ such that $\|U_t^s(\varepsilon, \omega) - I\| \le \varepsilon_1$ for $t - s < \delta$ and $t \le t_0$ (here $1 > \varepsilon_1 > \varepsilon$; otherwise, one can find the point u for which $\|U_u^{u-}(\varepsilon, \omega) - I\| \ge \varepsilon_1$). Thus, $U_t^s(\varepsilon, \omega)$ is invertible for sufficiently small $t - s$ and $\|(U_t^s(\varepsilon, \omega))^{-1}\| \le \frac{1}{1-\varepsilon_1}$ (since $\|U_t^s(\varepsilon, \omega) - I\| \le \varepsilon_1$). Using multiplicative property (3) of Theorem 19°, we obtain that $(U_t^s(\varepsilon, \omega))^{-1}$ exists for all $s < t$. The continuity of the operator inversion operation (in its domain of definition) implies the other statements of (e).

□

Remark 1. Statement (b) also yields the following equality, connecting the processes $Z_t(C)$ with the measure $\nu_t(A)$:

$$Z_t(C) = \int_C (Z - I) \nu_t(dZ) \tag{5.41}$$

(see 11°, Chapter 3).

Remark 2. The semigroup $U_t^s(\omega)$ can be expressed in terms of the semigroup $U_t^s(\varepsilon, \omega)$ and the process $Z_t(S_\varepsilon)$ as follows: If $\tau_1^\varepsilon < \tau_2^\varepsilon < \cdots < \cdots$ are jump times of the process $Z_t(S_\varepsilon)$, $Z_k^\varepsilon = Z_{\tau_k^\varepsilon}(S_\varepsilon) - Z_{\tau_k^\varepsilon -}(S^\varepsilon)$, then

$$U_t^s(\omega) = \begin{cases} U_t^s(\varepsilon, \omega), & \text{if } \sum I_{\{\tau_k \in [s,t]\}} = 0, \\ U^s_{\tau_{k+1}}(\varepsilon, \omega)(I + Z^\varepsilon_{k+1}) U^{\tau_{k+1}}_{\tau_{k+2}}(\varepsilon, \omega) \ldots (I + Z_l^\varepsilon) U_t^{\tau_l}(\varepsilon, \omega), \\ \text{if } \tau_k \le s < \tau_{k+1} \le \tau_l \le t < \tau_{l+1}. \end{cases}$$

Stochastic Semigroups with Bounded Jumps

Let us consider a semigroup $U_t^s(\varepsilon,\omega)$ for $\varepsilon < 1$. It is invertible in this case and

$$U_t^s(\varepsilon,\omega) = (U_s^0(\varepsilon,\omega))^{-1}U_t^0(\varepsilon,\omega).$$

Therefore, we can consider a multiplicative process on a group $GL(R^d)$ of invertible operators from $L(R^d)$. Distance in the group $GL(R^d)$ is defined by

$$r(A,B) = \frac{1}{2}(\|A-B\| + \|A^{-1} - B^{-1}\|),$$

and the norm is defined as follows:

$$N(A) = 1 + \frac{1}{2}(\|A-I\| + \|A^{-1} - I\|).$$

Then conditions (1)–(4), Section 5.1, are satisfied with $r_0 = 1$. Hence, the results of Section 5.1 are valid for stochastically continuous multiplicative processes, too. In particular, Lemma 10° implies that

$$\sup_{0\le s<t\le T} \mathbf{M}(\|U_t^s(\varepsilon,\omega)\|^m + \|(U_t^s(\varepsilon,\omega))^{-1}\|^m) < \infty \tag{5.42}$$

for all m and $T > 0$. Consider a family of operators in $L(R^d)$

$$A_t^s(\varepsilon) = \mathbf{M}U_t^s(\varepsilon,\omega)$$

(taking the mathematical expectation, we consider $U_t^s(\varepsilon,\omega)$ as a random element in the linear space $L(R^d)$). Equation (5.42) yields that $\|U_t^s(\varepsilon,\omega)\|$ is uniformly integrable, and therefore, $A_t^s(\varepsilon)$ is a continuous function of its arguments $s \le t$, $\lim_{t-s\to 0,\ t\le T} A_t^s(\varepsilon) = I$. Independence of $U_t^s(\varepsilon,\omega)$ and $U_u^t(\varepsilon,\omega)$ for $s < t < u$ implies that

$$A_t^s(\varepsilon)A_u^t(\varepsilon) = A_u^s(\varepsilon), \tag{5.43}$$

i.e., $A_u^s(\varepsilon)$ form a semigroup of nonrandom operators. $A_t^s(\varepsilon)$ is invertible for sufficiently small $t-s$; therefore, it is always invertible by virtue of (5.43). Consequently, the following operators are defined:

$$\widehat{U}_t^s(\varepsilon,\omega) = A_s^0(\varepsilon)U_t^s(\varepsilon,\omega)(A_t^0(\varepsilon))^{-1}, \quad 0 \le s < t. \tag{5.44}$$

Clearly, $\widehat{U}_t^s(\varepsilon,\omega)$ is an $\mathcal{F}_t^s$-measurable operator. These operators possess the multiplicative property

$$\widehat{U}_t^s(\varepsilon,\omega)\widehat{U}_u^t(\varepsilon,\omega) = \widehat{U}_u^s(\varepsilon,\omega), \quad 0 \le s < t < u.$$

Continuity of $A_s^0(\varepsilon)$ implies stochastic continuity of $\widehat{U}_t^s(\varepsilon,\omega)$. Thus, $\{\widehat{U}_t^s(\varepsilon,\omega),\ 0 \le s \le t\}$, is a stochastically continuous stochastic semigroup. It is regular if this is true for $U_t^s(\varepsilon,\omega)$. Since A_t^0 and $(A_t^0)^{-1}$ are continuous functions, they are locally bounded, and (5.42) is valid for $\widehat{U}_t^s(\varepsilon,\omega)$, too. We have

$$\mathbf{M}\widehat{U}_t^s(\varepsilon,\omega) = A_s^0(\varepsilon)A_t^s(\varepsilon)(A_t^0(\varepsilon))^{-1} = I.$$

22°. Theorem. *The limit*

$$\widehat{Z}_\varepsilon(t) = \lim_{n\to\infty} \sum_{k<2^n t} (\widehat{U}_{(k+1)/2^n}^{k/2^n}(\varepsilon,\omega) - I) \tag{5.45}$$

exists in the sense of mean square convergence. This limit is a stochastically continuous process with independent increments in $L(R^d)$ for which $\mathbf{M}\widehat{Z}_\varepsilon(t) = 0$ and $\mathbf{M}\|\widehat{Z}_\varepsilon(t)\|^2$ is locally bounded.

Proof. To prove the existence of the limit (5.45) in mean square, it suffices to show that the limit

$$\lim_{n\to\infty} \sum_{k<2^n t} (\widehat{U}_{(k+1)/2^n}^{k/2^n}(\varepsilon,\omega) - I)x, \tag{5.46}$$

exists in mean square for all $x \in R^d$, i.e.,

$$\lim_{n,m\to\infty} \mathbf{M}\Big| \sum_{k<2^n t} (\widehat{U}_{(k+1)/2^n}^{k/2^n}(\varepsilon,\omega) - I)x - \sum_{k<2^m t} (\widehat{U}_{(k+1)/2^m}^{k/2^m}(\varepsilon,\omega) - I)x\Big|^2 = 0. \tag{5.47}$$

Assuming that $n > m$, we can rewrite the difference of sums in (5.47) as follows:

$$\begin{aligned}
&\sum_{k=0}^{m_t-1} \Big[\sum_{k2^{n-m}\le i\le (k+1)2^{n-m}} (\widehat{U}_{(i+1)/2^n}^{i/2^n}(\varepsilon,\omega) - I) \\
&\qquad - \prod_{i=0}^{2^{n-m}-1} (\widehat{U}_{(k2^{n-m}+i+1)/2^n}^{(k2^{n-m}+i)/2^n}(\varepsilon,\omega) + I)\Big] x \\
&\qquad + \sum_{i=0}^{n_t - m_t 2^{n-m}} (\widehat{U}_{(i+1)/2^n}^{i/2^n}(\varepsilon,\omega) - I)x \\
&= \sum_{k=0}^{m_t-1} V_k(n,m)x + W_t(n,m)x,
\end{aligned}$$

Here m_t is an integral part of $2^m t$, n_t is an integral part of $2^n t$, and the product of operators is taken in increasing order of the indices. The operators

$V_k(n,m)$ and $W_t(n,m)$ are independent, and $\mathbf{M}V_k(n,m)=0$, $\mathbf{M}W_t(n,m)=0$. Therefore,

$$\mathbf{M}\Big|\sum_{k=0}^{m_t-1} V_k(n,m)x + W_t(n,m)x\Big|^2 = \sum_{k=0}^{m_t-1} \mathbf{M}|V_k(n,m)x|^2 + \mathbf{M}|W_t(n,m)x|^2.$$

Using independence of terms in the expression for $W_t(n,m)$ and the fact that their mathematical expectation equals zero, we get

$$\mathbf{M}|W_t(n,m)x|^2 = \sum_{i=m_t2^{n-m}}^{n_t} \mathbf{M}|(\widehat{U}^{i/2^n}_{(i+1)/2^n}(\varepsilon,\omega)-I)x|^2 = \sum_{i=m_t2^{n-m}}^{n_t} [\mathbf{M}|(\widehat{U}^{i/2^n}_{(i+1)/2^n}(\omega)x|^2 - |x|^2].$$

Now note that

$$0 \le \mathbf{M}|(\widehat{U}^s_t(\varepsilon,\omega)-I)x|^2 = \mathbf{M}|\widehat{U}^s_t(\varepsilon,\omega)x|^2 - |x|^2,$$
$$\mathbf{M}|\widehat{U}^s_t(\varepsilon,\omega)x|^2 \ge |x|^2. \tag{5.48}$$

Hence,

$$\begin{aligned}\mathbf{M}|\widehat{U}^s_{t+h}(\varepsilon,\omega)x - \widehat{U}^s_t(\varepsilon,\omega)x|^2 &= \mathbf{M}|\widehat{U}^s_{t+h}(\varepsilon,\omega)x|^2 - \mathbf{M}|\widehat{U}^s_t(\varepsilon,\omega)x|^2\\ &= \mathbf{M}|\widehat{U}^s_t(\varepsilon,\omega)[\widehat{U}^t_{t+h}(\varepsilon,\omega)-I]x|^2\\ &= \mathbf{M}\mathbf{M}(|\widehat{U}^s_t(\varepsilon,\omega)[\widehat{U}^t_{t+h}(\varepsilon,\omega)-I]x|^2/\mathcal{F}^t_{t+h})\\ &\ge \mathbf{M}|[\widehat{U}^t_{t+h}(\varepsilon,\omega)-I]x|^2\\ &= \mathbf{M}|\widehat{U}^t_{t+h}(\varepsilon,\omega)x|^2 - |x|^2.\end{aligned}$$

Thus, the following inequality holds:

$$\mathbf{M}|\widehat{U}^t_{t+h}(\varepsilon,\omega)x|^2 - |x|^2 \le \mathbf{M}|\widehat{U}^s_{t+h}(\varepsilon,\omega)x|^2 - \mathbf{M}|\widehat{U}^s_t(\varepsilon,\omega)x|^2 \tag{5.49}$$

for $s<t<t+h$. Using (5.49), we obtain

$$\mathbf{M}|W_t(n,m)x|^2 \le \mathbf{M}|\widehat{U}^0_{m_t+1/2^m}(\varepsilon,\omega)x|^2 - |x|^2$$

(we have taken into account that $\mathbf{M}|\widehat{U}^s_t(\varepsilon,\omega)x|^2$ increases with t; this follows from (5.48) and (5.49)).

For a random operator V such that $\mathbf{M}\|V\|^2<\infty$, we introduce a symmetric operator $D(V):(D(V)x,x)=\mathbf{M}|Vx|^2$.

In order to estimate $\mathbf{M}|V_k(n,m)x|^2$, we need the following lemma.

23°. Lemma. *Suppose that $V_1, V_2, \ldots, V_r$ are independent random operators in $L(R^d)$ for which $\mathbf{M}V_i = I$ and $M\|V_i\|^2 < \infty$. Then*

$$\mathbf{M}\left|\sum_{k=1}^{r}(V_k - I)x - \left(\prod_{k=1}^{r} V_k - I\right) x\right|^2$$
$$\leq \left\|D\left(\prod_{k=1}^{r} V_k - I\right)\right\| \sum_{k=1}^{r} \mathbf{M}|(V_k - I)x|^2.$$

Proof. We have

$$\prod_{k=1}^{r} V_k - I = V_1 \ldots V_{r-1}(V_r - I)$$
$$+ V_1 \ldots V_{r-2}(V_{r-1} - I) + \cdots + V_1 - I. \tag{5.50}$$

Hence,

$$\sum_{k=1}^{r}(V_k - I) - \prod_{k=1}^{r} V_k + I = \sum_{k=1}^{r}\left(I - \prod_{i=1}^{k-1} V_i\right)(V_k - I),$$

$$\mathbf{M}\left|\sum_{k=1}^{r}(V_k - I)x - \left(\prod_{k=1}^{r} V_k - I\right) x\right|^2$$
$$= \sum_{k=1}^{r} \mathbf{M}\left|\left(I - \prod_{i=1}^{k-1} V_i\right)(V_k - I)x\right|^2$$
$$= \sum_{k=1}^{r} \mathbf{M}\mathbf{M}\left(\left|\left(\prod_{i=1}^{k-1} V_i - I\right)(V_k - I)x\right|^2 / V_k\right)$$
$$\leq \sum_{k=1}^{r} \left\|D\left(\prod_{i=1}^{k-1} V_i - I\right)\right\| \mathbf{M}|(V_k - I)x|^2. \tag{5.51}$$

Eq. (5.50) yields

$$\mathbf{M}\left|\left(\prod_{k=1}^{r} V_k - I\right) x\right|^2 = \sum_{k=1}^{r-1} |V_1 \ldots V_{k-1}(V_k - I)x|^2,$$

and this expression increases with r. Therefore,

$$\left\|D\left(\prod_{i=1}^{k-1} V_i - I\right)\right\| \leq \left\|D\left(\prod_{i=1}^{r} V_i - I\right)\right\|.$$

Using this inequality in (5.51), we complete the proof of the lemma.

□

We now return to the theorem. By virtue of Lemma 23°, we have

$$\mathbf{M}|V_k(n,m)x|^2$$
$$\leq \|D(\widehat{U}^{k/2^m}_{(k+1)/2^m}(\varepsilon,\omega)-I)\| \sum_{i=2^{n-m}k}^{2^{n-m}(k+1)-1} \mathbf{M}|(\widehat{U}^{i/2^n}_{(i+1)/2^n}(\varepsilon,\omega)-I)x|^2.$$

In exactly the same way as in the estimation of $\mathbf{M}|W_t(n,m)x|^2$, we get

$$\sum_{i=2^{n-m}k}^{2^{n-m}(k+1)-1} \mathbf{M}|(\widehat{U}^{i/2^n}_{(i+1)/2^n}(\varepsilon,\omega)-I)x|^2$$
$$\leq \mathbf{M}|\widehat{U}^0_{(k+1)/2^m}(\varepsilon,\omega)x|^2 - \mathbf{M}|\widehat{U}^0_{k/2^m}(\varepsilon,\omega)x|^2.$$

Thus,

$$\mathbf{M}|\sum_{k=0}^{m_t-1} V_k(n,m)x + W_t(n,m)x|^2$$
$$\leq \max_{k<m_t} \|D(\widehat{U}^{k/2^m}_{(k+1)/2^m}(\varepsilon,\omega)-I)\|\mathbf{M}|\widehat{U}^0_{m_t/2^m}(\varepsilon,\omega)x|^2$$
$$+\mathbf{M}|\widehat{U}^0_{(m_t+1)/2^m}(\varepsilon,\omega)x|^2 - \mathbf{M}|\widehat{U}^0_{m_t/2^n}(\varepsilon,\omega)x|^2. \tag{5.52}$$

Since $\mathbf{M}\|\widehat{U}^s_t(\varepsilon,\omega)\|^m$ is bounded for all m, we have

$$\lim_{\substack{t-s\to 0\\ t\leq T}} \mathbf{M}\|\widehat{U}^s_t(\varepsilon,\omega)-I\|^2 = 0,$$

and therefore

$$\|D(\widehat{U}^s_t(\varepsilon,\omega)-I)\| \to 0$$

as $t-s\to 0$ uniformly on each finite interval. The function $\mathbf{M}|\widehat{U}^0_t(\varepsilon,\omega)x|^2$ is continuous, and thus the r.h.s. of (5.52) approaches zero as $m\to\infty$.

The existence of the limit (5.45) is proved.

Since $\widehat{Z}_\varepsilon(t)$ is a mean square limit of the processes with independent increments (given by the sums on the r.h.s. of (5.45)), it is also a process with independent increments and $\mathbf{M}\widehat{Z}_\varepsilon(t)=0$.

Using the inequality

$$\mathbf{M}|\sum_{k=1}^{n-1}(\widehat{U}^{t_k}_{t_{k+1}}(\varepsilon,\omega)-I)x|^2 \leq \mathbf{M}|\widehat{U}^{t_1}_{t_n}(\varepsilon,\omega)x - x|^2$$

for $t_1 < t_2 < \cdots < t_n$, we obtain that

$$\mathbf{M}|(\widehat{Z}_\varepsilon(t) - \widehat{Z}_\varepsilon(s))x|^2 \leq \mathbf{M}|\widehat{U}_t^s(\varepsilon,\omega)x - x|^2$$

for $s < t$. This implies that $\widehat{Z}_\varepsilon(t)$ is continuous in mean square.

□

Let us show how $\widehat{U}_t^s(\varepsilon,\omega)$ can be reconstructed in terms of $\widehat{Z}_\varepsilon(t)$.

Remark. Proving the theorem, we have not used the fact that $\widehat{U}_t^s(\varepsilon,\omega)$ has all the moments. We have only used the existence of the second moment and the condition

$$\lim_{t-s\to 0} \mathbf{M}\|\widehat{U}_t^s(\varepsilon,\omega) - I\|^2 = 0$$

for $t, s \leq T$, whatever T we take.

Hence, if $\widetilde{U}_t^s(\omega)$ is a stochastic semigroup such that

(1) $\mathbf{M}\|\widetilde{U}_t^s(\omega)\|^2 < \infty$,

(2) $\mathbf{M}\widetilde{U}_t^s(\omega) = I$,

(3) for all $T > 0$,

$$\lim_{h\to 0} \sup_{\substack{0<t-s\leq h\\ t\leq T}} \mathbf{M}\|\widetilde{U}_t^s(\omega) - I\|^2 = 0,$$

then the mean square limit

$$\widetilde{Z}_t = \lim_{n\to\infty} \sum_{k<2^n t} (\widetilde{U}_{(k+1)/2^n}^{k/2^n}(\omega) - I)$$

exists in $L(R^d)$, $\widetilde{Z}_t$ is a process with independent increments in $L(R^d)$ such that $\mathbf{M}\widetilde{Z}_t = 0$, $\mathbf{M}\|\widetilde{Z}_t\|^2 < \infty$, and for all $T > 0$,

$$\lim_{h\to 0} \sup_{\substack{0<t-s\leq h,\\ t\leq T}} \mathbf{M}\|\widetilde{Z}_t - \widetilde{Z}_s\|^2 = 0.$$

24°. Theorem. *If $\widehat{Z}_\varepsilon(t)$ is a process with independent increments constructed in Theorem 22°, then*

$$\widehat{U}_t^s(\varepsilon,\omega) = \lim_{n\to\infty} \prod_{2^n s\leq k<2^n t} \left(I + \widehat{Z}_\varepsilon\left(\frac{k+1}{2^n}\right) - \widehat{Z}_\varepsilon\left(\frac{k}{2^n}\right)\right). \tag{5.53}$$

Proof. Using Lemma 23°, we obtain for $s = t_1 < t_2 < \cdots < t_r = t$,

$$\mathbf{M}|[\widehat{U}_t^s(\varepsilon,\omega) - I - \sum_{i=1}^{r-1}(\widehat{U}_{t_{i+1}}^{t_i}(\varepsilon,\omega) - I)]x|^2$$
$$\leq \|D(\widehat{U}_t^s(\varepsilon,\omega) - I)\|(\mathbf{M}|\widehat{U}_t^s(\varepsilon,\omega)x|^2 - |x|^2).$$

Employing Theorem 22°, we get by means of the limit transition,

$$\begin{aligned}&\mathbf{M}|[\widehat{U}_t^s(\varepsilon,\omega) - I - (\widehat{Z}_\varepsilon(t) - \widehat{Z}_s(t))]x|^2\\ &\quad\le \|D(\widehat{U}_t^s(\varepsilon,\omega) - I)\|\mathbf{M}|(\widehat{U}_t^s(\varepsilon,\omega) - I)x|^2.\end{aligned}$$

Hence,

$$\begin{aligned}&\mathbf{M}\left|\left(\widehat{U}^{k/2^n}_{(k+1)/2^n}(\varepsilon,\omega) - I - \left(\widehat{Z}_\varepsilon\left(\frac{k+1}{2^n}\right) - \widehat{Z}_\varepsilon\left(\frac{k}{2^n}\right)\right)\right)x\right|^2\\ &\quad\le \|D(\widehat{U}^{k/2^n}_{(k+1)/2^n} - I)\|^2|x|^2.\end{aligned}$$

Taking into account the latter inequality, we can write

$$\begin{aligned}&\mathbf{M}\left|\left(\prod_{2^ns\le k<2^nt}\widehat{U}^{k/2^n}_{(k+1)/2^n}(\varepsilon,\omega) - \prod_{2^ns\le k<2^nt}\left(I + \widehat{Z}_\varepsilon\left(\frac{k+1}{2^n}\right) - \widehat{Z}_\varepsilon\left(\frac{k}{2^n}\right)\right)\right)x\right|^2\\ &= \mathbf{M}\left|\sum_{2^ns\le k<2^nt}\left(\prod_{2^ns\le i<k}\widehat{U}^{i/2^n}_{(i+1)/2^n}(\varepsilon,\omega)\left[\widehat{U}^{k/2^n}_{(k+1)/2^n}(\varepsilon,\omega) - I\right.\right.\right.\\ &\quad\left.\left.\left. - \widehat{Z}_\varepsilon\left(\frac{k+1}{2^n}\right) + \widehat{Z}_\varepsilon\left(\frac{k}{2^n}\right)\right]\prod_{k<j<2^nt}\left(I + \widehat{Z}_\varepsilon\left(\frac{j+1}{2^n}\right) - \widehat{Z}_\varepsilon\left(\frac{j}{2^n}\right)\right)\right)x\right|^2.\end{aligned}$$

This expression can be represented as a sum of the following terms: For $k < l$,

$$\begin{aligned}&\mathbf{M}\left(V_k\widehat{U}^{k/2^n}_{(k+1)/2^n}(\varepsilon,\omega)V_{kl}\right.\\ &\quad\times\left[\widehat{U}^{l/2^n}_{(l+1)/2^n}(\varepsilon,\omega) - I - \widehat{Z}_\varepsilon\left(\frac{l+1}{2^n}\right) + \widehat{Z}_l\left(\frac{l}{2^n}\right)\right]W_lx,\\ &\quad V_k\left[\widehat{U}^{k/2^n}_{(k+1)/2^n}(\varepsilon,\omega) - I - \widehat{Z}_\varepsilon\left(\frac{k+1}{2^n}\right) - \widehat{Z}_\varepsilon\left(\frac{k}{2^n}\right)\right]\\ &\quad\left.\times W_{kl}\left(I + \widehat{Z}_\varepsilon\left(\frac{l+1}{2^n}\right) - \widehat{Z}_\varepsilon\left(\frac{l}{2^n}\right)\right)W_lx\right)\\ &= \mathbf{M}\left(V_k\left[\widehat{U}^{k/2^n}_{(k+1)/2^n}(\varepsilon,\omega) - I\right]\right.\\ &\quad\times V_{kl}\left[\widehat{U}^{l/2^n}_{(l+1)/2^n}(\varepsilon,\omega) - I - \widehat{Z}_\varepsilon\left(\frac{l+1}{2^n}\right) + \widehat{Z}_\varepsilon\left(\frac{l}{2^n}\right)\right]W_lx,\\ &V_k\left[\widehat{U}^{k/2^n}_{(k+1)/2^n}(\varepsilon,\omega) - I - \widehat{Z}_\varepsilon\left(\frac{k+1}{2^n}\right)\right]W_{kl}\left.\left[\widehat{Z}_\varepsilon\left(\frac{l+1}{2^n}\right) - \widehat{Z}_\varepsilon\left(\frac{l}{2^n}\right)\right]W_lx\right).\end{aligned} \tag{5.54}$$

Here all the factors are independent and have bounded mathematical expectation of the squared norm. We note that

$$\|D(UV)\| \leq \|D(U)\|\,\|D(V)\|$$

for two independent operators V and U. Therefore, applying the Cauchy inequality $(a,b) \leq 1/2[(a,a)+(b,b)]$ to the variable (5.54), we obtain the following estimate for this expression

$$O(\|D(\widehat{U}^{k/2^n}_{(k+1)/2^n}(\varepsilon,\omega)-I)\| \cdot \|D(\widehat{U}^{l/2^n}_{(l+1)/2^n}(\varepsilon,\omega)-I)\|^2 + \|D(\widehat{U}^{k/2^n}_{(k+1)/2^n}(\varepsilon,\omega)-I)\|^2\|D(\widehat{U}^{l/2^n}_{(l+1)/2^n}(\varepsilon,\omega)-I)\|). \tag{5.55}$$

If $e_1,\dots,e_d$ is a basis in R^d, then using (5.49), we get

$$\sum_{k\leq t2^n} \|D(\widehat{U}^{k/2^n}_{(k+1)/2^n}(\varepsilon,\omega)-I)\|$$
$$\leq \sum_{k\leq t2^n}\sum_{r=1}^{d} \mathbf{M}|(\widehat{U}^{k/2^n}_{(k+1)/2^n}(\varepsilon,\omega)-I)e_r|^2$$
$$\leq \sum_{k\leq t2^n}\sum_{r=1}^{d} (\mathbf{M}|\widehat{U}^{0}_{(k+1)/2^n}(\varepsilon,\omega)e_r|^2 - \mathbf{M}|\widehat{U}^{0}_{k/2^n}(\varepsilon,\omega)e_r|^2)$$
$$\leq \sum_{r=1}^{d} \mathbf{M}|\widehat{U}^{0}_{(n_t+1)/2^n}(\varepsilon,\omega)e_r|^2$$

Here n_t is an integral part of $2^n t$. Since $\max_{k\leq t2^n}\|D(\widehat{U}^{k/2^n}_{(k+1)/2^n}(\varepsilon,\omega)-I)\| \to 0$ as $n\to\infty$, the proof of the theorem follows from (5.55) if we take the last inequality into account.

□

Remark 1. Using Lemma 23° just as in the theorem, we can prove the following statement: If $\widetilde{Z}_t$ is a process with independent increments in $L(R^d)$ for which (1) $\mathbf{M}\widetilde{Z}_t = 0$, (2) $\mathbf{M}\|\widetilde{Z}_t\|^2 < \infty$, (3) for all $T > 0$,

$$\lim_{h\to 0}\ \sup_{0<t-s\leq h} \mathbf{M}\|\widetilde{Z}_t - \widetilde{Z}_s\|^2 = 0,$$

then the mean square limit

$$\widetilde{U}^s_t(\omega) = \lim_{n\to\infty} \prod_{2^n s\leq k<2^n t} (I + \widetilde{Z}_{(k+1)/2^n} - \widetilde{Z}_{k/2^n}) \tag{5.56}$$

exists for all $0 \leq s < t$. Furthermore, $\widetilde{U}^s_t(\omega)$ is a stochastic semigroup such that (a) $\mathbf{M}\|\widetilde{U}^s_t(\omega)\|^2 < \infty$; (b) $\mathbf{M}\widetilde{U}^s_t(\omega) = I$; (c) for all $T > 0$,

$$\lim_{h\to 0}\ \sup_{\substack{0\leq t-s\leq h\\ t\leq T}} \mathbf{M}\|\widetilde{U}^s_t(\omega) - I\|^2 = 0;$$

and (d) the process $\widetilde{Z}_t$ is the mean square limit

$$\widetilde{Z}_t = \lim \sum_{k<2^n t} \widetilde{U}^{k/2^n}_{(k+1)/2^n}(\omega). \tag{5.57}$$

Remark 2. Since (5.56) and (5.57) give the one-to-one correspondence between processes with independent increments satisfying conditions (1)–(3) and stochastic semigroups satisfying (a)–(c), we can use Theorems 21°, 22°, and 24° to reduce the investigation of stochastically continuous stochastic semigroups to the analysis of stochastically continuous processes with independent increments in $L(R^d)$.

APPENDIX

Local Properties Of Sample Functions Of Homogeneous Processes In R

Since $\xi(t+s)-\xi(t)$ has the same distribution as $\xi(s)$ for all $t \geq 0$, it suffices to study the behavior of a homogeneous process at the point 0. We first investigate the behavior of the ratio $\xi(t)/t$ as $t \downarrow 0$.

Theorem 1. I. *If a process $\xi(t)$ has a bounded variation and if its cumulant $K(z)$ has the form*

$$K(z) = iaz + \int_{-\infty}^{\infty} (e^{izx} - 1)\Pi(dx), \tag{1}$$

then

$$\mathbf{P}\left\{\lim_{t\downarrow 0} \frac{\xi(t)}{t} = a\right\} = 1.$$

II. *If $\xi(t)$ has unbounded variation, then*

$$\overline{\lim_{t\downarrow 0}} \frac{\xi(t)}{t} = +\infty, \qquad \underline{\lim_{t\downarrow 0}} \frac{\xi(t)}{t} = -\infty.$$

with probability 1.

Proof. I. Let $\nu_t(A)$ be a Poisson measure constructed on the jumps of the process. Since

$$\sum \mathbf{M}\nu_{1/2^n}\left(\left\{x : |x| > \frac{1}{2^n}\right\}\right) \leq \sum \frac{1}{2^n}\Pi\left(\left\{x : |x| > \frac{1}{2^n}\right\}\right)$$
$$< 2\int_{-1}^{1} |x|\Pi(dx) + \Pi(\{x : |x| > 1\}),$$

the variables $\int_{|x|>2^{-n}} x\nu_{1/2^n}(dx)$ are zero beginning with a certain number. Hence, it suffices to show that the variables

$$\sup_{2^{-n}\le t\le 2^{-n+1}} 2^n\left|\int_{|x|\le 2^{-n}} x\nu_t(dx)\right| \le 2^n\int_{|x|\le 2^{-n}} |x|\nu_{2^{-n+1}}(dx) = \zeta_n$$

vanish with probability 1.

$$\mathbf{M}\zeta_n = 2^n\int_{|x|\le 2^{-n}} |x|2^{-n+1}\Pi(dx) = 2\int_{|x|\le 2^{-n}} |x|\Pi(dx) \to 0,$$
$$\mathbf{D}\zeta_n = 2^{2n}\int_{|x|\le 2^{-n}} |x|^2 2^{-n+1}\Pi(dx) = 2^{n+1}\int_{|x|\le 2^{-n}} x^2\Pi(dx)$$

and

$$\sum_{n=1}^{\infty}\mathbf{D}\zeta_n \le \int_{|x|\le 1} x^2 \sum_{2^n\le 1/|x|} 2^{n+1}\Pi(dx) \le 4\int_{|x|\le 1}|x|\Pi(dx) < \infty.$$

Therefore, $\sum_{n=1}^{\infty}(\zeta_n - \mathbf{M}\zeta_n)^2 < \infty$, $\zeta_n \to 0$ with probability 1. The statement I is proved.

II. Let us first consider the process with negative jumps only. Without loss of generality we can assume that the cumulant of this process has the form

$$K(z) = iaz - \frac{bz^2}{2} + \int_{-1}^{0}(e^{izx} - 1 - izx)\Pi(dx), \tag{2}$$

where $a > 0$. Then for all $y > 0$ the period of time τ_y, during which the process reaches the value y for the first time, is finite; τ_y is a monotonic homogeneous process with independent increments with respect to y. Let $K_-(\lambda) = K(-i\lambda)$. It was shown in 32°, Chapter 4, that $\mathbf{M}e^{-\lambda\tau_y} = e^{-yB(\lambda)}$, where $B(\lambda)$ is a solution of the equation $\lambda = K_-(B(\lambda))$. The function $B(\lambda)$ has the form

$$B(\lambda) = a_1(\lambda) + \int_0^{\infty}(e^{-\lambda x} - 1)\Pi_1(dx) \qquad (a_1 > 0). \tag{3}$$

Clearly, $a_1 = 0$ if $B(\lambda)$ remains bounded when $\lambda \to +\infty$. Assume that $B(\lambda) \to -\infty$ as $\lambda \to \infty$. Then using (3), we find

$$a_1 = \lim_{\lambda\to\infty} B'(\lambda) = \lim_{\lambda\to\infty}\frac{1}{K'_-(-B(\lambda))} = \lim_{\lambda\to\infty}\frac{1}{K'_-(\lambda)}$$
$$= \lim_{\lambda\to\infty}\left[a + b\lambda + \int_{-1}^{0}(e^{\lambda x} - 1)x\Pi(dx)\right]^{-1} = \frac{1}{+\infty} = 0.$$

Hence, $a_1 = 0$ and $\mathbf{M}e^{iz\tau_y} = \exp\left\{y\int_0^\infty (e^{izx}-1)\Pi_1(dx)\right\}$. Thus, $\mathbf{P}\left\{\lim_{y\downarrow 0}\frac{\tau_y}{y}=0\right\} = 1$ by virtue of statement I. Consequently, $\mathbf{P}\left\{\lim_{y\downarrow 0}\frac{\xi(\tau_y)}{\tau_y}=+\infty\right\}=1$, i.e., $\mathbf{P}\left\{\overline{\lim}_{t\downarrow 0}\frac{\xi(t)}{t}=+\infty\right\}=1$.

In order to show that for the process considered

$$\mathbf{P}\left\{\underline{\lim}_{t\downarrow 0}\frac{\xi(t)}{t}=-\infty\right\}=1, \tag{4}$$

it is sufficient to prove that

$$\mathbf{P}\left\{\underline{\lim}_{t\downarrow 0}\frac{\xi(t)}{t}\le -v\right\}=\mathbf{P}\left\{\underline{\lim}_{t\downarrow 0}\frac{\xi(t)+vt}{t}\le 0\right\}=1$$

for all $v>0$. The last equality holds if $P\{\inf_{t\le\delta}[\xi(t)+vt]<0\}=1$ for every $\delta>0$. Thus (4) is valid if $P\{\inf_{0\le s\le t}\xi(s)<0\}=1$ for the process with the cumulant (2), whatever $a>0$ is taken.

The following equality was obtained in the proof of Theorem 32°, Chapter 4,

$$\int_{-\infty}^0 e^{zx}d_xq_-(\lambda,x)=\frac{\lambda}{\lambda-\left[az+\frac{bz^2}{2}+\int_{-1}^0(e^{zx}-1-zx)\Pi(dx)\right]}\cdot\frac{B(\lambda)-z}{B(\lambda)},$$

where $q_-(\lambda,x)=\lambda\int_0^\infty e^{-\lambda t}\mathbf{P}\{\inf_{s\le t}\xi(s)<x\}dt$ and $B(\lambda)$ is defined by (3). Let us find $1-q_-(\lambda,-0)$:

$$\begin{aligned}1-q_-(\lambda,-0)&=\lim_{z\to\infty}\int_{-\infty}^0 e^{zx}d_xq_-(\lambda,x)\\&=\lim_{z\to\infty}\frac{\lambda z(B(\lambda))^{-1}}{\lambda-\left[az+\frac{bz^2}{2}+\int_{-1}^0(e^{zx}-1-zx)\Pi(dx)\right]}.\end{aligned}$$

Since

$$\begin{aligned}\lim_{z\to\infty}\frac{1}{z}\int_{-1}^0(e^{zx}-1-zx)\Pi(dx)&=\lim_{z\to\infty}\int_{-1}^0\left(\frac{e^{zx}-1}{z}-x\right)\Pi(dx)\\&\ge\lim_{z\to+\infty}\int_{-1}^{-\varepsilon}\left(\frac{e^{zx}-1}{z}-x\right)\Pi(dx)\\&=-\int_{-1}^{-\varepsilon}x\Pi(dx)\end{aligned}$$

where the last expression tends to $+\infty$ as $\varepsilon\to 0$, we have $1-q_-(\lambda,-0)=0$. Therefore, $\int_0^\infty e^{-\lambda t}\mathbf{P}\{\inf_{0\le s\le t}\xi(s)>0\}\,dt=0$. Our statement is proved.

□

Now suppose that

$$\int_{-1}^{1} |x|\Pi(dx) = +\infty, \qquad \int_{0}^{1} x\Pi(dx) < \infty.$$

Then the process $\xi(t)$ can be represented as $\xi(t) = \xi_1(t)+\xi_2(t)$, where $\xi_1(t)$ is a process with bounded variation, $\xi_2(t)$ has no positive jumps, and its variation is unbounded. Using Theorem 1 we obtain

$$\overline{\lim_{t\downarrow 0}}(\xi(t)/t) = \lim_{t\downarrow 0}(\xi_1(t)/t) + \overline{\lim_{t\downarrow 0}}(\xi_2(t)/t) = +\infty,$$
$$\underline{\lim}_{t\downarrow 0}(\xi(t)/t) = \underline{\lim}_{t\downarrow 0}(\xi_2(t)/t) = -\infty.$$

Changing the sign of $\xi(t)$, we establish validity of the theorem in the case when

$$\int_{-1}^{1} |x|\Pi(dx) = +\infty, \qquad \int_{-1}^{0} |x|\Pi(dx) < \infty.$$

Finally, consider the case of

$$\int_{-1}^{0} |x|\Pi(dx) = +\infty, \qquad \int_{0}^{1} |x|\Pi(dx) = +\infty.$$

Then the process $\xi(t)$ can be represented as follows: $\xi(t) = \xi_1(t) + \xi_2(t)$, where $\xi_1(t)$ has no positive jumps, $\xi_2(t)$ has no negative jumps, $\xi_1(t)$ and $\xi_2(t)$ are independent, and they both have unbounded variations. Without loss of generality we can take $\mathbf{M}\xi_1(t) = 0$. Since $\overline{\lim}_{t\downarrow 0} \frac{\xi_2(t)}{t} = +\infty$ with probability 1, then in order to prove that

$$\mathbf{P}\left\{\overline{\lim_{t\downarrow 0}} \frac{\xi_1(t) + \xi_2(t)}{t} = +\infty\right\} = 1,$$

it suffices to show that

$$\mathbf{P}\left\{\overline{\lim_{k\to\infty}} \frac{\xi_1(t_k)}{t_k} \geq 0\right\} = 1 \tag{5}$$

for every sequence $t_k \downarrow 0$. Let us choose a sequence t_{n_k} so that

$$\mathbf{P}\left\{\lim_{k\to\infty} \frac{1}{t_{n_k}}\xi_1(t_{n_{k+1}}) = 0\right\} = 1.$$

Then (5) holds if

$$\mathbf{P}\left\{\overline{\lim_{k\to\infty}} \frac{1}{t_{n_k}}(\xi_1(t_{n_k}) - \xi_1(t_{n_{k+1}})) \geq 0\right\} = 1. \tag{6}$$

The variables $\xi_1(t_{n_k}) - \xi_1(t_{n_{k+1}})$ are independent, and thus (6) is valid (by virtue of the Borel–Cantelli theorem) if $\sum \mathbf{P}\{\xi_1(t_{n_k}) - \xi_1(t_{n_{k+1}}) \geq 0\} = +\infty$. The last relation follows from the next lemma, which is of independent interest.

Lemma. *If $\xi_1(t)$ is a homogeneous process with independent increments without positive jumps, for which $\mathbf{M}\xi_1(t) = 0$, then*

$$\mathbf{P}\{\xi_1(t) > 0\} \geq 1/16. \tag{7}$$

Proof. Let $\mathbf{M}e^{z\xi_1(t)} = e^{tv(z)}$, where

$$v(z) = \left[\frac{bz^2}{2} + \int_{-\infty}^{0} (e^{zx} - 1 - zx)\Pi(dx)\right].$$

Then,

$$\begin{aligned} v(2z) &= 2bz^2 + \int_{-\infty}^{0} (e^{2zx} - 1 - 2zx)\Pi(dx) \\ &= 2bz^2 + 2\int_{-\infty}^{0} (e^{zx} - 1 - zx)\Pi(dx) + \int_{-\infty}^{0} (e^{zx} - 1)^2\Pi(dx) \\ &\leq 2bz^2 + 4\int_{-\infty}^{0} (e^{zx} - 1 - zx)\Pi(dx) = 4v(z) \end{aligned}$$

since $(e^u - 1)^2 \leq 2(e^u - 1 - u)$ for $u \leq 0$. Using the Cauchy inequality we get

$$[\mathbf{M}e^{z\xi_1(t)}I_{(0,\infty)}(\xi_1(t))]^2 \leq \mathbf{M}e^{2z\xi_1(t)}\mathbf{P}\{\xi_1(t) > 0\}.$$

Hence,

$$\mathbf{P}\{\xi_1(t) > 0\} \geq \frac{(\mathbf{M}e^{z\xi_1(t)}I_{(0,\infty)}(\xi_1(t)))^2}{\mathbf{M}e^{2z\xi_1(t)}} \geq \frac{(e^{tv(z)} - 1)^2}{e^{4tv(z)}}.$$

Since $v(0) = 0$ and $v(+\infty) = +\infty$, one can find z such that $e^{tv(z)} = 2$. Inserting this z into the last inequality we obtain (7).

A local growth of arbitrary processes can be estimated with the help of the theorem given below.

Theorem 2. *Suppose that $\xi(t)$ is a homogeneous process with independent increments and $\varphi(t)$ is a continuous nonnegative function increasing on $[0,1]$ that satisfies the following conditions:*

(a) $\lim_{u\downarrow 1} \sup_t \left|\frac{\varphi(ut)}{\varphi(t)} - 1\right| = 0$;

(b) *for every $\varepsilon > 0$, there exists $a_\varepsilon > 0$ such that*

$$\mathbf{P}\{\xi(t) < -\varepsilon\varphi(t)\} \leq 1 - a_\varepsilon.$$

Then,

(1) *if* $\int_0^1 \frac{1}{t}\mathbf{P}\{\xi(t) > \varphi(t)\}dt < \infty$, *then* $\mathbf{P}\left\{\lim_{t\downarrow 0} \frac{\xi(t)}{\varphi(t)} \leq 1\right\} = 1$;

(2) *if* $\int_0^1 \frac{1}{t}\mathbf{P}\{\xi(t) > \varphi(t)\}dt = \infty$, *then* $\mathbf{P}\left\{\lim_{t\downarrow 0} \frac{\xi(t)}{\varphi(t)} \geq 1\right\} = 1$.

Proof. Using 7°, Chapter 1, separability of the process $\xi(t)$, and the condition (b), we conclude that for $a < 1$,

$$\mathbf{P}\{\sup_{0\leq s\leq a^k} \xi(s) > (1+2\varepsilon)\varphi(a^k)\} \leq \frac{1}{a_\varepsilon}\mathbf{P}\{\xi(t) > (1+\varepsilon)\varphi(a^k)\}$$

if $a^{k+1} < t < a^k$. Choosing a so close to 1, that $(1+\varepsilon)\varphi(a^k) > \varphi(a^{k-1})$, we obtain

$$\mathbf{P}\{\sup_{0\leq s\leq a^k} \xi(s) > (1+2\varepsilon)\varphi(a^k)\} \leq \frac{1}{a_\varepsilon(1-a)}\int_{a^k}^{a^{k-1}} \frac{1}{t}\mathbf{P}\{\xi(t) > \varphi(t)\}dt.$$

Consequently, $\sum_{k=1}^{\infty} \mathbf{P}\{\sup_{0\leq s<a^k} \xi(s) > (1+2\varepsilon)\varphi(a^k)\} < \infty$. Hence, beginning with a certain k, $\xi(t) \leq (1+2\varepsilon)\varphi(a^k) \leq (1+2\varepsilon)(1+\varepsilon)\varphi(t)$ for $a^{k+1} \leq t \leq a^k$ with probability 1. Therefore, $\overline{\lim}_{t\downarrow 0} \frac{\xi(t)}{\varphi(t)} \leq (1+2\varepsilon)(1+\varepsilon)$ with probability 1. Since this is valid for arbitrary $\varepsilon > 0$, the statement (1) is proved.

Now we prove statement (2). First, we establish that for every $\varepsilon > 0$, one can find $\delta > 0$ such that

$$\sum_{k=1}^{\infty} \mathbf{P}\{\xi(a^k) > (1-\varepsilon)\varphi(a^k)\} = +\infty \tag{9}$$

for $0 < 1 - a < \delta$.

Let $a^{k+1} < t < a^k$. If a is such that $\varphi(a^k) - \varphi(a^{k+1}) \leq \frac{\varepsilon}{2}\varphi(a^k)$, then

$$\begin{aligned}
&\mathbf{P}\{\xi(a^k) > (1-\varepsilon)\varphi(a^k)\}\\
&\quad\geq \mathbf{P}\{\xi(t) > \varphi(t)\}\mathbf{P}\{\xi(a^k) - \xi(t) > (1-\varepsilon)\varphi(a^k) - \varphi(t)\}\\
&\quad\geq \mathbf{P}\{\xi(t) > \varphi(t)\}\mathbf{P}\{\xi(a^k) - \xi(t) > (1-\varepsilon)\varphi(a^k) - \varphi(a^{k+1})\}\\
&\quad\geq \mathbf{P}\{\xi(t) > \varphi(t)\}\\
&\qquad\times \mathbf{P}\left\{\xi(a^k) - \xi(t) > -\frac{[\varepsilon\varphi(a^k) - (\varphi(a^k) - \varphi(a^{k+1}))]}{\varphi(a^k - a^{k+1})}\varphi(a^k - t)\right\}\\
&\quad\geq \alpha_{\varepsilon/2}\mathbf{P}\{\xi(t) > \varphi(t)\};
\end{aligned}$$

$$\begin{aligned}
&\alpha_{\varepsilon/2}\frac{1}{a^k - a^{k+1}}\int_{a^{k+1}}^{a^k} \mathbf{P}\{\xi(t) > \varphi(t)\}dt\\
&\quad\geq \alpha_{\varepsilon/2}\frac{a}{1-a}\int_{a^k}^{a^{k+1}} \frac{1}{t}\mathbf{P}\{\xi(t) > \varphi(t)\}dt.
\end{aligned}$$

Hence,

$$\sum_{k=1}^{\infty} \mathbf{P}\{\xi(a^k) > (1-\varepsilon)\varphi(a^k)\} \geq \frac{\alpha_{\varepsilon/2} a}{1-a} \int_0^1 \frac{1}{t} \mathbf{P}\{\xi(t) > \varphi(t)\} dt = \infty,$$

i.e., (9) is valid.

It follows from (9) that for every N there exists l such that

$$\sum_{k=1}^{\infty} \mathbf{P}\{\xi(a^{l+Nk}) > (1-\varepsilon)\varphi(a^{l+Nk})\} = \infty.$$

Since

$$\mathbf{P}\left\{\xi\left(\frac{a^{l+Nk}}{1-a^N}\right) - \xi\left(\frac{a^{l+Nk+N}}{1-a^N}\right) > (1-\varepsilon)\varphi(a^{l+Nk})\right\}$$
$$= \mathbf{P}\{\xi(a^{l+Nk}) > (1-\varepsilon)\varphi(a^{l+Nk})\}$$

and since the events

$$\left\{\xi\left(\frac{a^{l+Nk}}{1-a^N}\right) - \xi\left(\frac{a^{l+Nk+N}}{1-a^N}\right) > (1-\varepsilon)\varphi(a^{l+Nk})\right\}$$

are independent for different k, an infinite number of these events happen with probability 1, i.e., one can find with probability 1 the sequence k_n such that

$$\xi\left(\frac{a^{l+Nk_n}}{1-a^N}\right) - \xi\left(\frac{a^{l+Nk_n+N}}{1-a^N}\right) > (1-\varepsilon)\varphi(a^{l+Nk_n}). \tag{10}$$

The inequality (10) implies one of the two inequalities:

$$\xi(a^{l+Nk_n}/(1-a^N)) > (1-2\varepsilon)\varphi(a^{l+Nk_n}), \tag{11}$$
$$\xi(a^{l+Nk_n+N}/(1-a^N)) < -\varepsilon\varphi(a^{l+Nk_n}). \tag{12}$$

Let us show that we can choose k_n approaching infinity sufficiently fast so that the opposite of inequality (12) holds for an infinite number of indices k_n. This means that (11) is valid for these indices. We take k_n such that

$$\lim_{n\to\infty} \frac{1}{\varphi(a^{l+Nk_n})} \xi\left(\frac{a^{l+Nk_{n+1}+N}}{1-a^N}\right) = 0$$

with probability 1 (this is possible since $\lim_{t\downarrow 0} \xi(t) = 0$). Then beginning with some n

$$\left|\xi\left(\frac{a^{l+Nk_{n+1}+N}}{1-a^N}\right)\right| < \frac{\varepsilon}{2}\varphi(a^{l+Nk_n}). \tag{13}$$

Consider events

$$\left\{\xi\left(\frac{a^{l+Nk_n+N}}{1-a^N}\right)-\xi\left(\frac{a^{l+Nk_{n+1}+N}}{1-a^N}\right)\geq-\frac{\varepsilon}{2}\varphi(a^{l+Nk_n})\right\}. \tag{14}$$

They are independent and

$$\mathbf{P}\left\{\xi\left(\frac{a^{l+Nk_n+N}}{1-a^N}\right)-\xi\left(\frac{a^{l+Nk_{n+1}+N}}{1-a^N}\right)\geq-\frac{\varepsilon}{2}\varphi(a^{l+Nk_n})\right\}$$

$$=\mathbf{P}\left\{\xi\left(a^{l+Nk_n}\frac{a^N}{1-a^N}\right)(1-a^{N(k_n-k_{n+1})})\geq-\frac{\varepsilon}{2}\varphi(a^{l+Nk_n})\right\}\geq\alpha_{\varepsilon/2}>0,$$

provided that N is such that $\frac{a^N}{1-a^N}\leq 1$ (by virtue of condition (b)). Thus, an infinite number of the events (14) take place with probability 1. But for all k_n for which events (13) and (14) happen, we have

$$\xi\left(\frac{a^{l+Nk_n}}{1-a^N}\right)>-\varepsilon\varphi(a^{l+Nk_n}),$$

i.e., the event opposite to (12) happens. Hence, we have proved that there exists (with probability 1) an infinite sequence of indices k_n, for which (11) holds. The condition (a) of the theorem implies that there exists N such that

$$(1-2\varepsilon)\varphi(a^{l+Nk_n})\geq(1-3\varepsilon)\varphi\left(\frac{a^{l+Nk_n}}{1-a^N}\right).$$

If N is chosen so, the inequality (11) yields

$$\overline{\lim_{n\to\infty}}\left(\varphi\left(\frac{a^{l+Nk_n}}{1-a^N}\right)\right)^{-1}\xi\left(\frac{a^{l+Nk_n}}{1-a^N}\right)\geq 1-3\varepsilon.$$

And this implies statement (2) since $\varepsilon>0$ is arbitrary.

□

Remark 1. Condition (b) is satisfied automatically for symmetric processes since $P\{\xi(t)<0\}\leq 1/2$ in this case.

Remark 2. It is clear that we can consider a function $\varphi(t)$ given on an arbitrary interval $[0,\delta]$ and satisfying the conditions of the theorem on it. The only difference is that we must substitute integrals from 0 to δ for the integrals in statements (1) and (2).

Theorem 3. (Local log–log law). *If $\xi(t)$ is a Wiener process and $D\xi(t)=bt$, then*

$$\mathbf{P}\left\{\overline{\lim_{t\downarrow 0}}\frac{\xi(t)}{\sqrt{2bt\ln\ln\frac{1}{t}}}=1\right\}=1.$$

Proof. It suffices to examine the case $b = 1$, $a = \mathbf{M}\xi(1) = 0$. Taking $\varphi(t) = (1+\varepsilon)\sqrt{2t\ln\ln\frac{1}{t}}$, we find

$$\mathbf{P}\{\xi(t) > \varphi(t)\} = \frac{1}{\sqrt{2\pi}}\int_{\varphi(t)/\sqrt{t}}^{\infty} e^{-x^2/t}dx \le \frac{1}{\sqrt{2\pi}}\cdot\frac{\sqrt{t}}{\varphi(t)}\cdot e^{-\varphi^2(t)/(2t)}$$
$$= \frac{1}{\sqrt{4\pi(1+\varepsilon)\ln\ln\frac{1}{t}}}\left(\ln\frac{1}{t}\right)^{-(1+\varepsilon)^2}.$$

Thus, $\int_0^\delta \frac{1}{t}\mathbf{P}\{\xi(t) > \varphi(t)\}dt < \infty$ for sufficiently small $\delta > 0$. Further, we have $\mathbf{P}\{\xi(t) < -\varepsilon\varphi(t)\} \le \frac{t}{\varepsilon^2\varphi(t)} \to 0$ as $t \to 0$. Therefore,

$$\mathbf{P}\left\{\overline{\lim_{t\downarrow 0}}\frac{\xi(t)}{\sqrt{2(1+\varepsilon^2)t\ln\ln\frac{1}{t}}} \le 1\right\} = 1.$$

This is valid for all $\varepsilon > 0$, and thus

$$\mathbf{P}\left\{\overline{\lim_{t\downarrow 0}}\frac{\xi(t)}{\sqrt{2\ln\ln\frac{1}{t}}} \le 1\right\} = 1.$$

Let us estimate

$$\mathbf{P}\left\{\xi(t) > \lambda\sqrt{2t\ln\ln\frac{1}{t}}\right\}$$

from below for $\lambda < 1$:

$$\frac{1}{\sqrt{2\pi}}\int_{\lambda\sqrt{2\ln\ln\frac{1}{t}}}^{\infty} e^{-x^2/2}\,dx \ge \frac{1}{\sqrt{2\pi}}\int_{\lambda\sqrt{2\ln\ln\frac{1}{t}}}^{\Delta+\lambda\sqrt{2\ln\ln\frac{1}{t}}} e^{-x^2/2}\,dx$$
$$\ge \frac{\Delta}{\sqrt{2\pi}}\exp\left\{-\frac{1}{2}\left(\Delta+\lambda\sqrt{2\ln\ln\frac{1}{t}}\right)^2\right\}$$
$$\ge \frac{\Delta}{\sqrt{2\pi}}\exp\left\{-\lambda^2(1+\Delta)^2\ln\ln\frac{1}{t}\right\},$$

provided that $\lambda\sqrt{2\ln\ln\frac{1}{t}} \ge 1$. Choosing Δ so that $\lambda^2(1+\Delta)^2 = \gamma < 1$, we get

$$\mathbf{P}\left\{\xi(t) > \lambda\sqrt{2t\ln\ln\frac{1}{t}}\right\} \ge \frac{C}{(\ln\frac{1}{t})^{\nu}};$$

and therefore,

$$\int_0^\delta \frac{1}{t}\mathbf{P}\left\{\xi(t) > \lambda\sqrt{2t\ln\ln\frac{1}{t}}\right\}dt = +\infty.$$

Hence,

$$\mathbf{P}\left\{\overline{\lim_{t\downarrow 0}}\frac{\xi(t)}{\sqrt{2t\ln\ln\frac{1}{t}}} \geq \lambda\right\} = 1$$

for all $\lambda < 1$. This implies the required result.

□

We will now establish the theorem of A. Ya. Khinchin.

Theorem 4. *If $\xi(t)$ is a homogenous process with independent increments without a Gaussian component, then*

$$\mathbf{P}\left\{\lim_{t\downarrow 0}\frac{\xi(t)}{\sqrt{t\ln\ln\frac{1}{t}}} = 0\right\} = 1.$$

Proof. Without loss of generality we can assume that $\xi(t)$ has a finite dispersion. Denote $\varphi(t) = \sqrt{t\ln|\ln t|}$. Since

$$\mathbf{P}\{|\xi(t)| > \varepsilon\varphi(t)\} \leq \frac{\mathbf{D}\xi(t)}{\varepsilon^2\varphi^2(t)} \to 0,$$

by virtue of Theorem 3 it suffices to show that

$$\int_0^c \frac{1}{t}\mathbf{P}\{|\xi(t)| > \varepsilon\varphi(t)\}dt < \infty$$

for $c > 1$ and all $\varepsilon > 0$.

Let $\xi'(t)$ be a process that is independent of $\xi(t)$ but has the same distribution. Then

$$\begin{aligned}\mathbf{P}\{|\xi(t) - \xi'(t)| > \varepsilon\varphi(t)\} &\geq \mathbf{P}\{|\xi(t)| > 2\varepsilon\varphi(t)\}\mathbf{P}\{|\xi'(t)| \leq \varepsilon\varphi(t)\}\\ &\geq \mathbf{P}\{|\xi(t)| > 2\varepsilon\varphi(t)\}\left(1 - \frac{\mathbf{D}\xi(t)}{\varepsilon^2\varphi^2(t)}\right).\end{aligned}$$

Therefore, we can assume that $\xi(t)$ has a symmetric distribution. Let

$$\mathbf{M}e^{iz\xi(t)} = \exp\{tK(z)\},$$

where

$$K(z) = \int_0^\infty (\cos zx - 1)\Pi(dx).$$

Representing $\xi(t)$ as a sum of independent random variables ($\xi(t) = \xi_1(t) + \xi_2(t)$) with characteristic functions

$$\mathbf{M}e^{iz\xi_1(t)} = \exp\left\{t \int_0^\delta (\cos zx - 1)\Pi(dx)\right\},$$

$$\mathbf{M}e^{iz\xi_2(t)} = \exp\left\{t \int_\delta^\infty (\cos zx - 1)\Pi(dx)\right\},$$

where $\delta = \delta(t) = 1/\varphi(t)$, we obtain

$$\mathbf{P}\{|\xi(t)| > \varepsilon\varphi(t)\} \leq \mathbf{P}\left\{|\xi_1(t)| > \frac{\varepsilon}{2}\varphi(t)\right\} + \mathbf{P}\left\{|\xi_2(t)| > \frac{\varepsilon}{2}\varphi(t)\right\}.$$

Let us show that

$$\int_0^c \frac{1}{t}\mathbf{P}\left\{|\xi_k(t)| > \frac{\varepsilon}{2}\varphi(t)\right\} dt < \infty, \qquad k = 1, 2.$$

Setting $z = \Gamma\varphi(t)/t$ we get for $k = 1$,

$$\begin{aligned}
\mathbf{P}\left\{|\xi_1(t)| > \frac{\varepsilon}{2}\varphi(t)\right\} &= 2\mathbf{P}\left\{\xi_1(t) > \frac{\varepsilon}{2}\varphi(t)\right\} \leq e^{-\frac{\varepsilon}{2}z\varphi(t)}\mathbf{M}e^{z\xi_1(t)} \\
&= \exp\left\{-\frac{\varepsilon}{2}z\varphi(t) + t\int_0^\delta (\operatorname{ch} zx - 1)\Pi(dx)\right\} \\
&\leq \exp\left\{-\frac{\varepsilon}{2}\Gamma\frac{\varphi^2(t)}{t} + t\int_0^\delta \left(\operatorname{ch}\Gamma\frac{\varphi(t)}{t}x - 1\right)\Pi(dx)\right\} \\
&\leq \exp\left\{-\frac{\varepsilon}{2}\Gamma\frac{\varphi^2(t)}{t} + \Gamma^2\frac{\varphi^2(t)}{t}\int_0^\delta x^2\Pi(dx)\right\}.
\end{aligned}$$

Taking Γ such that $\varepsilon\Gamma/2 = \alpha > 1$, we obtain

$$\mathbf{P}\left\{|\xi_1(t)| > \frac{\varepsilon}{2}\varphi(t)\right\} = O\left(\left[\ln\frac{1}{t}\right]^{-\alpha}\right).$$

Hence,

$$\int_0^c \frac{1}{t}\mathbf{P}\left\{|\xi_1(t)| > \frac{\varepsilon}{2}\varphi(t)\right\} dt < \infty.$$

Further,

$$\begin{aligned}
\mathbf{P}\left\{|\xi_2(t)| > \frac{\varepsilon}{2}\varphi(t)\right\} &= 2\mathbf{P}\left\{\xi_2(t) > \frac{\varepsilon}{2}\varphi(t)\right\} \\
&\le \frac{4}{\varepsilon\varphi(t)} \int_0^{\frac{\varepsilon}{2}\varphi(t)} \mathbf{P}\{\xi_2(t) > x\}dx \\
&= \frac{4}{\varepsilon\varphi(t)} \int \frac{1-\cos z\frac{\varepsilon}{2}\varphi(t)}{z^2}(1 - \mathbf{M}e^{iz\xi_2(t)})dz \\
&\le \frac{\varepsilon\varphi(t)}{4t} \int \frac{1-\cos z\frac{\varepsilon}{2}\varphi(t)}{z^2} \int_{x>\delta} (1-\cos zx)\Pi(dx) \\
&= \frac{4t}{\varepsilon\varphi(t)} \int_\delta^\infty \Pi(dx) \int \frac{(1-\cos zx)\left(1-\cos z\frac{\varepsilon}{2}\varphi(t)\right)}{z^2}dz.
\end{aligned}$$

Using the equality

$$\int \frac{(1-\cos az)(1-\cos bz)}{z^2}dz = \pi \min[|a|, |b|],$$

we get

$$\begin{aligned}
\mathbf{P}\left\{|\xi_2(t)| > \frac{\varepsilon}{2}\varphi(t)\right\} &\le \frac{4t}{\varepsilon\varphi(t)} \int_\delta^\infty \Pi(dx)\min\left[|x|, \frac{\varepsilon}{2}\varphi(t)\right] \\
&= \frac{4t}{\varepsilon\varphi(t)} \int_\delta^{\frac{\varepsilon}{2}\varphi(t)} x\Pi(dx) + 2t\int_{\frac{\varepsilon}{2}\varphi(t)}^\infty \Pi(dx).
\end{aligned}$$

It remains to note that

$$\begin{aligned}
\int_0^c \frac{1}{\varphi(t)} \int_{\delta(t)}^1 x\Pi(dx)dt &\le \frac{1}{2}\int_0^c \int_{\delta(t)}^1 x\Pi(dx)d\delta(t) \\
&\le \frac{1}{2}\int_0^1 du \int_u^1 x\Pi(dx) = \frac{1}{2}\int_0^1 x^2\Pi(dx),
\end{aligned}$$

$$\int_0^c \int_{\frac{\varepsilon}{2}\varphi(t)}^1 \Pi(dx)dt \le \int_0^1 \int_{\sqrt{t}}^1 \Pi(dx)dt = \iint_{0<t<x^2<1} dt\Pi(dx) = \int_0^1 x^2\Pi(dx).$$

□

NOTES

Processes with independent increments, namely the process of Brownian motion, were considered for the first time (as a mathematical subject) by Bachelier in 1900 [1]. The rigorous theory of the Brownian motion was constructed on the basis of measure theory by Wiener (1923) [1]. The process constructed by him is now called the Wiener process, and his scheme became the basis on which Kolmogorov constructed the general notion of a random process [6] (1930). Later on, Finetti [1] (1929), Kolmogorov [5] (1932), and Lévy [2] (1934) found the general form for a characteristic function of a process with independent increments.

We note that the theory of processes with independent increments is closely connected with limit theorems for sums of independent random variables and with infinitely divisible distributions. These problems were considered with different degrees of generality in the books by Khinchin [1,2], Gnedenko and Kolmogorov [1,2], Ibragimov and Linnik [1], Feller [1], Loev [1], Zolotaryov [3], and Lévy [3].

Chapter 0

For proofs of the results presented here, see courses of the probability theory and theory of random processes.

Chapter 1

§1.1 Inequalities for sums of independent random variables appear, first, in the works by Khinchin and Kolmogorov [1], Kolmogorov [1,2,3], and Lévy [1]. Further, various modifications of these inequalities were obtained by different authors.

§1.2. The renewal theory is included to this book, because the renewal process is the simplest example of a discrete process with independent increments. The history of the problem and its contemporary state are presented more completely in the review by Sevastyanov [1].

§1.3, 1.4. Random walks are homogeneous processes with independent increments with discrete time. The most complete presentation of the theory is given in the book by Spitzer [2].

Chapter 2

Results of this chapter are mainly of a methodical character. Usually the results concerning real-valued processes are known, and they are used to determine characteristics of random measures. Here we act vice versa and use the results concerning measures on finite and countable algebras to construct the theory of processes with independent increments. Random measures and problems of extension of measures up to countably additive ones were studied by Prekopa [1]. Stochastic integrals with respect to a Poisson measure were defined by Itô [2].

Chapter 3

Decomposition of a process into nonrandom, discrete, and stochastically continuous components belongs to Lévy [2]. For the study of stochastically continuous processes, we have used the approach by Itô [1,3], in which the jump Poisson part and the continuous Wiener part are separated. Similar decompositions have been constructed up to now for a wide class of random processes (semimartingales); for details see Dellacheria, and Meyer [1]. Equations for locally homogeneous processes are in fact Kolmogorov's [4] equations.

Chapter 4

§§4.1, 4.2. Homogeneous processes with independent increments are homogeneous Markov processes. A semigroup approach to studying these processes was suggested by Feller and developed in the book by Dynkin [1]. Additive functionals of Markov processes have also been described in this book. Their representation, that is used here, can be found in the book by Gikhman and Skorokhod [2]; see also Skorokhod [2].
§§4.3, 4.4. Boundary functionals of homogeneous processes with independent increments (including composed Poisson processes and semicontinuous processes) were studied by different authors. The essential role was played by Spitzer's work [1]. We also mention works by Baxter and Donsker [1], Rogozin [1], Zolotaryov [2], Borovkov [1], Korolyuk [1], Gusak [1], Gusak and Korolyuk [1], and Shurenkov [1].
§4.5 The growth of processes with independent increments was studied in the works by Kolmogorov [1,2], Khinchin [3], Gnedenko [1,2], Zolotaryov [1], Rogozin [4], Shtatland [1].

Chapter 5

§§1.1, 1.2. Fundamental problems of the theory of processes with independent increments on groups are considered in the book by Grenander [1].
§1.3. Operator processes with independent multiplicative increments were introduced by Skorokhod [1]; and stochastic semigroups were introduced by Skorokhod [3] and Butsan [1]. Here we have cited the results of Butsan (see also Skorokhod [4]).

References

Bachelier, P.

[1] "Théorie de la spéculation," *Ann. Sci. École Norm. Sup.* **17** (1900), 21–86.

Baxter, J., Donsker, M.

[1] "On the distribution of the supremum functional for processes with stationary, independent increments," *Trans. Amer. Math. Soc.* **85**, (1957) 73.

Borovkov, A. A.

[1] "On the time of the first passage for one class of processes with independent increments," *Teor. Ver. i Prim* **X** (1965), 360–364 (in Russian).

Butsan, E. P.

[1] *Stochastic Semigroups.* Kiev, Naukova Dumka, 1977 (in Russian).

Dellacheria, C., Meyer, P.-A.

[1] *Probabilities and Potential.* North-Holland, Amsterdam, 1982.

Doob, J. L.

[1] *Stochastic Processes.* New York, 1953.

Dynkin, E. B.

[1] *Foundations of the Theory of Markov Processes.* Moscow, Fizmatgiz, 1959 (in Russian).

Feller, W.

[1] *An Introduction to Probability Theory and its Applications*, Vol. 2, Moscow, Mir, 1967 (Russian translation).

Finetti, B.

[1] "Sulla funzioni a incremento aleatorio," *Rend. Acad. Noz. Lincei. Cl. Sci. Fis. Math. Natur.* **10**, (1829), 163–168.

Gnedenko, B. V.

[1] "On the growth of homogeneous random processes with independent increments," *Izv. Acad. Sci. USSR, Ser. Math.* **7** (1947), 89–110 (in Russian).

[2] "On the theory of growth of homogeneous random processes with independent increments," *Trans. Inst. Math. Acad. Sci. Ukrainian SSR* **10**, (1948) 60–82 (in Russian).

Gnedenko, B. V., Kolmogorov, A. N.
[1] *Limit Distributions for Sums of Independent Random Variables.* Moscow, Gostekhizdat, 1949 (in Russian).

Grenander, U.
[1] *Probabilities on Algebraic Structures.* Moscow, Mir, 1965 (Russian translation).

Gusak, D. V.
[1] "On the joint distribution of time and value of the first passage for homogeneous processes with independent increments," *Teor. Ver. i Prim.* **XIV** (1969), 15–23 (in Russian).

Gusak, D. V., Korolyuk, V. S.
[1] "On the time of the first passage through the given level for processes with independent increments," *Teor. Ver. i Prim.* **XIII** (1968), 471–478 (in Russian).
[2] "On the joint distribution of a process with standard increments and its maximum," *Teor. Ver. i Prim.* **XIV** (1969), 421–430 (in Russian).

Ibragimov, I. A., Linnik, Yu. V.
[1] *Independent and Stationary Dependent Variables.* Moscow, Nauka, 1965 (in Russian).

Itô, K.
[1] " On stochastic processes," *Japan J. Math* **18** (1942), 261–301.
[2] "Stochastic processes," *Moscow, Foreign Lit.*, (1960), iss. 1 (Russian translation).
[3] "On stochastic differential equations," *Mem. Amer. Math. Soc.*, (1951), 4, p. 1–51.

Khinchin, A. Ya.
[1] *Asymptotic Laws of Probability Theory.* Moscow, ONTI, 1936 (in Russian).
[2] "Two theorems on stochastic processes with one-type increments," *Matem. Sb.* **3** (1938), 577–584 (in Russian).
[3] "On the local growth of homogeneous stochastic processes without aftereffect," *Izv. Acad. Sci. USSR, Ser. Math.* **5–6** (1939), 487–508 (in Russian).

Kolmogorov, A. N.
[1] "Über die Summen durch den Zufall bestimmter unabhängiger Grössen," *Math. Ann.* **99** (1928), 309–319.
[2] Ibid. **102**, (1929), 484–488.
[3] "Über das Gesetz des iterierten Logarithmus," *Math. Ann.* **101** (1929) 126–135.
[4] "On analytic methods in probability theory," *Usp. Mat. Nauk* **5** (1938), 5–41 (in Russian).
[5] "Sulla forma generale di un processo stocastico omogeneo," *Atti Accad. Naz. Lincei* **15** (1932), 805–808, 866–869.

[6] *Fundamental Notions of Probability Theory*. Moscow, ONTI, 1936 (in Russian).

Korolyuk, V. S.

[1] *Boundary Problems for Composed Poisson Processes*. Kiev, Naukova Dumka, 1975 (in Russian).

Lévy, P.

[1] *Calcul des Probabilitiés*. Paris, 1925.

[2] "Sur les intégrales dont les éléments sont des variables aléatoires indépendents," *Ann. Scuola Norm. Super. Pisa, Sci. Fis. e Mat.* **2** (1934), 337–366; **4** (1935), 217–218.

[3] *Théorie de l'Addition des Variables Aléatoires*. Paris, 1937.

Loev, M.

[1] *Probability Theory*. Moscow, Foreign Lit., 1962 (Russian translation).

Prekopa, A.

[1] "On the stochastic set function," *Acta Math. Acad. Scient. Hung.* **7** (1956), 215–263; **8** (1957), 337–400.

Rogozin, B. A.

[1] "On the distribution of the first passage value," *Teor. Ver i Prim.* **IX** (1964), 498–515 (in Russian).

[2] "On some classes of processes with independent increments," *Teor. Ver. i Prim.* **X** (1965), 527–531 (in Russian).

[3] "On the distributions of some functionals arising in boundary value problems for processes with independent increments," *Teor. Ver. i Prim.* **XI** (1966), 656–670 (in Russian).

[4] "On the local behaviour of processes with independent increments," *Teor. Ver. i Prim.* **XIII** (1968), 507–512 (in Russian).

Sevastyanov, B. A.

[1] "Renewal Theory," *Itogi Nauki i Tekhn. Ser. Teor. Ver. Mat. Stat., Cybernetics* **11** (1973/1974), 99–128 (in Russian).

Shtatland, E. S.

[1] "On local properties of processes with independent increments," *Teor. Ver. i Prim.* **X** (1965), 344–350 (in Russian).

Shurenkov, V. M.

[1] *Ergodic Theorems and Adjacent Problems of the Theory of Random Processes*. Kiev, Naukova Dumka, 1981 (in Russian).

Skorokhod, A. V.

[1] "Multiplicative matrix random processes," *Works of the 7th Soviet Conf. Prob. Theor. Math. Stat., Tbilisi* (1963), 81–85 (in Russian).

[2] "Nonnegative additive functionals of a process with independent increments," *In: Prob. Theory Math. Stat., Kiev, Kiev Univ.* **4**, (1971) (in Russian).

[3] "Martingales and stochastic semigroups," *Theory of Random Processes* **4** (1976), 86–94 (in Russian).
[4] "Operator stochastic differential equations," *Usp. Mat. Nauk* **37** (1982), 157–183 (in Russian).

Smith, R. L.
[1] "Regenerative stochastic processes," *Proc. Roy. Soc. Edinburgh* **A 232**, (1955), 6–31.

Spitzer, F.
[1] "A combinatorial lemma and its application to probability theory," *Trans. Amer. Math. Soc.* **82** (1956), 323–339.
[2] *Principles of Random Walk*. D. Van Nostrand, Princeton, New Jersey, 1964.

Wiener, N. D.
[1] "Differential space," *J. Math. Phys. Mass. Inst. Technology* **2** (1923), 131–174.

Zolotaryov, V. M.
[1] "An analogue of log–log law for semi-continuous stable processes," *Teor. Ver. i Prim.* **IX** (1964), 566 (in Russian).
[2] "The time of the first passage through the level and behaviour at infinity of one class of processes with independent increments," *Teor. Ver. i Prim.* **III** (1969), 724–733 (in Russian).
[3] *One-dimensional Stable Distributions*. Moscow, Nauka, 1983 (in Russian).

SUBJECT INDEX